T0184906

# Communications
# in Computer and Information Science 1025

*Commenced Publication in 2007*
Founding and Former Series Editors:
Phoebe Chen, Alfredo Cuzzocrea, Xiaoyong Du, Orhun Kara, Ting Liu,
Krishna M. Sivalingam, Dominik Ślęzak, Takashi Washio, Xiaokang Yang,
and Junsong Yuan

## Editorial Board Members

More information about this series at http://www.springer.com/series/7899

Abdullah Bin Gani · Pradip Kumar Das ·
Latika Kharb · Deepak Chahal (Eds.)

# Information, Communication and Computing Technology

4th International Conference, ICICCT 2019
New Delhi, India, May 11, 2019
Revised Selected Papers

 Springer

*Editors*
Abdullah Bin Gani
University of Malaya
Kuala Lumpur, Malaysia

Latika Kharb
Department of IT
Jagan Institute
of Management Studies
New Delhi, India

Pradip Kumar Das
Department of Computer Science
and Engineering
Indian Institute
of Technology Guwahati
Guwahati, India

Deepak Chahal
Department of IT
Jagan Institute
of Management Studies
New Delhi, India

ISSN 1865-0929          ISSN 1865-0937 (electronic)
Communications in Computer and Information Science
ISBN 978-981-15-1383-1          ISBN 978-981-15-1384-8 (eBook)
https://doi.org/10.1007/978-981-15-1384-8

This Springer imprint is published by the registered company Springer Nature Singapore Pte Ltd.
The registered company address is: 152 Beach Road, #21-01/04 Gateway East, Singapore 189721, Singapore

# Preface

The International Conference on Information, Communication, and Computing Technology (ICICCT 2019) was held on May 11, 2019, in New Delhi, India. ICICCT 2019 was organized by the Department of Information Technology, Jagan Institute of Management Studies (JIMS) Rohini, New Delhi, India. The conference received 315 submissions and after rigorous reviews, 24 papers were selected for this volume. The acceptance rate was around 18.1%. The contributions came from diverse areas of Information Technology categorized into two tracks – Next Generation Networking & Communication Systems and Evolutionary Computing through Machine Learning.

The aim of ICICCT 2019 was to provide a global platform for researchers, scientists, and practitioners from both academia and industry to present their research and development activities in all the aspects of Communication & Network Systems and Evolutionary Computing through Machine Learning.

We thank all the members of the Advisory Committee and the Program Committee for their hard work. We are very grateful to Prof. Mohamad Habli (Aldhar University College, Dubai, UAE) as the keynote speaker, Prof. Bharat Bhasker (Director at IIM Raipur, India) as session chair for Track 1, Prof. Ajai Jain (Department of Computer Science Engineering, IIT Kanpur, India) as Session Chair for Track 2.

We thank all the Technical Program Committee members and referees for their constructive and enlightening reviews on the manuscripts. We thank Springer for publishing the proceedings in the *Communications in Computer and Information Science* (CCIS) series. We thank all the authors and participants for their great contributions that made this conference possible.

July 2019

Pradip K. Das
Abdullah Bin Gani
Latika Kharb
Deepak Chahal

# Organization

## General Chair

Pradip K. Das            IIT Guwahati, India

## Program Chair

Abdullah Bin Gani        University of Malaya, Malaysia

## Conference Secretariat

Praveen Arora           Jagan Institute of Management Studies, India

## Session Chair for Track 1

Bharat Bhasker        Director at Indian Institute of Management, Raipur, India

## Session Chair for Track 2

Ajai Jain             IIT Kanpur, India

## Convener

Latika Kharb         Jagan Institute of Management Studies, India
Deepak Chahal       Jagan Institute of Management Studies, India

## Technical Program Committee

Siddhivinayak Kulkarni    MIT World Peace University, India
Malti Bansal           Delhi Technology University, India
Mohd. Faizal Bin Abdollah   Fakulti Teknologi Maklumat dan Komunikasi, Malaysia
Ahmad Khan          COMSATS University Islamabad, Pakistan
Mohd Abdul Ahad     Jamia Hamdard, India
Shahab Shamshirband    Iran University Science and Technology, Iran
Atul Gonsai Gosai      Saurashtra University, India
Shamimul Qamar      King Khalid University, Saudi Arabia
P. Subashini           Avinashilingam University for Women, India
Partha Pakray        National Institute of Technology, Assam, India
Azurah              Universiti Teknologi Malaysia, Malaysia
Anazida            Universiti Teknologi Malaysia, Malaysia

| | |
|---|---|
| Chan Weng Howe | Universiti Teknologi Malaysia, Malaysia |
| R. Sujatha | UKF College of Engineering and Technology, India |
| C. Shoba Bindu | JNTUA College of Engineering, India |
| S. Pallam Setty | Andra University, India |
| K. Madhavi | JNTUA College of Engineering, India |
| Janaka Wijekoon | Sri Lanka Institute of Information Technology, Sri Lanka |
| Hanumanthappa J. | University of Mysore, India |
| K. Thabotharan | University of Jaffna, Sri Lanka |
| Kamal Eldahshan | Al-Azhar University, Egypt |
| Tony Smith | University of Waikato, New Zealand |
| Abdel-Badeeh Salem | Ain Shams University, Egypt |
| Khalid Nazim Sattar Abdul | Majmaah University, Saudi Arabia |
| H. S. Nagendraswamy | University of Mysore, India |
| S. R. Boselin Prabhu | Anna University, India |
| S. Rajalakshmi | Sri Chandrasekharendra Saraswathi Viswa Mahavidyalaya, India |
| Anastasios Politis | Technological and Educational Institute of Central Macedonia, Greece |
| Subhash Chandra Yadav | Central University of Jharkhand, India |
| Uttam Ghosh | Vanderbilt University, USA |
| Wafaa Shalash | King Abdulaziz University Jeddah, Saudi Arabia |
| Etimad Fadel | King Abdulaziz University Jeddah, Saudi Arabia |
| Oleksii Tyshchenko | University of Ostrava, Czech Republic |
| Hima Bindu Maringanti | North Orissa University, India |
| Froilan D. Mobo | Philippine Merchant Marine Academy, The Philippines |
| Latafat A. Gardashova | Azerbaijan State Oil Academy, Azerbaijan |
| Wenjian Hu | Dynamic Ads Ranking, Facebook, USA |
| Muhammad Umair Ramzan | King Abdulaziz University Jeddah, Saudi Arabia |
| Areej Abbas Malibary | King Abdulaziz University Jeddah, Saudi Arabia |
| Dilip Singh Sisodia | National Institute of Technology Raipur, India |
| P. R. Patil | PSGVP Mandal's D.N. Patel College of Engineering, India |
| Jose Neuman Souza | Federal University of Ceara, Brazil |
| Nermin Hamza | King Abdulaziz University Jeddah, Saudi Arabia |
| R. Chithra | K. S. Rangasamy College of Technology, India |
| Homero Toral Cruz | University of Quintana Roo, Mexico |
| J. Viji Gripsy | PSGR Krishnammal College for Women, India |
| Boudhir Anouar Abdelhakim | Abdelmalek Essaâdi University, UAE |
| Muhammed Ali Aydin | Istanbul Cerrahpaşa University, Turkey |
| Suhair Alshehri | King Abdulaziz University Jeddah, Saudi Arabia |
| Dalibor Dobrilovic | University of Novi Sad, Serbia |
| A. V. Petrashenko | National Technical University of Ukraine, Ukraine |
| Ali Hussain | Sri Sai Madhari Institute of Science and Technology, India |

| | |
|---|---|
| A. NagaRaju | Central University Rajasthan, India |
| Cheng-Chi Lee | Fu Jen Catholic University, Taiwan |
| Apostolos Gkamas | University Ecclesiastical Academy of Vella of Ioannina, Greece |
| M. A. H. Akhand | Khulna University of Engineering & Technology, Bangladesh |
| Saad Talib Hasson | University of Babylon, Iraq |
| Valeri Mladenov | Technical University of Sofia, Bulgaria |
| Kate Revoredo | Departamento de Informática Aplicada, Brazil |
| Dimitris Kanellopoulos | University of Patras, Greece |
| Samir Kumar Bandyopadhyay | University of Calcutta, India |
| Baljit Singh Khehra | BBSBEC, India |
| Nitish Pathak | BVICAM, India |
| Md Gapar Md Johar | Management Science University, Malaysia |
| Kathemreddy Ramesh Reddy | Vikrama Simhapuri University, India |
| Shubhnandan Singh Jamwa | University of Jammu, India |
| Surjeet Dalal | SRM University, Delhi-NCR, India |
| Manoj Patil | North Maharashtra University, India |
| Rahul Johari | GGSIPU, India |
| Adeyemi Ikuesan | University of Pretoria, South Africa |
| Pinaki Chakraborty | Netaji Subhas University of Technology, India |
| Subrata Nandi | National Institute of Technology Durgapur, India |
| Vinod Keshaorao Pachghare | College of Engineering, Pune, India |
| A. V. Senthil Kumar | Hindusthan College of Arts and Science, India |
| Khalid Raza | Jamia Milia Islamia, India |
| Parameshachari B. D. | TGSSS Institute of Engineering and Technology for Women, India |
| E. Grace Mary Kanaga | Karunya University, India |
| Subalalitha C. N. | SRM University Kanchipuram, India |
| Niketa Gandhi | Machine Intelligence Research Labs, USA |
| T. Sobha Rani | University of Hyderabad, India |
| Zunnun Narmawala | Nirma University, India |
| Aniruddha Chandra | National Institute of Technology Durgapur, India |
| Ashwani Kush | Kurukshetra University, India |
| Manoj Sahni | Pandit Deendayal Petroleum University, India |
| Promila Bahadur | Maharishi University of Management, USA |
| Gajendra Sharma | Kathmandu University, Nepal |
| Rabindra Bista | Kathmandu University, Nepal |
| Renuka Mohanraj | Maharishi University of Management, USA |
| Eduard Babulak | Institute of Technology and Business, Czech Republic |
| Zoran Bojkovic | University of Belgrade, Serbia |
| Pradeep Tomar | Gautam Buddha University, India |
| Arvind Selwal | University of Jammu, India |
| Yashwant Singh | University of Jammu, India |

# Contents

## Evolutionary Computing Through Machine Learning

# Next Generation Networking and Communication Systems

# Building Dominating Sets to Minimize Energy Wastage in Wireless Sensor Networks

Nilanshi Chauhan$^{(\boxtimes)}$ and Siddhartha Chauhan

Computer Science and Engineering Department, National Institute
of Technology Hamirpur, Hamirpur, Himachal Pradesh, India
{nilanshi,sid}@nith.ac.in

**Abstract.** Quality of wireless sensor networks (WSNs) can be very well judged with the help of two key metrics viz. coverage and connectivity. In the process of enhancing these two metrics, it is very crucial to ensure minimum energy wastage in energy critical WSNs. This study constructs disjoint dominating sets to activate them in a rotation for minimizing the energy wastage in the deployed network. The proposed methodology utilizes the unique properties of hexagon tessellations to generate the maximal number of dominating sets. Sensor spatial density of the activated dominating set has also been minimized to reach an optimized scenario. The simulation results attained in the present study prove the superiority of the proposed technique, over the existing works.

**Keywords:** Coverage · Connectivity · Energy · Dominating sets · Wireless sensor networks

## 1 Introduction

Recently, wireless sensor networks (WSNs) have become a dire need for many event critical applications such as health monitoring, habitat monitoring, battlefield monitoring, etc. Such a network is composed of a large number of very small low powered sensing devices, called sensor nodes. These nodes collect data from their surroundings, process it and route to a distant base station, called sink, for further processing. One of the fundamental issues in WSN is regarding the quality of coverage provided by the sensor nodes. Coverage can be categorized in three ways, namely, Point Coverage, Target Coverage and Area Coverage [1]. Connectivity is another important issue for WSNs. Usually, coverage is provided in such a way that every point in the target area, boundary and every target point covered by multiple nodes. Such coverage is referred to as k-coverage, where k is the degree of coverage provided to an area, boundary or point. Being energy critical, WSNs must ensure minimum energy wastage without compromising with the requirements of the application, it serves. In the present study, a randomly deployed WSN has been considered for area-based coverage. If all the deployed nodes are kept active at all times then such a situation will increase the chances of collisions amongst them. This will be followed by a number of retransmissions, thereby increasing the communication overhead. Both collisions and communication overhead are major sources of energy wastage. Hence instead of keeping all the nodes active, only a subset of total nodes is kept active at a time.

© Springer Nature Singapore Pte Ltd. 2019
A. B. Gani et al. (Eds.): ICICCT 2019, CCIS 1025, pp. 3–11, 2019.
https://doi.org/10.1007/978-981-15-1384-8_1

The subsets are also known as dominating sets and are kept active in rotation to save energy of the other nodes. This dominating set rotation technique is one of the most effective techniques to prolong the lifetime of a WSN. A solution to the k-coverage problem with minimum sensor spatial density has been presented in this paper. This research is based on a novel concept of choosing nodes from a smaller area to cover a larger area which yields an optimal value for the SSD. The larger area to be covered is defined over the smaller area in such a way that any node deployed in the smaller area shall completely cover it.

Hence the unique hexagon tessellation is utilized to minimize the sensor spatial density. Rest of this paper is structured as follows. A literature survey has been presented in Sect. 2. Network model with some important definitions has been described in Sect. 3. Coverage strategy has been illustrated in Sect. 4, along with the proposed algorithms. Simulation results and discussion are presented in Sect. 5. Finally conclusion is drawn in Sect. 6.

## 2    Literature Review

The focus of the literature review is to know about the various dominating sets generation techniques for enhancing the WSN lifetime. Each dominating set algorithm aims to generate as many dominating sets as possible, of minimal average size while maintaining the required degree of coverage and connectivity.

In [1] the problem of determining the number of sensor nodes to ensure four connectivity and full coverage with respect to different ratios of the transmission range of sensor nodes to their sensing range has been dealt with. A diamond pattern which is said to consist of six triangular patterns and can regenerate itself to a square grid pattern has been used. The key metric, coverage to ensure the quality of service provided by a wireless sensor network has been explored [2]. Two important aspects of WSNs viz., minimum node degree and k-connectivity have been given in [3]. Cover sets generated for efficient area monitoring have been activated in a round-robin fashion in [4].

Coverage problem has been formulated as a decision problem with a goal to cover the target area with at least k sensors. This decision problem has been termed as k-unit disc coverage (K-UC) problem and k-non unit disc coverage (K-NUC) problem. In both the cases, insufficiently covered areas have been detected using polynomial time algorithms [5]. The point based coverage to monitor a set of targets with known locations has been addressed [6]. The problem of optimal set generation while preserving the network connectivity and eliminating the redundant nodes has been addressed [7]. Distributed algorithms have been developed to determine and control the levels of coverage provided by the deployed nodes in the target area [8]. Reference [9] presents and compares various coverage and connectivity techniques. Such algorithms have been divided into three categories viz. exposure-based, node mobility based and integrated coverage and connectivity based algorithms. Two transition phases viz. sensing coverage phase transition (SCPT) which denotes the abrupt change of small fragmented areas to a large covered area, and network connectivity Phase Transition (NCPT) which denotes the change from a number of connected components to a fully connected WSN are defined in [10]. Coverage and connectivity have been considered together. Further, percolation theory of Boolean models has been used to estimate the coverage area in WSNs. In [11] dominating set problem has been abstracted as backbone rotation problem. A random geometric graph G(n, r) with 'p' as the minimum degree has been used to represent a WSN and p + 1 backbones were selected as independent dominating subsets of G. Graph coloring algorithms have been implemented to find suboptimal solutions.

The problem of k-coverage in WSNs has been modeled as the scheduling of cover-sets in each round of scheduling so that every point in the target area is k-covered and all the active set of nodes are connected [12]. The optimum sensor spatial density has been analyzed for one and two-dimensional WSNs which have been designed in such a way that the mean square error distortion between original and reconstructed spatial correlated signal is minimal [13]. A cell structure is reported to derive a constant factor approximation algorithm referred to as distributed nucleus algorithm (DNA), for the domatic partition problem [14]. Two distributed approaches for connected dominating sets (CDSs) generation in WSNs are given in Ref. [15].

Thus it is seen from the literature review that minimization of energy wastage is of paramount importance in wireless sensor networks. Hence the same has been dealt with in the present research work. The present paper uses the very unique concentric hexagon tessellations to partition the target area. k-coverage to the whole area is ensured by ensuring the coverage to each outer hexagon partition by choosing nodes from its concentric hexagonal region. For each of the concentric hexagons, the side has been chosen as radius/$\Theta$, where $\Theta$ is a positive integer. The value of $\Theta$ has been selected such that the sensor spatial density is optimal for every dominating set. Centralized approach based algorithm has been proposed to generate disjoint dominating sets.

## 3  Network Model

Network model has been introduced in this section along with the assumptions and definitions of various parameters used in this study. All sensor nodes deployed in a target field F are static and all of them are similar in terms of sensing, processing, and data routing capabilities. Hence these have the same sensing radius and the communication radius. Each node has its own unique identification (ID) number in the WSN and their locations and IDs are known to the sink node. Circular sensing disk model has been assumed. This states that the sensing region of a node comprises of a circular area of radius $r$, where $r$ is the sensing radius of that node. Similarly, a circular transmission disk model has been assumed which takes the transmission region in a circular shape of radius equal to the transmission radius of sensor nodes.

A point $p$ in the target field F is covered by a sensor node $s$ if and only if

$$d(s, p) \leq r \tag{1}$$

where function d calculates the Euclidean distance between two points. Similarly, a sensor node $s_1$ can communicate with another node $s_2$ if and only if

$$d(s_1, s_2) \leq R \tag{2}$$

where $R$ is the communication radius.

In case $s_1$ and $s_2$ are heterogeneous and both have different communication radii, $R_1$ and $R_2$ respectively then both of them can communicate with each other if and only if

$$d(s_1, s_2) \leq min(R_1, R_2) \tag{3}$$

where the function min finds a minimum of two values. The minimum value required for communication radius must not be too high because the energy consumed by the sender directly depends on the distance between the sending and the receiving nodes.

SSD of a region 'a' is defined as the ratio of total active nodes in 'a' to the area of 'a'. This study aims to minimize its value, provided total nodes active in target field F provide coverage to every point in F. Present research work requires k nodes to cover every partition.

In order to optimize the value of SSD, efforts have been made to cover as much area as possible with k sensor nodes and for that, it is very crucial to decide the region from where k nodes are chosen. Present work depends on the novel concept of choosing nodes from the as much small area as possible to cover as much large area as possible. Further, if the covered area could be converted to one of the three popular tessellations then minimum overlap area can also be minimized, which is illustrated in the next sections.

### 3.1    The Concentric Hexagonal Tessellations

Concentric hexagonal tessellations have been introduced in this section. The outer hexagon is constructed by condensation of the area of coverage provided by the sensing ranges of the nodes deployed inside the smaller hexagon concentric to it, as shown in Fig. 1. Outer hexagon is also used to partition the target field F and is covered by choosing k nodes from the inner hexagon. This way the idea of choosing nodes from a smaller area to cover a larger area has been implemented. In order to be able to provide k coverage to outer hexagon using the inner hexagon, at least k nodes must be available inside the inner hexagon. The present study considers the random deployment of nodes in the target area and, as the size of inner hexagon reduces the chances of nodes available in it decrease. The area $A_1$ as shown in Fig. 2, is the maximum area that the nodes in small hexagon $H_s$ can cover while $a_1$ denotes the area of $H_s$. $A_1$ is also called the area of coverage contribution (AOCC). It is very important to note that the coverage area i.e. $A_1$, contributed by nodes deployed in $H_1$, also include the area of $H_1$. Area of coverage contribution and $H_1$ share a special relation which can be described in two parts as follows:

  i. Any node inside $H_1$ will completely cover its AOCC and
 ii. Any node inside AOCC/$H_1$ will completely cover $H_1$.

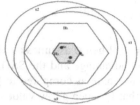
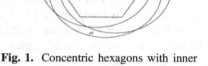

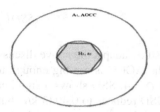

**Fig. 1.** Concentric hexagons with inner hexagon nodes providing coverage to outer hexagon.

**Fig. 2.** AOCC of inner hexagon tessellation.

The first part of the relationship is exploited in the present study for hexagonal tessellation. The second part has been exploited for the triangular tessellation in Ref. [14].

### 3.2 The Reuleaux Triangle

It is important to notice that the hexagon is actually made up of six triangular tessellations or six overlapping reuleaux triangles. A reuleaux triangle rt(m) is formed by the intersection of three circles of equal radii where 'm' denotes the width of the triangle, which is defined as the distance between two parallel lines which bound the triangle.

### 3.3 Sensor Spatial Density Analysis

This section presents the sensor special density (SSD) analysis for the hexagonal tessellation when the size of its side is varied in terms of the sensing radius r. If the size is taken greater than r, then the area of hexagon generated, can't be covered completely with the sensing range of a sensor node. Hence the maximum value that can be assigned to the side of the hexagon is r. The case wh7ich has been opted for this study is discussed as follows:

**Hexagonal Tessellation of Side s Equal to One-Fourth of the Sensing Radius**
In this case, side of the hexagonal tessellation is taken as r/$\theta$, where $\theta$ = 4. Area is calculated in terms of the sensing radius of the deployed nodes. Again $A_1$ denotes the area of each triangular tessellation, T and $A_2$ denotes the area of each arc.

$$Area\left(AOCC\left(\frac{r}{4}\right)\right) = 54A_1 + 6A_2 \tag{4}$$

No matter what the sample value for r is taken, total triangles and arcs constituting the AOCC for a particular value of $\theta$ for a particular tessellation shall be the same.

$$Area\left(AOCC\left(\frac{r}{4}\right)\right) = (1.7694)r^2 \tag{5}$$

$$SSD\left(\frac{r}{4}\right) = (0.5651)\frac{k}{r^2} \tag{6}$$

This study adapts the above discussed case for $\theta = 4$. When $\theta$ assumes a value of 1 and 2, the AOCC is not big enough to be taken as an advantage and the SSD is very high. At $\theta = 4$, SSD shows a significant decrease in its value as compared to [2]. For $\theta = 5$, SSD reduces to 0.4973 $k/r^2$ but it is analysed that for such a low value of SSD, the highest degree of coverage that could be achieved for a good number of deployed nodes as 1000, is just 2. This is because the size of the inner tessella becomes too small and the probability of finding k number of nodes for $k \geq 3$ becomes close to zero. Hence the best case considered for implementation of proposed cover-set/dominating set generating algorithm is for $\theta = 4$.

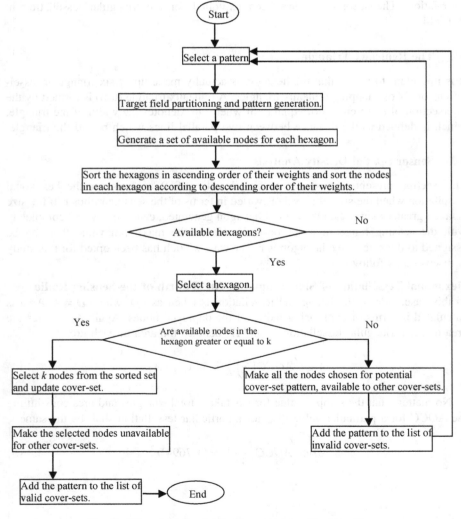

**Fig. 3.** Flow-chart for dominating set generation algorithm.

# 4   Coverage Strategy

A hexagonal tessellation of side equal to one-fourth of the sensing radius is considered to select nodes deployed on or inside its boundary. The area of coverage contribution is condensed to form a bigger concentric hexagon with side equal to three-fourths of sensing radius and three times the side of inner hexagonal tessellation. The area of coverage contribution itself consists of this outer hexagon and the arcs around its edges.

Figure 3 illustrates the flow-chart of the proposed algorithm for dominating sets generation. A pattern $p_i$ is considered a valid dominating set only if every hexagon $h_j \in H_i^{outer}$ in this pattern is k-covered. If at least k nodes are present in every $h_j \in H_i^{outer}$, then all the hexagons in $H_i^{outer}$ will be k covered. Such a pattern $p_i$ is added to the set of valid cover-sets 'DS'. Nodes are arranged in the descending order of their weights with respect to each hexagon $h_j$. Unused with higher value of residual energy, are given a fair chance to provide their services to the network.

# 5   Results and Discussion

This section presents the results generated for the dominating sets algorithms and the effect of a decrease in the value of SSD, on the number of dominating sets generated and time complexity of sorting the inner hexagons in ascending order of their weights. This section presents a comparative analysis of the results generated by the proposed algorithm with some of the works reported in literature. Different scenarios have been considered regarding the target field size, sensing radius, and transmission radius. The number of dominating sets generated and the average cardinality of each such dominating set, for a particular number of total deployed nodes in the target area have been compared with Refs. [2, 14, 15].

**Table 1.** Comparison of proposed algorithm with Ref [2] with respect to number of DS

| N | $|D|$, k = 1 [2] | $|D|$, k = 1 [Present] | $|D|$, k = 2 [2] | $|D|$, k = 2 [Present] | $|D|$, k = 3 [2] | $|D|$, k = 3 [Present] | $|D|$, k = 4 [2] | $|D|$, k = 4 [Present] |
|---|---|---|---|---|---|---|---|---|
| 1000 | 0 | 2 | 0 | 2 | 0 | 1 | 0 | 0 |
| 2000 | 6 | 15 | 3 | 12 | 2 | 3 | 1 | 3 |
| 3000 | 12 | 28 | 5 | 20 | 3 | 12 | 3 | 7 |
| 4000 | 15 | 45 | 9 | 25 | 5 | 16 | 4 | 12 |

Table 1 shows a comparison between the present work and the results presented in Ref. [2]. Number of dominating sets produced has been compared for different network sizes, varying from 1000 deployed nodes to 4000. $|D|$ represents the number of dominating sets (DS) produced. k = 1, 2, 3, 4 represents the degree of coverage. As it is evident from the comparison results that DS generating technique presented in [2] does not produce any DS of degree 1, 2, 3, 4 for a network size of thousand while the proposed approach successfully generates DS up to a degree of 3 for the same network size. A target field of $100 \times 100$ m$^2$ was considered for this case and a sensing range of 25 m assumed. Table 2 shows a comparison between the number of DS produced by the proposed approach and the one presented in Ref. [14] with respect to two network sizes of 400, and 800 for different transmission ranges ($T_{tx}$). Degree of coverage has been assumed 1 and a target field of size 50 × 50 m$^2$ has been taken for the purpose of simulations. It is evident from the table that the number of DS produced in present study is greater as compared to [14].

**Table 2.** Comparison of the performance of the proposed algorithm with Ref. [14] with respect to number of DS. |D| represents the number of DS produced.

| N | $R_{tx}$(m) | |D|, k = 1 [14] | |D|, k = 1 [Present] |
|---|---|---|---|
| 400 | 15 | 7 | 7 |
| 800 | 15 | 18 | 20 |
| 400 | 20 | 17 | 17 |
| 800 | 20 | 31 | 35 |
| 400 | 25 | 19 | 22 |
| 800 | 25 | 34 | 36 |

**Table 3.** Performance comparison of the proposed algorithm with Ref. [15] with respect to average size of DS

| N | Average size of each DS [15] | Average size of each DS [Present] |
|---|---|---|
| 500 | 27 | 15 |
| 600 | 28 | 15 |
| 700 | 27 | 15 |
| 800 | 27 | 15 |
| 900 | 26 | 16 |
| 1000 | 28 | 15 |

Table 3 shows the results obtained regarding the average DS size. These results have been compared with the ones presented in Ref. [15]. Target area of size 100 × 100 m$^2$ has been considered and a sensing range of 20 m was assumed. Degree of coverage has been chosen as 1. Average size produced in present study is less as compared to the average size of generated in Ref. [15]. Hence, the same degree of coverage can be provided with lesser number of nodes in the dominating set, with the proposed technique.

## 6   Conclusion

A randomly deployed wireless sensor network has been dealt with in the present research work. A solution to k-coverage problem has been presented and the active sensor spatial density has been optimized. Hexagon tessellation has been opted due to the properties of its area of coverage contribution which have been utilized to minimize the energy wastage in the sensor network. Furthermore, the hexagon tessellation minimizes the sensor spatial density by nearly 9% compared to triangular one. Results have been compared with the other research works reported in literature, regarding the dominating sets generation. In future, distributed algorithms for dominating sets generation and sleep/wake-up scheduling shall be explored.

# References

1. Bai, X., Yun, Z., Xuan, D., Lai, T.H., Jia, W.: Optimal patterns for four connectivity and full coverage in wireless sensor networks. IEEE Trans. Mob. Comput. **9**(3), 435–448 (2010)
2. Yu, J., Wan, S., Cheng, X., Yu, D.: Coverage contribution area-based k-coverage for wireless sensor networks. IEEE Trans. Veh. Technol. **66**(9), 8510–8523 (2017)
3. Bettstetter, C.: On minimum node degree and connectivity of a wireless multi-hop network. In: Proceeding of 3rd International Symposium on Mobile Ad Hoc Networking and Computing, Lausanne, Switzerland, vol. 3, pp. 80–91. ACM (2002)
4. Abrams, Z., Goel, A., Plotkin, S.: Set k-cover algorithms for efficient monitoring in wireless sensor networks. In: 3rd International Symposium on Information Processing in Sensor Networks, Berkeley, USA, vol. 3, pp. 424–432, IEEE (2004)
5. Huang, C., Tseng, Yu.: The coverage problem in a wireless sensor network. Mob. Netw. Appl. **10**(4), 519–528 (2005)
6. Cardei, M., Du, D.: Improving wireless sensor networks lifetime through power aware organization. Wirel. Sens. Netw. Lifetime **11**, 333–340 (2005)
7. Tezcan, N., Wang, W.: Effective coverage and connectivity preserving in wireless sensor networks. In: Proceedings of IEEE Wireless Communications and Networking, Kowloon, China, pp. 3388–3393. IEEE (2007)
8. Huang, C., Tseng, Y., Lu-Wu, H.: Distributed protocols for ensuring both coverage and connectivity of a wireless sensor network. ACM Trans. Sens. Netw. **3**(1), 1–22 (2007)
9. Ghosh, A., Das, S.K.: Coverage and connectivity issues in wireless sensor networks: a survey. Pervasive Mob. Comput. **4**, 303–334 (2008)
10. Ammari, H.M., Das, S.K.: Integrated coverage and connectivity in wireless sensor networks: a two dimensional percolation problem. IEEE Trans. Comput. **57**(10), 1423–1434 (2008)
11. Mahjoub, D., Matula, D.W.: Employing (1-$\epsilon$) dominating set partitions as backbones in wireless sensor networks. In: 12th Workshop of Algorithm Engineering and Experiments, Austin, Texas, pp. 98–111. SIAM Publications (2011)
12. Ammari, H.M., Das, S.K.: Centralized and clustered k-coverage protocols for wireless sensor networks. IEEE Trans. Comput. **61**(1), 118–133 (2012)
13. Wu, J., Sun, N.: Optimum sensor spatial density in distortion-tolerant wireless sensor networks. IEEE Trans. Wirel. Commun. **11**(6), 2056–2064 (2012)
14. Yu, J., Zhang, Q., Yu, D., Chen, C., Wang, G.: Domatic partition in homogeneous wireless sensor networks. J. Netw. Comput. Appl. **37**, 186–193 (2014)
15. Al-Nabham, N., Al-Rhodhan, M., Al-Dhelan, A.: Distributed energy-efficient approaches for connected dominating set construction in wireless sensor networks. Int. J. Distrib. Sens. Netw. **10**(6), 1–11 (2014)

# Design and Implementation of Low Noise Amplifier in Neural Signal Analysis

Malti Bansal$^{(\boxtimes)}$ (iD) and Diksha Singh

Department of Electronics and Communication Engineering,
Delhi Technological University (DTU), Delhi 110042, India
maltibansal@gmail.com

**Abstract.** LNA is an important component of transceivers and is widely used in neural signal analysis. In this paper, we review the different topologies and configurations used for LNA in neural applications. We compare the different topologies and conclude which one is the best topology among the ones studied on basis of certain parameters that govern the performance of a LNA for neural applications. According to our analysis, CMOS bipotential amplifier is the most appropriate neural amplifier in terms of all design parameters taken into consideration for use of LNA in neural applications.

**Keywords:** LNA · SNR · Neural · Feedback amplifier · Differential · Impedance

## 1 Introduction

Nowadays, diseases are being diagnosed on the basis of analysis of the signals produced as a result of chemical reactions in our body. Brain activities are also studied by analyzing the neural signals produced in our brain. These neural signals produced are very weak signals and noise gets added as they are passed through various components which are used in their analysis. Noise degrades the signal to noise ratio of these neural signals and as a consequence of addition of noise, it becomes difficult to study brain activity pertaining to these signals. To analyze these signals in an appropriate manner, they need to be amplified up to a certain threshold level without much noise being added to them. Low Noise Amplifiers are best suited to analyze these weak signals and achieve this function. LNA amplifies the signal received at the input; and very less noise is added due to its components to the received signal. Hence, the SNR of the signal is maintained at an appropriate level which is suitable for analysis of neural signals. LNA can be utilized for neural signal analysis with different design techniques to improve the performance of LNA. In this paper, we review the different topologies and configurations used for LNA in neural applications. We compare the different topologies and conclude which one is the best topology among the ones studied on basis of certain parameters that govern the performance of a LNA for neural applications. Section I of the paper is an introduction to the topic, section II briefly reports

© Springer Nature Singapore Pte Ltd. 2019
A. B. Gani et al. (Eds.): ICICCT 2019, CCIS 1025, pp. 12–24, 2019.
https://doi.org/10.1007/978-981-15-1384-8_2

about an LNA, section III reviews the use of LNA for neural applications, section IV compares the different topologies of LNA utilized for neural applications and finally section V presents conclusion and future scope of this area.

## 2 Low Noise Amplifier

Amplitude of signals decreases as they pass through environment and additional noise also gets added to them. Amplifiers amplify the signals received at the input but they also amplify the noise along with the signal. As an amplifier is an active device, its components also add additional noise to the signal and the signal to noise ratio (SNR) of the signal is decreased. Amplifier should be chosen such that it adds very little noise to the signal which passes through it. Low Noise Amplifier serves this purpose and it is designed such that, noise added due to its configuration is very low. LNA is placed at the initial stage of the receiver or the transceiver. Placing the LNA at the initial stage of the receiver reduces the cascaded effect of noise due to other components of the system which may include mixer, oscillator etc. Noise performance of the LNA governs the performance of the complete receiver system. LNA should be designed such that it has a very high gain and improved noise efficiency factor (NEF). LNA need to be connected to other components of the receiver system, so input and output impedance should be high enough to prevent loading effect on other components connected. Loading effect tend to decrease the gain of the amplifier as effective load resistance of the amplifier decreases as a consequence of low resistance of the adjacent components. A proper matching network needs to be used for complete transmission of signals without loss [1]. However, many other parameters also govern the performance of a LNA. For example, power consumption for an LNA should be low, CMRR should be high, chip area being utilized should be as minimum as possible and various other design parameters should also be considered. Imai et al. [2] reported various steps involved in the design of LNA which includes biasing, proper transistor choice, frequency stabilization, etc. It stated that the very first step involves the proper selection of the transistor to be used. We should check if the components of the amplifier are stable at the frequency of operation, if not then techniques should be used to make the operation stable at the required frequency. Proper biasing needs to be done to optimize the parameters like noise, gain, power dissipation etc.

## 3 Use of LNA for Neural Applications

Detailed examination of neural signals produced in our brain helps us to understand the brain activities. Detailed analysis also provides a better understanding of disorders related to brain i.e. Parkinson's disease, Alzheimer's disease and various others [3, 4]. Electrodes are used for interfacing these with the tissues present in the brain [5]. Multiple sites can also be observed at the same point of tine i.e., activities in different parts of brain can be studied by using multichannel neural amplifiers. Multichannel LNA are capable of observing multiple neuron activities at the same instance but the area occupied by these amplifiers is increased because multiple amplifiers are to be fabricated on a single chip. However, many multi-channel amplifiers reported so far were also found to use a very small chip area.

Neural signals lie in the range of few millivolts and are therefore considered as very weak signals and they may get degraded if more noise gets added to them. LNA is suitable for analysis of these signals because the noise added by this type of amplifier is very less and does not degrade the SNR of the signal. Power consumption is a very crucial factor in case of neural amplifiers as high power may result in increase of temperature and this may damage the brain tissues. An important requirement for the use of LNA for neural applications is that they should be designed in such a way that power consumed by them for operation is very low. Neurons may generate a rhythmic pattern of action potentials or spikes and if they generate action potentials in sync, this gives rise to local field potentials. Neural signal can also be considered as superimposition of action potentials or spikes on low frequency local field potentials. Frequency range for these signals is different. Action potentials lie in higher frequency range (300–6000 Hz) in comparison to local field potentials (1–200 Hz) [6]. LNA's are specially designed to study these signals separately in their respective frequency band. Some configurations focus on action potential analysis and some on local field potential analysis; because of the different frequency range. However some amplifiers used filters which filtered out the desired frequencies to handle Action potentials as well as local field potential simultaneously. In any circumstances, the gain of the LNA should be high because the signals associated are always weak signals. Noise that affects the LNA in neural signal analysis can be thermal noise, flicker noise, background noise, etc. The frequency of the signals is very low hence the flicker noise is more as compared to other types of noise in case of these signals. Amplitude of neural signals is weak and frequency is also not much large so these signals can easily be sensed from skin surface [7]. Electrodes acquire the required signals generated from the neurons. The position of the electron from the neuron also governs the background noise added to the signal [8]. We analyze the different neural amplifiers developed so far and study their performance with respect to different parameters such as noise, gain, power consumption, chip area, etc.

A basic LNA for neural applications always uses an operational transconductance amplifier (OTA) along with additional circuitry. Additional circuitry with an OTA may be a feedback circuit or it may differ in type of load connected. Current mirrors may be used as load in some configurations. Cascade configuration is also used in some LNA's which may further be modified as folded cascode to improve the performance of the LNA. AC coupling is always provided to the neural amplifier to block DC voltages developed. This can be achieved by using a capacitor in the circuit which requires a larger chip area. Telescopic OTA were used which provided better performance than other OTA structures available. These were observed to provide low noise and power consumption in comparison to other architectures [8–10]. Telescopic OTA can only be used for signals which have low swing amplitude. However, these were used in neural signal analysis with some modification because neural signals have weak amplitudes. Bulk Driven method was used which was observed to improve the output swing [11].

(i) Feedback Configuration: Using feedback in amplifier significantly improves the noise figure (NF) of the amplifier. Feedback can be provided by using a capacitor in general. However, pseudo resistors using MOS along with the capacitor were also used in some feedback circuits. Capacitive feedback is most widely used in case of LNAs. We discuss the different types of capacitive feedback topologies employed for LNA in neural applications. Kim et al. [12] reported a low power, low noise neural LNA for implantable biomedical devices. This configuration used an OTA block which was provided with capacitive-feedback. The feedback provided using a capacitor helped to obtain an accurate gain. Use of capacitive feedback also improved the gain and in this configuration, it was found to depend on the ratio of the capacitances $C_{IN}$ and $C_F$ used in the circuit of this amplifier. Gain obtained in this topology was around 46 dB which was obtained on the basis of capacitance values $C_{IN}$ and $C_F$ used in this configuration. In general, the frequency range of the amplifier should be sufficient enough to include all types of signals produced by the neuron activities. In this topology, the higher cutoff frequency was obtained by using the value of the capacitor placed at the input terminal of OTA. The lower cutoff frequency was obtained by expression: $G_M/(A_M.C_L)$ where $G_M$ is the transconductance of the OTA used. AC coupling was also used to remove the large offset DC voltage which was develops due to chemical reactions at the neuron interface. AC coupling can be achieved by placing a capacitor at the input pin of the OTA block. A capacitor acts as short circuit for the AC signals and it acts as an open circuit for DC signals and blocks them. The configuration was as shown in the Fig. 1. Positive terminal of the OTA was grounded and the input was applied at the negative terminal. A capacitor $C_{IN}$ was also placed at the input negative terminal to provide AC coupling because the capacitor blocks the DC signals. $C_F$ is the feedback capacitor and it provides a negative feedback to the amplifier. A capacitor is also connected to the non-inverting terminal of the OTA along with additional circuit. The architecture of the OTA used in the amplifier is held responsible for the performance of the LNA. This neural amplifier used a two stage architecture for the OTA and it helped to obtain a large gain and a large output swing. The first stage of the OTA was designed to obtained better noise performance mainly focused on flicker noise. Current mirrors used as load were implemented using source degeneration resistors which provided a better noise performance. Miller capacitor was also used to improve the bandwidth of the OTA. This configuration was found to consume very low power; and other parameters were also found to be in sufficient range to analyze the neural signals. This amplifier was designed on a 0.18 μm CMOS process and occupied a chip area of 0.136 mm$^2$. A NEF of 2.6 was obtained which was found to be better than pervious works done in the similar configuration. Dwivedi et al. [13] reported a neural amplifier for recording local field potentials and consumed low power. Local field potentials are related to the spikes formed during neuron interaction. This amplifier was specifically designed for analyzing the local field potential developed due to neuron interactions. This amplifier reported also used an OTA along with a capacitive feedback (Fig. 2). The architecture uses a capacitor which is connected at the inverting terminal of the OTA and a reference voltage source connected to the positive terminal of the OTA. The feedback circuit uses a capacitor and MOS pseudo resistor elements $M_{a,b,c}$ which are biased such that they exhibit a high resistive value. The neural signals are connected to the input terminal via a capacitor which blocks unwanted DC signals rising due to the electrode and tissue interaction. The use of pseudo resistors decreases the chip area because elements with large values of resistance can be implemented in a small chip area by using MOS.

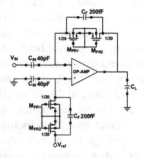

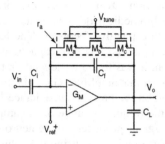

**Fig. 1.** Basic diagram of a neural amplifier (Reproduced from [12])

**Fig. 2.** Schematic of neural amplifier (Reproduced from [13])

Gain of the amplifier can be calculated by the ratio of the capacitance values used; one at the input terminal of OTA (Fig. 3) and other as feedback element. Using a capacitor at the input also helps to remove DC signals and filter out the irrelevant signals. The high cutoff frequency was determined by the value of the capacitor connected in the feedback circuit and the equivalent resistance of the MOS elements which were used in the feedback circuit. The lower cutoff frequency was obtained by the capacitor $C_L$ connected at the load terminal. The capacitor values should be chosen such that the cutoff frequencies are able to cover up the frequency range of the signals produced by neurons. Noise analysis for this configuration can be done by using the circuit for OTA being used in this neural amplifier. Only the transistors in the architecture which lie along the signal path added noise to the signal. Others do not add noise because of common mode cancellation. This configuration used 0.18 µm CMOS technology at 27 °C. This amplifier occupied a chip area of about 0.10 mm$^2$ which was found to be very much area efficient. This amplifier used a very low power for its operation. However noise efficiency was not improved significantly in this configuration.

(ii) Cascode Configuration: Cascode configuration is obtained by connecting a common emitter stage in cascade with a common base stage in case of bipolar junction transistors. This configuration may improve the bandwidth of the amplifier, input and output impedance or the gain of the amplifier. Telescopic and folded cascade amplifiers were used where low noise, large DC gain, high unity gain frequency were required as the important characteristics [14]. Telescopic architecture provided better power efficiency and folded cascade was seen to provide higher output swing. Further changes were done in these architectures to improve the various performance parameters. AC path and DC path were separated and the transistors were split to ensure proper matching [15]. Many other modifications were done but those cascode configurations were not intended to be utilized in neural applications. Cerida et al. [16] reported a low-noise amplifier which was based on fully differential recycling cascode architecture. This amplifier was based on recycling architecture [17] (Fig. 4) which provided a better gain and bandwidth efficiency.

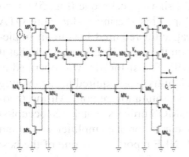

**Fig. 3.** Architecture of OTA used (Reproduced from [13])

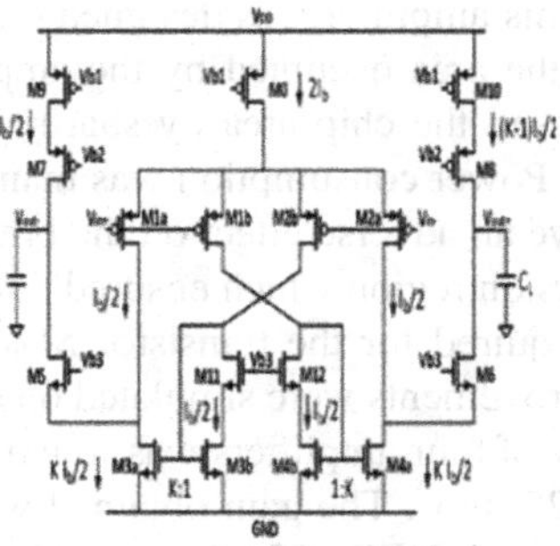

**Fig. 4.** Fully differential recycling folded cascode amplifier (Reproduced from [17])

The architecture of this amplifier consisted of two current mirrors M3a–M3b and M4a–M4b in which currents were reused; which was not implemented in any of the earlier architectures. These current mirrors improved the gain of this amplifier and bandwidth was also improved. The architecture was observed to have symmetry on both the sides of the amplifier and hence all the parameters were observed to be the same. The symmetry was a result of fully differential architecture of the amplifier. The gain of the amplifier was found to be dependent on the transconductance and output impedance. Parameters for this architecture also depend on the value of K which was termed as the current gain in the recycling path. Depending on the overdrive voltage, transistors operate in weak, moderate and strong inversion. In this architecture, M3a–M3b, M4a–M4b, M9, M10 were required to operate in strong inversion and a large $V_{ov}$ was maintained to do so. Operating these transistors in strong inversion reduces the thermal noise. Through analysis, they found that value of K should be as small as possible but greater than one. For reducing flicker noise, it was found that value of K should be smaller than one. So K was chosen to be greater and close to 1. The differential pair transistors were operated in weak inversion by keeping the $V_{OV}$ small. These were operated in weak inversion for increasing the transconductance which resulted in increase in gain. The current mirrors used in the recycling path which was realized using the transistors M3a–M3b; M4a–M4b may add a pole-zero pair near to the frequencies relevant. Hence, they may tend to generate stability problems. The pole-zero pair should have been such that it is far from frequencies relevant. This architecture was simulated on a 0.35 µm CMOS process and offered a gain of around 42 dB. Power consumption was found to be 66 µW and noise efficiency factor of 2.58 was obtained. This work was compared with the previous works and the NEF was improved; however the power consumption was higher than others.

Majidzadeh et al. [18] reported an amplifier intended for recording activities of multiple neurons at a single point of time. Brain activities can only be studied by analyzing signals from multiple neurons so amplifier designed should be efficient when implemented in multichannel mode. Therefore chip area occupied by an amplifier becomes an important factor in implementing multichannel amplifier for neural signal analysis. This amplifier was designed on partial OTA sharing technique (Fig. 5) which decreased the area occupied by the amplifier on chip. This sharing architecture OTA also decreased the chip area by sharing the bulk capacitor required for circuit implementation. Power consumption was maintained within the prescribed limits such that it did not have an adverse effect on the brain tissues [19]. Transistor $M_{c0}$ was operated in weak inversion region which ensured that no extra circuitry was required for generating voltages required for the transistor $M_{d0}$ to operate in saturation region. All the theoretical improvements were simulated on a 0.18 µm CMOS process. Total area occupied by an array of four amplifiers was found and effective area for one amplifier was found to be 0.0625 mm². The gain obtained was 39.4 dB. NEF reported was one of the best values obtained. NEF of 3.35 was achieved for an array of four amplifiers. However, a better NEF could be achieved by reducing the flicker noise contribution by size of the devices. Since it was a multichannel amplifier, crosstalk between adjacent channels was also taken into consideration and efforts were made to reduce it for effective observations. CMRR was measured to be 70.1 dB which was within the desired limits. PSRR of 63.8 dB was obtained for this amplifier.

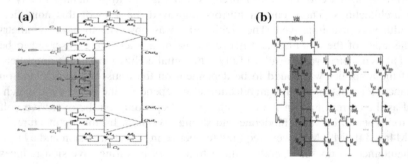

**Fig. 5.** (a): Partial sharing OTA architecture (Reproduced from [18]). (b): Circuit of partial sharing OTA (Reproduced [18])

Lopez et al. [20] reported an amplifier which was multi-channel and consisted of 16 channels and each channel had a low noise fully differential amplifier (Fig. 6). Each channel had a AC coupling and was using fourth order filter which helped to attenuate the out of band frequency noise and also large DC input offsets were rejected. This amplifier was designed to operate in two modes: one mode was to operate all channels at the same point of time to observe activities of multiple neurons. Other mode was to operate a single channel at a time using a multiplexer. In single channel mode, the channels that were not in use were switched off to save power. Noise analysis of this amplifier considered all the sources in the circuit that could contribute to the noise of the amplifier. Electrodes used were also a source of noise and the electronic circuitry also adds noise to the signal produced by the neurons. Thermal and flicker noise was kept lower than electrode noise, this helped to reduce contribution of overall noise. Design of the amplifier involved various parameters like power consumption, chip area, etc. In this architecture, a single channel was using a fully differential folded cascade OTA. Gain of the amplifier was calculated by the ratio of value of the capacitor placed in the feedback circuit and the value of capacitor placed at the input terminal. The differential pair in this amplifier used PMOS, which were operated in weak inversion. Weak inversion region operation of the transistors ensured that the transconductance was high and hence, it reduced the effect of noise [21]. M4–M5 and M10–M11 were operated in strong inversion to reduce the effect of noise. Flicker noise was also minimized by increasing the size of the transistors. Pseudo resistors were effective in decreasing the chip area because they can realize high resistance values [21].

Ghaderi et al. [22] reported LNA which used a bulk driven cascade current mirror as load. This used a new telescopic OTA which was found to have a better output swing than other telescopic OTA. Output swing could be increased by reducing the voltage drop on the load. Bulk driven cascade current mirror was used along with telescopic OTA to increase the output swing in this amplifier reported. Gate driven cascade current mirror was also used in some architectures but it had certain limitations which restricted its use to only single stage telescopic amplifiers [23]. BD cascade current mirror (BDCCM) (Fig. 7) was reported to decrease the threshold voltage of the transistors used, as a result of which output swing was found to increase. Architectures with BDCCM as a load had a better noise efficiency and gain as compared to architectures without BDCCM as a load. Noise aspects of the amplifier depended on the transistors used and their transconductance values. The input noise was found to be given as:

$$V_{n,in,eff} = \int_{fl}^{fH} (\frac{C_1 + C_2 + C_{in}}{C_1})^2 V_{n,in,OTA.df}^2 \tag{1}$$

Where $C_1, C_2, C_{in}$ were the capacitors used in the circuit of the amplifier. These parts of noise were minimized to increase the noise efficiency of the amplifier. Flicker noise was reduced by using PMOS in the circuit and the thermal noise was minimized by adjusting transconductance values of the transistors used in OTA. This amplifier had a gain equal to 38.9 dB, power consumption of 6.9 µw, NEF of 2.2 and output swing of 0.7 V. This was capable of amplifying all types of neural signals produced in body.

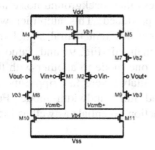

**Fig. 6.** Fully differential folded cascade input amplifier (Reproduced from [20])

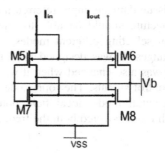

**Fig. 7.** BDCCM (Reproduced from [22])

(iii) Bipotential Configuration: Yang et al. [24] reported an array of neural amplifiers which were found to be suitable for large scale integration. Array of neural amplifiers is capable of analyzing activities of multiple neurons simultaneously and it provides a better understanding of the associated brain disorder if any. This particular configuration used two stage amplifier (Fig. 8) in combination with Current Reuse Complimentary Input (CRCI) technique [25] to achieve the desired performance of the neural amplifier. The first stage used an amplifier which was found to use a capacitive feedback and CRCI technique. This stage achieved good noise efficiency but the Power Supply Rejection Ratio was found to be below desired value. Hence, a second stage was used to improve the value of the PSRR. The second stage used fully differential OTA along with capacitive feedback. A reference amplifier was also used which was common to all the channels and the architecture of this reference amplifier was responsible for significant reduction in power dissipation. Noise was equally coupled to the first stage and the reference amplifier was suppressed by the second stage as a common mode signal and it improved the noise performance of this LNA. Similar to other configurations, large DC offsets should be removed which was done with the help of MOS-bipolar pseudo resistors with high resistance; and on-chip capacitors were also employed. The first stage amplifier and the reference amplifier were matched to each other in terms of gain which helped to achieve a good PSRR and CMRR. Mismatch between the first stage and the reference amplifier gave rise to a poor value of PSRR. Capacitive feedback was used in the first stage and the reference amplifier, which helped to achieve a good gain, accuracy and linearity. Since it was a multichannel amplifier, crosstalk was a major factor which affected the performance and this was minimized using a careful layout. This configuration was found to achieve a NEF of 1.93 which is convincing and a PSRR of 50 dB which is sufficient for typical use. Each channel occupied a chip area of 0.137 mm$^2$. NEF of this configuration could be further improved by making certain changes to the reference amplifier, which could be shared between more number of channels.

(iv) Reconfigurable configuration: Valtierra et. al [26] reported a 4 mode reconfigurable low noise amplifier. Two types of signals are produced in the brain: action potentials and local field potentials. These signals have different amplitudes and frequency range, so this amplifier was capable of analyzing both the signals produced in the brain. Bandwidth could be selected to analyze these signals separately or at the same point of time. Analog Front End amplifier should have low power consumption to prevent tissue damage, input referred noise should be less than background noise and the architecture should be able to reject DC offsets produced. This amplifier was capable of selecting different modes, which could select different neural signal frequency ranges. It was designed such that its circuit changed (Fig. 9) and value of capacitor was also changed when it was switched from one mode to another with the help of control signals. The four modes which were available were fast ripple mode, action potential mode, local field potential mode and full bandwidth mode. Full bandwidth mode covered both the frequency bands of action potentials as well as local field potentials. Local field potential mode covered the frequency range of action potentials. Local field potential mode covered frequency range of local field potentials. Fast ripple mode covered the high pass pole of action potential range and low pass pole of local field potential range. The differential pairs used were operated in weak inversion region. Noise which was considerable in this architecture was produced by differential pair of the OTA. This was considered to consume the lowest power among the similar reported circuits. If small channel length transistors are used, power consumption can be further reduced. Gain obtained was different for the four modes, which were available in this architecture.

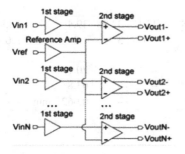

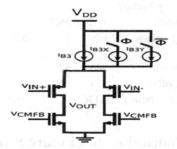

**Fig. 8.** Multichannel bipotential amplifier (Reproduced from [24])

**Fig. 9.** Simple programmable OTA (Reproduced from [26])

## 4  Comparison of Architectures

After analysis of different architecture discussed so far, we compare the architectures with the help of different parameters such as NEF, power, gain, bandwidth, etc. Out of the different architectures studied, we try to find out the best architecture for use in neural applications. As discussed so far, neural signals are weak signals and hence, the gain of the architecture chosen should be high enough to increase the amplitude to a threshold level which makes these signals suitable for analysis. However, the best architecture could be different depending on the real time requirements. Some may require a better noise efficiency, some may require a particular bandwidth depending on the signals to be analyzed (action potentials and local filed potentials), etc. We have considered 10 parameters to analyze the performance of the amplifiers and represented the data in tabular form (Table 1). In this discussion, NEF and gain are considered as dominant aspects. Hence, best architecture has been primarily finalized on the basis of these parameter values.

**Table 1.** Comparison of LNA architectures for neural amplifiers

| Parameter | [12] | [24] | [13] | [26 (a)] | [26 (b)] | [16] | [20] | [18] | [22] |
|---|---|---|---|---|---|---|---|---|---|
| CMOS Technology (μm) | 0.18 | 0.09 | 0.18 | 0.18 | 0.18 | 0.35 | 0.35 | 0.18 | 0.18 |
| Bandwidth (Hz) | 13800 | 4900–10500 | 3–164 | 0.129–590 | 224–6800 | 6023 | 6000 | 7200 | 89–7000 |
| Input-referred Noise (μVrms) | 5 | 3.04 | 5.62 | 4.2 | 4.1 | 1.16 | 1.9 | 3.5 | 2.45 |
| Power (μW) | – | – | – | 0.454 | 0.454 | 66.03 | 66 | 7.92 | 6.9 |
| NEF (Noise Efficiency Factor) | 2.6 | 1.93 | 6.33 | 4.5 | 1.31 | 2.58 | 1.9 | 3.5 | 2.2 |
| Gain (dB) | 46.2 | 58.7 | 31.7 | 48 | 48 | 42.1 | 33.98 | 39.4 | 38.9 |
| Supply voltage (V) | 1.2 | 1 | 0.8 | 1 | 1 | 3.3 | 3.3 | 1.8 | ±0.9 |
| CMRR (dB) | >66 | >45 | – | – | – | – | – | 70.1 | – |
| PSRR (dB) | >74 | >50 | 54 | – | – | – | – | 63.8 | – |
| Chip area (mm$^2$) | 0.136 | – | 0.098 | – | – | – | – | 0.0625 mm$^2$ | – |

## 5    Conclusion and Future Scope

All the topologies studied so far emphasized on certain fixed parameters which were to be optimized. Future research in this particular area will also focus only on those parameters. Further improvements can be done in these topologies to improve their performance and increase their efficiency in real time application for signal recording and analysis. The discussion can be endless as there are many different architectures that have been suggested so far for implementing LNA in neural signal analysis. LNA in neural applications can be extended by making changes in the existing architectures and simulating them to see their performance variations. LNA should mainly have a very low noise efficiency factor. As the name suggests, LNA should have low noise; and also it should have a considerable gain, sufficient for neural signal analysis. As per the above discussion, Bipotential amplifier proves to be the best LNA for use in neural applications. It was reported to have a noise efficiency factor of 1.9 and a gain of 58.7 which are the best figures as observed by us.

## References

1. Bhalani, H.V., Prabhakar, N.M.: Rudimentary study and design process of low noise amplifier at Ka band. IJ Publ. 3(2), 1181–1183 (2015)
2. Imai, Y., Tokumitsu, M., Minakawa, A.: Design and performance of low-current GaAs MMIC's for L-band front-end applications. IEEE Trans. Microw. Theory Tech. 39(2), 209–215 (1991)
3. Mussa-Ivaldi, F.A., Miller, L.E.: Brain-machine interfaces: computational demands and clinical needs meet basic neuroscience. Trends Neurosci. 26(6), 329–334 (2003)

4. Wise, K.D.: Silicon microsystems for neuroscience and neural prostheses. IEEE Eng. Med. Biol. Mag. **24**(5), 22–29 (2005)
5. Butson, C.R., McIntyre, C.C.: Role of electrode design on the volume of tissue activated during deep brain stimulation. J. Neural Eng. **3**(1), 1–8 (2006)
6. Logothetis, N.K.: The neural basis of the blood-oxygen-level-dependent functional magnetic resonance imaging signal. Philos. Trans. Roy. Soc. Lond. B Biol. Sci. **357**(1424), 1003–1037 (2002)
7. Gosselin, B.: Recent advances in neural recording microsystems. Sensors **11**, 4572–4597 (2011)
8. Chaturvedi, V., Amrutur, B.: An area efficient noise-adaptive neural amplifier in 130 nm CMOS technology. IEEE J. Emerg. Sel. Top. Circuits Syst. **1**, 536–545 (2011)
9. Ng, K.A., Xu, Y.P.: A compact, low input capacitance neural recording amplifier. IEEE Trans. Biomed. Circuits Syst. **7**, 610–620 (2013)
10. Saberhosseini, S.S.: A micro-power low-noise amplifier for neural recording microsystems. In: ICEE 2012 - 20th Iranian Conference on Electrical Engineering, pp. 314–317 (2012)
11. Blalock, B.J., Allen, P.E., Rincon-Mora, G.A.: Designing 1-V Op amps using standard digital CMOS technology. IEEE Trans. Circuits Syst. II Analog Digit. Signal Process. **45**, 769–780 (1998)
12. Kim, H.S., Cha, H.-K.: A low power, low-noise neural recording amplifier for implantable devices. In: Proceedings of International SoC Design Conference (ISOCC), pp. 275–276 (2016)
13. Dwivedi, S., Gogoi, A.K.: Local field potential measurement with low-power area-efficient neural recording amplifier. In: Proceedings of IEEE International Conference on Signal Processing, Informatics, Communication and Energy Systems (SPICES), pp. 1–5 (2015)
14. Ahmed, M., Shah, I., Tang, F., Bermak, A.: An improved recycling folded cascode amplifier with gain boosting and phase margin enhancement. In: Proceedings of IEEE International Symposium on Circuits and Systems (ISCAS), pp. 2473–2476 (2015)
15. Li, Y.L., Han, K.F., Tan, X., Yan, N., Min, H.: Transconductance enhancement method for operational transconductance amplifiers. Electron. Lett. **46**(19), 1321–1323 (2010)
16. Cerida, S., Raygada, E., Silva, C., Monge, M.: Low noise differential recycling folded cascade neural amplifier. In: Proceedings of IEEE 6th Latin America Symposium on Circuits and Systems (LASCAS), pp. 1–4 (2015)
17. Assaad, R.S., Silva-Martinez, J.: The recycling folded cascode: a general enhancement of the folded cascode amplifier. IEEE J. Solid State Circuits **44**(9), 2535–2542 (2009)
18. Majidzadeh, V., Schmid, A., Leblebici, Y.: Energy efficient low-noise neural recording amplifier with enhanced noise efficiency factor. IEEE Trans. Biomed. Circuits Syst. **5**(3), 262–271 (2011)
19. IEEE Standard for Safety Levels With Respect to Human Exposure to Radio Frequency Electromagnetic Fields, 3 kHz to 300 GHz. IEEE Std. C95.1-2005 (2006)
20. Lopez, C.M., et al.: A multichannel integrated circuit for electrical recording of neural activity with independent channel programmability. IEEE Trans. Biomed. Circuits Syst. **6**(2), 101–110 (2012)
21. Harrison, R.R., Charles, C.: A low-power low-noise CMOS amplifier for neural record ing applications. IEEE J. Solid-State Circuits **38**(6), 958–965 (2003)
22. Ghaderi, N., Kazemi-Ghahfarokhi, S.-M.: A low noise neural amplifer using bulk driven cascode current mirror load. In: Proceedings of 9th International Conference on Electrical and Electronics Engineering (ELECO), pp. 76–80 (2015)
23. Razavi, B.: Design of Analog CMOS Integrated Circuits. McGraw-Hill, New York (2001)
24. Yang, T., Hollemann, J.: An ultralow-power low noise CMOS bipotential amplifier for neural recording. IEEE Trans. Circuits Syst. II Express Briefs **62**(10), 927–931 (2015)

25. Holleman, J., Otis, B.: A sub-microwatt low-noise amplifier for neural recording. In: Proceedings of 29th Annual International Conference on IEEE Engineering in Medicine and Biology Society, pp. 3930–3933 (2007)
26. Valtierra, J.L., Rodríguez-Vázquez, Á., Delgado-Restituto, M.: 4 mode reconfigurable low noise amplifier for implantable neural recording channels. In: 12th Conference on PhD Research in Microelectronics and Electronics (PRIME), pp. 1–4 (2016)

# Apache Hadoop Based Distributed Denial of Service Detection Framework

Nilesh Vishwasrao Patil[1]([⊠]) [ID], C. Rama Krishna[1] [ID],
and Krishan Kumar[2] [ID]

[1] National Institute of Technical Teachers Training & Research,
Chandigarh, India
patil.nilesh38@gmail.com
[2] University Institute of Engineering and Technology,
Panjab University, Chandigarh, India

**Abstract.** Distributed Denial of Service (DDoS) attack is one of the most powerful and immense threats to internet-based services. It hinders the victim services within a short duration of time by overwhelming with the huge amount of attack traffic. A sophisticated attacker closely follows the current research of DDoS defense, perform a sophisticated attack by compromising millions of unsecured devices, and send a huge amount of attack traffic (Big Data) to destroy a victim. The attack volume size pattern is shifted to Terabits per second (Tbps) from Gigabits per second (Gbps). When a large amount of traffic is processed by the defense system to identify attack traffic, seldom defense system itself can become a victim of DDoS attack. Therefore, there is a demand to implement DDoS defense system which can efficiently process a massive amount of network traffic and immediately distinguish attack traffic. In this paper, we propose a victim-end Hadoop based DDoS defense framework to identify an attack using the MapReduce programming model based on information theory metric. Further, we have implemented Hadoop based DDoS testbed and validated proposed framework using real datasets, such as, MIT Lincoln LLDDoS1.0, CAIDA and live traffic generated using testbed. The experimental result of proposed framework shows higher detection accuracy (average detection accuracy is 97%).

**Keywords:** Network security · Denial of Service (DoS) · Distributed Denial of Service (DDoS) · Big Data · Hadoop · MapReduce

## 1 Introduction

In the present era, every organization has moved their services online for accessible 24/7 to grow business and revenue. When the Internet was designed, the main objectives were fast data transfer, fast processing, and identification of packet tampering [1]. Everyday Internet users and Internet of Things (IoT) devices are exponentially multiplying because of easy access and decentralized nature of the Internet. Denial of Service (DoS) is an attack to submerge the victim service and denied access to genuine users. DoS attack is launched easily using a single device that continuously

A. B. Gani et al. (Eds.): ICICCT 2019, CCIS 1025, pp. 25–35, 2019.
https://doi.org/10.1007/978-981-15-1384-8_3

forwards random traffic to a victim service. However, it can be undoubtedly identified, infiltrated and trace-back immediately to take legal action because of a single source [1–3]. A Distributed Denial of Service (DDoS) is an attack which completely deprives the performance of a victim service or seldom it may be unavailable [3]. The DDoS attack can be arisen by compromising multiple devices, launch in a coordinated manner, and sending unnecessary traffic through bots towards a victim service. Therefore, it is a challenging job to detect DDoS attack with greater accuracy in real time.

As per Kaspersky Lab report [4], it has remarked that in a first part (Q1) of 2018, significant growth in number of attack occurrences as well as the span of attack when linked with last part (Q4) of 2017. It presents the incidence of DDoS attacks increasing because of various causes such as exponential increase of the non-secure IoT devices, user-friendly attack tools, and security defects in the network. The DDoS attacks volume size is constantly increasing each year despite effective and powerful detection, mitigation and trace back mechanism have introduced by fellow researchers. Figure 1 shows how the volume size of DDoS attack increases every year.

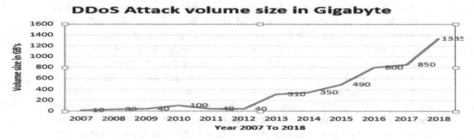

**Fig. 1.** Year-wise DDoS attack volume size from the year 2007 to 2018.

The first DDoS attack was witnessed in June-July 1999, filed in August 1999 which is a target on a single computer system of the University of Minnesota with the help of 227 compromised systems [5]. It was sustained for almost two days and an attack was launched using DDoS Trinoo [6]. And this onwards in 2018, Github has experienced the highest DDoS attack in the records which is around 1.35 Terabits per seconds (Tbps). However, Github was recovered from this attack within 8 min [7].

Peng et al. [8] categorized the DDoS defense system into four comprehensive categories such as Prevention, Detection, Traceback and Mitigation. Further, DDoS attack can be deployed at victim-end (destination-end), source-end, intermediate (core-network) and distributed [2, 9]. Bhuyan et al. [10] analyzed each deployment location of the defense system and presented victim-end DDoS defense system which is better. The reasons are: (i) It was deployed near to victim, hence closely watched network traffic, (ii) Victim-end deployment is quite simple cost effective (iii) It gets aggregated network traffic for analysis which improve detection accuracy and lessen false positive rate. However, victim-end defense system needs to process a large amount of network traffic flows and sometimes the system can itself become a victim of DDoS attack. Therefore, there is a demand to implement systems which can exploit the benefits of victim-end defense system and efficiently analyze massive amount of network traffic to discriminate DDoS attacks from legitimate traffic on a cluster of nodes.

Apache Hadoop [11] is an open source, reliable, scalable and distributed framework. It is one of the most powerful frameworks to store and process a huge amount of data i.e., Big Data on a cluster of nodes. In this paper, we implemented a victim-end Hadoop based DDoS defense framework which detects DDoS attack traffic using MapReduce programming model [17] and validated using real datasets (MIT Lincoln LLDDoS1.0, CAIDA) and live traffic generated using proposed testbed.

The rest of paper is organized as follows, Sect. 2 discuss existing literature in the field of DDoS using Hadoop framework, Sect. 3 proposed Hadoop based detection system, and Sect. 4 present the methodology. Section 5 presents details of our experimental setup, Sect. 6 we present the performance results followed by remarks in Sect. 7.

## 2 Related Work

In this section, outlined existing literature presented by fellow researchers to combat against a DDoS attack based on a Hadoop framework. The fellow researchers proposed numerous powerful solutions to fight against DDoS attack and to address volume based detection in a Hadoop framework. However, after these attack incidents are increasing linearly. The modern attacker generates low rate DDoS attacks by compromising millions of devices which can undoubtedly circumvent the volume based detection system.

Lee and Lee [12] proposed a Hadoop based DDoS attack system. They implemented a counter based detection algorithm to perform detection using the MapReduce programming model and performed implementation in testbed. However, they validated the defense system using offline batch processing only. According to performance evaluation parameters, the proposed system requires approximate 25 min for 500 GB and 47 min for 1 TB of network traffic. It implies that approximate 5 to 10 min is enough to crash victim service and to refuse access to legitimate users. Khattak et al. [13] proposed a Hadoop based DDoS forensics framework using the MapReduce programming model. They applied "horizontal threshold" and "vertical threshold" inside the distinct time window. They verified a defense system using MIT Lincoln LLS-DDoS-1.0 [23] real datasets and efficiently detected high rate DDoS (HR-DDoS) attacks. However, a defense system is validated only using offline batch processing and low rate DDoS (LR-DDoS) attack easily circumvent the system. Zhao et al. [14] proposed Hadoop and HBase based DDoS detection framework using a neural network. They implemented a testbed setup on cloud platform comprises of a victim web server, attacker nodes, and defense system. However, a defense system demands more time for training and testing phase.

Dayama et al. [15] proposed a Hadoop-based DDoS protection framework. They used a MapReduce programming model to implement a detection algorithm based on threshold value (count number of requests) to discriminate DDoS attack and genuine traffic flows. However, if a sophisticated attacker performs LR-DDoS attack, then it surely circumvents defense system where as in case of a flash event [16], genuine users can be treated as attack traffic.

Hameed et al. [18] proposed Hadoop based framework to combat against DDoS attack. They designed an algorithm for DDoS detection to detect attack four influential attacks such as ICMP, UDP, TCP-SYN, HTTP-GET using MapReduce programming model and extended their own work [19] by proposing a HADEC framework to detect HR-DDoS attack within fair time. They generated attack traffic using Mausezahn tool [20] and added legitimate traffic. A HADEC framework is comprised of traffic capturing server, detection server (Namenode) and data nodes (ranges from 2 to 10). A threshold value (500 & 1000) is used to discriminate between attack traffic & genuine network traffic. However, almost 77% time is demanded by traffic capturing server of total detection time and because of a threshold value (500 & 1000) LR-DDoS attack can easily circumvent the defense system and can be treated as legitimate traffic. Chhabra et al. [21] presented a Hadoop based forensics analytic system for DDoS and implemented using a supervised machine learning algorithm. They have validated framework using CAIDA dataset and claimed 99.34% detection accuracy. However, the proposed system requires more time for training and testing phase. Also, they validated proposed system only using real datasets.

Utmost of the existing literature has widely used volume based detection method to discriminate DDoS attack from legitimate network traffic. Nowadays sophisticated attacker is compromising millions of unsecure devices, originate LR-DDoS attack from each device, and consequently the tremendous amount of useless network traffic target towards the victim server. In this paper we proposed a victim-end Hadoop based DDoS defense system using Information theory metric i.e. Shannon entropy [22].

## 3  Proposed Hadoop Based DDoS Detection Framework

In this section, we proposed a victim-end Hadoop based DDoS defense system by employing Shannon entropy. The detection framework consists of two phases (i) Network traffic sniffing phase and (ii) Detection process phase. In sniffing phase, live traffic is captured by using Wireshark network traffic sniffer tool and stored in the Hadoop Distributed File System (HDFS). In the detection process phase, the resources are allocated with the help of Yet Another Resource Allocator (YARN) to perform the detection job using the MapReduce programming model. The architecture of the proposed system is depicted in Fig. 2.

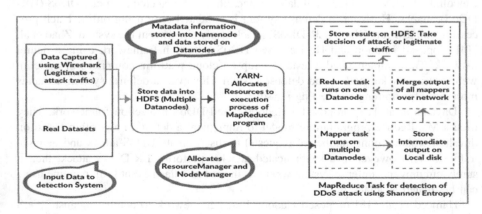

**Fig. 2.** Logical architecture of proposed framework.

Figure 2 consists of three phases, (i) Captures live traffic from legitimate and attacker nodes, (ii) store captured traffic into HDFS and YARN allocates resources for analyzed network traffic flows and, (iii) Using MapReduce programming model traffic to analyze and store result on HDFS and decide whether it is a DDoS attack or normal traffic.

## 4  Methodology

Information theory plays a significant role in the domain of mathematics, physics, statistics, mechanical engineering, civil engineering, computer science & engineering, and many more areas. Information theory based detection metric is often used in the anomaly detection research from the past several years because it is offering notable divergence between an anomaly and legitimate packet. However, in the case of a DDoS detection based on Hadoop framework, the information theory metric is seldom used to detect attack traffic.

### 4.1  Shannon Entropy

Shannon entropy can be defined mathematically as,

$$SE = \sum_{i=1}^{m} \frac{Pi}{S} \log \frac{Pi}{S} \tag{1}$$

where $P_i$ is total number of request with the $i^{th}$ source IP in time window. And $S$ can be defined as

$$S = \sum_{i=1}^{m} Pi \tag{2}$$

For our detection framework used information theory detection metric such as Shannon entropy to discriminate DDoS attack from legitimate traffic flows, for that we defined T is the sampling period in which incoming packets are $X_1, X_2, X_3 \ldots X_n$ and time window is set to 1 to analyze network traffic flows. Where $n$ – total number of packets, $t$ – total number of time window, $m$ – total number of packets in each time window and its value may be different for each time window or may be zero (if no incoming packet). Hence value of $n = m_1 + m_2 + m_3 + \ldots + m_t$. It is require for Eq. (1).

### 4.2  Dataset Used

The proposed framework is validated using different live network traffic scenarios as depicted in Table 1. Also real dataset CAIDA is used to validate proposed framework.

**Table 1.** Live traffic scenario.

| Case | No. of time slots | No. of nodes used | No. of packets |
|---|---|---|---|
| Legitimate traffic | 120 | 10 | 3718 |
| Attack traffic (LR-DDoS) | 120 | 42 | 22069 |
| Legitimate traffic | 360 | 10 | 9327 |
| Attack traffic (LR-DDoS) | 360 | 42 | 50423 |

The real datasets such as MIT Lincoln and live traffic (i.e. no attack scenario) are used to form baseline behavior of our proposed framework to get average value (μ), and standard deviation (σ).

### 4.3  Detection Algorithm

In the Hadoop framework, data processing job consists of a couple of parts, such as Mapper and Reducer job. Each network traffic block (default size is 128 MB) is processed by one mapper that implies if our network traffic data file splits into 10 blocks then concurrently 10 mappers are executed on a cluster of nodes (datanodes) to execute this job. A reducer job is performed by one datanode which is decided by YARN manager.

**Algorithm 4.1: Performed by each Mapper (# of mappers = # of blocks):**

1. Set

      *ip []* ← store source IP of all packets.

      *T*← sampling time (120)

      *time []* ← store arrival time of all packets in seconds

      *tw*← 1(time window is set to 1)

      *entropy[]* ← store entropy of each time window

      *m*← store number of packets (require for equation 1)

2. Extract from packet header *srcIP, arrivalTime, dstIP, protocol*

3. Probability distribution of each unique source IP

      *Key* = distinct Source IP in *tw*

      *Value* = frequency of IP in *tw*

4. Calculate entropy using equation (1).

5. Pass the result to reducer algorithm (4.2)← stored result in local disk (called as intermediate output)

6. Exit

**Algorithm 4.2: Performed by Reducer**
1. Takes the input from each mapper (Algorithm 4.1, step 5 above).
2. Combine the result of each mapper.
3. Compare entropy value of each time window with threshold value (*th* = $\mu \pm$ *k\*σ as shown in Table 2 in our case baseline network threshold range is - 2.59 to +2.59*).
   a) If incoming traffic entropy value lies between above mentioned range, declare legitimate traffic flow.
   b) Else, attack traffic flow (If not lies between above mentioned range).
4. Store the result into HDFS ← To take action, i.e., attack or normal traffic
5. Exit

# 5  Experimental Setup

In this section, we explain the details of our experimental proposed testbed. In Fig. 3, Hadoop based testbed consists of one sniffing node (victim), one namenode (master), three datanodes (slaves), and multiple traffic generators (legitimate and attacker) nodes. Figure 3 shows the experimental testbed of proposed Hadoop based detection framework.

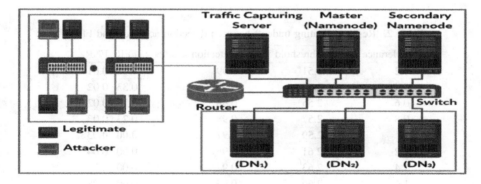

**Fig. 3.** Testbed: Hadoop based detection framework.

In Fig. 3, multiple attackers and legitimate systems generates live network traffic flows and send towards capturing server (victim) node (live traffic captured scenario depicted in Table 1, Sect. 4.2). The job of namenode is to only monitoring and metadata management of data blocks stored in different datanodes. The role of datanodes (DN1, DN2, and DN3) is to process the mapper and reducer job to discriminate legitimate and DDoS attack traffic. SecondaryNamenode is used to provide backup in case of failure of Namenode.

# 6   Results and Discussion

To validate a proposed framework, live traffic is generated using a testbed, the details of generated traffic are discussed in Table 1, Sect. 4.2. We have calibrated the threshold value using the following Eq. 3, and get average value ($\mu$), and standard deviation ($\sigma$) values from baseline behavior (i.e. no attack scenario).

$$th = \mu \pm k * \sigma \tag{3}$$

where $\mu$ - entropy mean of each time window, k-tolerance factor and $\sigma$-standard deviation of entropies value. To measure the performance of proposed framework, we have used detection accuracy, false positive rate and false negative rate which defined in Eqs. (4), (5) and (6). Table 2 shows results of each performance parameter. Four important parameters of confusion matrix are True Positive (TP), True Negative (TN), False Positive (FP) and False Negative (TN), which are required to calculate performance metrics such as Detection Accuracy, FPR and FNR. Detection accuracy can be calculated using a fraction of attack events detected correctly. False Positive Rate (FPR) is the percentage of normal traffic reported as attack traffic. False Negative Rate (FNR) is the percentage of attack traffic stated as legitimate traffic. The value of tolerance factor k is chosen in such a way where False Positive Rate (FPR) and False Negative Rate (FNR) is crossing each other (in our use case value of k is 1.0). This provides tradeoff between the detection accuracy and false positive rate.

$$DetectionAccuracy = \frac{TP}{TP + FN} \tag{4}$$

**Table 2.** Result indicating tradeoff between detection accuracy and FPR.

| Tolerance factor | Threshold value | Detection accuracy | FPR | FNR |
|---|---|---|---|---|
| 0.2 | 2.51 | 0.99 | 0.68 | 0.01 |
| 0.4 | 2.53 | 0.98 | 0.33 | 0.02 |
| 0.6 | 2.55 | 0.98 | 0.28 | 0.02 |
| 0.8 | 2.57 | 0.98 | 0.23 | 0.03 |
| **1** | **2.59** | **0.97** | **0.00** | **0.03** |
| 1.2 | 2.61 | 0.96 | 0.00 | 0.04 |
| 1.4 | 2.63 | 0.94 | 0.00 | 0.06 |
| 1.6 | 2.66 | 0.93 | 0.00 | 0.08 |
| 1.8 | 2.68 | 0.92 | 0.00 | 0.08 |
| 2 | 2.70 | 0.90 | 0.00 | 0.10 |
| 2.2 | 2.72 | 0.88 | 0.00 | 0.12 |
| 2.4 | 2.74 | 0.88 | 0.00 | 0.13 |
| 2.6 | 2.76 | 0.86 | 0.00 | 0.14 |
| 2.8 | 2.79 | 0.83 | 0.00 | 0.17 |
| 3.0 | 2.81 | 0.79 | 0.00 | 0.21 |

$$FPR = \frac{FP}{TN + FP} \tag{5}$$

$$FNR = \frac{TN}{TN + FP} \tag{6}$$

Threshold calibrated (tolerance factor) is done as shown in Fig. 4. Tolerance factor value is calculated in such a way where False Positive Rate (FPR) value and False Negative Rate (FNR) value is crossing each other (in our case is k = 1.0).

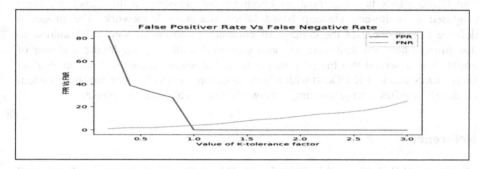

**Fig. 4.** Threshold calibration.

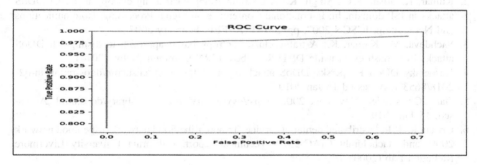

**Fig. 5.** ROC Curve.

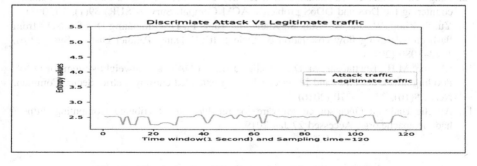

**Fig. 6.** Discrimination DDoS attack and legitimate traffic.

Figure 5 shows Receiver operating characteristic (ROC) curve between the detection accuracy and false positive rate. Entropy values are calculated for attack traffic, and legitimate traffic flows as shown in Fig. 6. It shows that low rate DDoS attack entropies values are higher compared to legitimate traffic due to large number hosts are sending traffic to the victim server which helps us to discriminate attack traffic from legitimate traffic flows.

# 7  Conclusions

The DDoS attack is a big threat to Internet-based services. In this paper, we have proposed a victim-end Hadoop based DDoS detection framework. The proposed defense framework uses the concept of computing entropy of source IP address to discriminate between legitimate and attack network traffic by employing a cluster of nodes. It is observed that the proposed defense framework recognizes the application layer DDoS attack (LR-DDoS) with a high detection rate (97%). The proposed system efficiently handles a large amount of network traffic with quick response.

# References

1. Bhuyan, M.H., Kashyap, H.J., Bhattacharyya, D.K., Kalita, J.K.: Detecting distributed denial of service attacks: methods, tools and future directions. Comput. J. **57**(4), 537–556 (2013)
2. Kumar, K., Joshi, R.C., Singh, K.: A distributed approach using entropy to detect DDoS attacks in ISP domain. In: International Conference on Signal Processing, Communications and Networking, ICSCN 2007, pp. 331–337. IEEE, February 2007
3. Sachdeva, M., Kumar, K.: A traffic cluster entropy based approach to distinguish DDoS attacks from flash event using DETER testbed. ISRN Commun. Netw. (2014)
4. Kaspersky DDoS Kaspersky DDoS attack report. https://securelist.com/ddos-report-in-q2-2018/86537/. Accessed 15 Jan 2019
5. Gary, C.: Kessler, November 2002. https://www.garykessler.net/library/ddos.html. Accessed: 15 Jan 2019
6. Criscuolo, P.J.: Distributed denial of service: Trin00, tribe flood network, tribe flood network 2000, and stacheldraht CIAC-2319. Technical report, California University Livermore Radiation Lab (2000)
7. Github survived Biggest DDoS attack ever recorded wired.com 03/01/2018. https://www.wired.com/story/github-ddos-memcached/. Accessed 24 Jan 2019
8. Peng, T., Leckie, C., Ramamohanarao, K.: Survey of network-based defense mechanisms countering the DoS and DDoS problems. ACM Comput. Surv. (CSUR) **39**(1), 3 (2007)
9. Yu, S., Zhou, W., Jia, W., Guo, S., Xiang, Y., Tang, F.: Discriminating DDoS attacks from flash crowds using flow correlation coefficient. IEEE Trans. Parallel Distrib. Syst. **23**(6), 1073–1080 (2012)
10. Bhuyan, M.H., Bhattacharyya, D.K., Kalita, J.K.: E-LDAT: a lightweight system for DDoS flooding attack detection and IP traceback using extended entropy metric. Secur. Commun. Netw. **9**(16), 3251–3270 (2016)
11. Apache Hadoop: Open-source software for reliable, and distributed Computing. https://hadoop.apache.org/. Accessed 27 Jan 2019

12. Lee, Y., Lee, Y.: Detecting DDoS attacks with Hadoop. In: Proceedings of the ACM CoNEXT Student Workshop, p. 7. ACM (2011)
13. Khattak, R., Bano, S., Hussain, S., Anwar, Z.: DOFUR: DDos forensics using MapReduce. In: Frontiers of Information Technology (FIT), pp. 117–120. IEEE (2011)
14. Zhao, T., Lo, D.C.T., Qian, K.: A neural-network based DDoS detection system using hadoop and HBase. In: 2015 IEEE 17th International Conference on High Performance Computing and Communications (HPCC), 2015 IEEE 7th International Symposium on Cyberspace Safety and Security (CSS), 2015 IEEE 12th International Conference on Embedded Software and Systems (ICESS), pp. 1326–1331. IEEE (2015)
15. Dayama, R.S., Bhandare, A., Ganji, B., Narayankar, V.: Secured network from distributed DOS through HADOOP. Int. J. Comput. Appl. **118**(2), 20–22 (2015)
16. Behal, S., Kumar, K.: Detection of DDoS attacks and flash events using novel information theory metrics. Comput. Netw. **116**, 96–110 (2017)
17. MapReduce programming model. https://www.tutorialspoint.com/map_reduce/. Accessed 24 Feb 2019
18. Hameed, S., Ali, U.: Efficacy of live DDoS detection with Hadoop. In: IEEE/IFIP Network Operations and Management Symposium (NOMS), pp. 488–494. IEEE (2016)
19. Hameed, S., Ali, U.: HADEC: Hadoop-based live DDoS detection framework. EURASIP J. Inf. Secur. **2018**(1), 11 (2018)
20. Mausezahn- fast traffic generator. http://man7.org/linux/man-pages/man8/mausezahn.8.html. Accessed 24 Jan 2019
21. Chhabra, G.S., Singh, V., Singh, M.: Hadoop-based analytic framework for cyber forensics. Int. J. Commun Syst **31**(15), e3772 (2018)
22. Shannon, C.E.: A mathematical theory of communication. Bell Syst. Tech. J. **27**(3), 379–423 (1948)
23. MIT LLDOS 1.0 - Scenario One. https://www.ll.mit.edu/r-d/datasets/2000-darpa-intrusion-detection-scenario-specific-datasets. Accessed 24 Jan 2019
24. CAIDA DDoS dataset. https://www.caida.org/data/passive/ddos-20070804_dataset.xml. Accessed 24 Jan 2019

# Quality Assessment Models for Open Source Projects Hosted on Modern Web-Based Forges: A Review

Jaswinder Singh[1]([✉]) [ID], Anu Gupta[1] [ID], and Preet Kanwal[2] [ID]

[1] Panjab University, Chandigarh, India
jaswinder.davc@gmail.com, anugupta@pu.ac.in
[2] Sri Guru Gobind Singh College, Chandigarh, India
pk.sggs@yahoo.com

**Abstract.** Selection and adoption of appropriate open source application is quite critical for an organization as the open source projects exhibit a quality range from very low to very high. There exist various open source software quality assessment models that aid organizations in measuring the quality of a software project and to select a particular software application based upon some quality score. Commercial involvement and usage of social networking features by modern web-based open source project hosting sites has given rise to new challenges for existing open source software quality assessment models. This paper verifies that whether the existing open source software quality assessment models are able to address these challenges or not. Answering the formulated research questions, a summary of findings we obtained is that existing open source quality assessment models do not consider some of the factors that influence the quality of open source projects hosted on modern web-based forges. Hence there is a need for a quality assessment model addressing the shortcomings of existing models.

**Keywords:** Open source software · Social networking ·
Quality assessment models · Commercial involvement

## 1 Introduction

Open source software (OSS) follows collaborative software development model, where self-motivated volunteers distributed all over the world publicly contribute to the software projects in the form of source code, documentation, peer review, bug reports and patches etc. The communication among OSS project communities regarding the development of the software project also occurs in an open manner [1]. In the last 2 decades, OSS has scaled great heights. One can see the presence of OSS in variety of software domains like operating systems, database applications, software libraries and network infrastructure applications [2]. This distinctive open development approach has even motivated major proprietary software organizations to own and manage hundreds of OSS projects [3]. To make it mandatory for government institutions to use OSS, Ministry of Communication and Information Technology, Government of India

A. B. Gani et al. (Eds.): ICICCT 2019, CCIS 1025, pp. 36–47, 2019.
https://doi.org/10.1007/978-981-15-1384-8_4

has proposed "Policy on Adoption of Open Source Software for Government of India" under its Digital India program [4].

Rapid OSS adoption and its publicly available development history motivated various software engineering researchers and organizations to develop quality assessment models for OSS projects. These models analyze OSS product and development process along with associated community for a complete project quality assessment. Most of the OSS quality assessment models are hierarchical in nature, where the overall software quality is divided into high-level quality characteristics, sub-characteristics and corresponding metrics. It has been observed that IT practitioners don't use the existing OSS quality models too often [5]. Instead of applying the quality models for OSS evaluation, acquaintance or recommendation by fellows is taken up as the main criteria for OSS selection. It has been asserted that lack of proper documentation and ambiguity can be the reasons behind non-adoptability of existing OSS quality assessment models [6].

The aim of this paper is to study and compare various existing OSS quality assessment models in order to discover their strengths and limitations. IT professionals can make use of this study to choose among the existing quality assessment models which fulfill their needs. Moreover, the study will guide researchers in addressing the shortcomings of the existing models and thereby lead to development of new models. The remaining paper has been organized in the following manner. Section 2 presents the background on which the current study is based. Section 3 details the research method used. Section 4 compares the selected OSS models and Sect. 5 discusses the comparison results. Section 6 presents the concluding remarks.

## 2  Background

In the past, researchers have attempted to compare the OSS quality assessment models from various dimensions. Deprez et al. [7] have compared two OSS quality assessment models i.e. Qualification and Selection of Open Source Software (QSOS) and Business Readiness Rating for Open Source (OpenBRR). The comparison was based on definition of quality characteristics, availability of requisite raw data, evaluations steps involved and analysis of the scoring procedure for the defined quality evaluation criteria. Petrinja et al. [8] have conducted a controlled experiment to compare three OSS quality assessment models namely QSOS, OpenBRR and Open Source Maturity Model (OMM). The comparison was carried out to identify the strengths and limitations of quality and usability of the selected assessment models by applying them on 2 open source web browser projects: Mozilla firefox and Google chromium. Stol et al. [9] have identified twenty OSS quality assessment models from literature. Further, they have formulated a comparison framework named as FOCOSEM (Framework fOr Comparing Open Source software Evaluation Methods) to compare the identified models on the basis of model context, model users, model process and model evaluation method. Haaland et al. [10] have divided the OSS quality assessment models into two categories: First generation quality models and second generation quality models. They have compared a first generation quality model (OpenBRR) with a second generation quality model (QualOSS) by assessing the quality of Asterisk OSS project to find the improvements in second generation models over first generation models. Adewumi et al. [2] have reviewed six OSS quality assessment models based upon the origin of the model, online availability of the evaluation results and tool support. Miguel et al. [11] have compared fourteen traditional as well as OSS quality assessment models to ascertain the inclusion of common quality characteristics in their evaluation process.

International Organization for Standardization and International Electrotechnical Commission (ISO/IEC) have jointly developed two software quality standards named as ISO/IEC 9126 [12] and ISO/IEC 25010 [13] in 1991 and 2011 respectively. ISO/IEC 9126 model has defined 6 quality characteristics namely Functionality, Reliability, Usability, Efficiency, Maintainability and Portability which has further been refined into 24 sub-characteristics. ISO/IEC 25010 has defined 8 quality characteristics: Functional Suitability, Reliability, Performance Efficiency, Operability, Security, Compatibility, Maintainability and Transferability which have been refined into 31 sub-characteristics. Capability Maturity Model Integration (CMMI) [14] is another quality assessment standard developed in 2002. The model focuses on the quality of software development process. There are 5 CMMI maturity levels (Initial, Managed, Defined, Quantitatively Managed and Optimizing) according to the effectiveness of the development process.

Present study is different from existing studies as it mainly focuses on the effective applicability of OSS quality assessment models in the current scenario. These days OSS projects are being developed on web-based distributed version control environments that employ various social networking features to attract the communities and popularize the OSS project. Moreover various commercial organizations are also actively participating in OSS development. Keeping in view these advancements in OSS development, there is a need to carry out a detailed study and comparison of existing OSS quality assessment models to test their applicability on OSS projects hosted on modern web-based distributed version control environments.

## 3  Research Method

The current study follows an exploratory research methodology wherein a comparison of existing OSS quality assessment models has been carried out through in-depth literature survey. From literature, it has been observed that first OSS quality assessment model emerged in 2003 [10]. Hence the literature from 2003 till date has been explored to identify the various existing OSS quality assessment models. For the purpose of comparison, 8 models that explicitly define the quality characteristics, their corresponding metrics and also outline a detailed evaluation process have been selected. A brief description of the selected models is as follows.

Capgemini Open Source Maturity Model (OSMM) [15] enables organizations and OSS practitioners to choose the most suitable OSS option based on software product's maturity. The model makes use of product and application indicators to evaluate a software system. Product indicators determine the inherent characteristics of the software whereas application indicators are devised by taking into account user needs and reviews. Navicasoft Open Source Maturity Model (OSMM) [16] evaluates the available software alternatives in phases. First phase is to select open source products that have to be evaluated. In the second phase, weighting factors are assigned to key aspects of the software. In third phase, each of these categories is decomposed into metrics which are scored in accordance with the method specified by OSMM. In the final phase, the scores within each category are multiplied with their weights to produce a final assessment score. SpikeSource (Carnegie Mellon University) and Intel Corporation have jointly developed Business Readiness Rating for Open Source (OpenBRR) model [17] that defines 12 categories for the assessment of software product. The model first formulates a list of viable open source products. Depending upon the user requirements, the assessment categories are assigned weights and mapped to metrics for measurement. The metrics are computed and assessment categories are quantified. Final business readiness rating score for the OSS project is obtained by aggregating the assessment categories in accordance with their stated weights.

Atos Origin has proposed Qualification and Selection of Open Source (QSOS) model [18] defining the quality evaluation process in 4 interdependent steps - Define, Evaluate, Qualify and Select. The first step involves defining 3 aspects of the software product namely type of software, type of license and type of community. In the second step, information is collected from various sources and defined quality attributes are scored in a range from 0–2. In the third step, various filters are defined in accordance with the needs and constraints set by the user. Last step involves final selection of the OSS that successfully realizes user's requirements. Taibi et al. have proposed Open Business Quality Rating (OpenBQR) model [19], as an extension and integration of OpenBRR and QSOS. OpenBQR considers several indicators divided into 5 different areas namely Functional requirement analysis, Target usage assessment, Internal quality, External quality and Likelihood of support in the future. Final quality score is produced by dividing these 5 categories into sub-categories and evaluating each of them. Samoladas et al. have proposed Software Quality Observatory for Open Source Software (SQO-OSS) [20] with focus on automation for OSS quality evaluation. The model is hierarchical in nature with Source code quality and Community quality at the top layer of hierarchy. Petrinja et al. have proposed Quality Platform for Open Source Software-Open Maturity Model (QualiPSo-OMM) [21] for quality and maturity assessment of OSS. The model has been motivated by the Capability Maturity Model Integration (CMMI). The model computes the trustworthiness and quality of an OSS project on the basis of 10 factors. Aversano et al. have proposed Evaluation Framework for Free/Open Source Projects (EFFORT) [22], a framework for selection and adoption of OSS for small and medium enterprises. The framework is designed specially to evaluate open source Customer Relationship Management (CRM) and Enterprise Resource Planning (ERP) systems based on product quality, community trustworthiness and product attractiveness.

A set of 7 research questions has been framed after an extensive literature study of the selected eight quality assessment models in order to establish their effectiveness in the quality evaluation of OSS projects.

### 3.1 Research Questions

To accomplish the aim of the current research work, following research questions have been formulated. Research questions have been framed to determine the

- Strengths and limitations of existing OSS quality models for quality evaluation of OSS projects.
- Applicability of existing OSS quality assessment models on modern web-hosted OSS projects by ascertaining whether existing OSS quality models are considering the technological advancements adopted by modern OSS hosts for effective software development.

The findings and subsequent discussion will help to better understand the quality evaluation process of existing OSS models and to determine whether there is a need for a newer quality model addressing the gaps of existing approaches.

**RQ1.**   Does the model follow an existing quality standard for defining and mapping OSS quality characteristics and metrics?

**RQ2.**   Does the model involve subjectivity and require expert opinion in its evaluation process?

**RQ3.**   Does the model consider the social networking features provided by modern web-based OSS project hosting sites in its evaluation process?

**RQ4.**   Does the model provide sufficient tool support and documentation for its application?

**RQ5.**   Has the model been validated on a wide spectrum of OSS projects?

**RQ6.**   Does the model consider the impact of commercial involvement on the quality of an OSS project?

**RQ7.**   How does the model assign weights to quality characteristics/metrics?

## 4   Comparison

This section presents a summary of the comparison of selected eight OSS quality assessment models. The comparison has been made on the basis of seven research questions formulated in the previous section. The summarized comparison results are presented in Table 1.

**Table 1.** Comparison of identified quality assessment models.

| OSS quality assessment model | RQ1 | RQ2 | RQ3 | RQ4 | RQ5 | RQ6 | RQ7 |
|---|---|---|---|---|---|---|---|
| Capgemini Open Source Maturity Model (2003) | Partially Follows ISO/IEC 9126 | Involves subjectivity and requires expert opinion | No | No | No proof of model validation found from existing literature | No | According to the need of the user or organization |
| Navicasoft Open Source Maturity Model (2003) | Does not follow any software quality standard | Involves subjectivity and requires expert opinion | No | No | No proof of model validation found from existing literature | No | According to what the software is to be used for and by whom |
| Open Business Readiness Rating (OpenBRR) (2005) | Partially Follows ISO/IEC 9126 | Involves subjectivity and requires expert opinion | No | No | No proof of model validation found from existing literature | No | According to metric's importance within Each evaluation category |
| Qualification and Selection of Open Source Software (QSOS) (2006) | Does not follow any software quality standard | Involves subjectivity and requires expert opinion | No | Yes | No proof of model validation found from existing literature | Only checks for sponsor availability | According to importance of each evaluation criterion based on the particular context |
| Open Business Quality Rating (OpenBQR) (2007) | Extended from ISO/IEC 9126 | Involves subjectivity and requires expert opinion | No | No | Validated on 3 content management systems | Takes into account some indirect effect | According to metric's importance in the evaluation process |
| Software Quality Observatory for Open Source Software (SQO-OSS) (2008) | Extended from ISO/IEC 9126 | No subjectivity and no expert opinion required | No | Yes | Validated for very few OSS projects | No | All metrics are equally important, so weighting is not recommended |
| Quality Platform for Open Source Software-Open Maturity Model (QualiPSo-OMM) (2009) | Inspired from CMMI Model | Involves subjectivity and requires expert opinion | No | No | Validated for very few OSS projects | Checks for contribution from software companies | According to the evaluator's point of view |
| Evaluation Framework for Free/Open Source Projects (EFFORT) (2010) | Extended from ISO/IEC 9126 | Involves subjectivity and requires expert opinion | No | No | Validated for 5 open source ERP systems | Only checks for sponsor availability | According to the metric's importance in the evaluation of a goal |

# 5  Discussion

A detailed discussion of the results depicted in Table 1, regarding the comparison of OSS quality assessment models, is presented in this section.

**RQ1.**  Does the model follow an existing quality standard for defining and mapping OSS quality characteristics and metrics?

To answer this question, OSS quality assessment models, selected for this research study, need to be compared with three standard quality assessment models namely ISO/IEC 9126, ISO/IEC 25010 and CMMI by matching the quality characteristics defined in selected models with the quality characteristics present in the standard models. But as ISO/IEC 25010 was introduced after all the selected OSS quality assessment models were developed, hence it is left out from the comparison.

Navicasoft OSMM and QSOS do not follow any of the software quality standards as none of the quality characteristics defined in these models correspond to any of the quality characteristics present in ISO/IEC 9126 or CMMI. Capgemini OSMM and OpenBRR models partially follow ISO/IEC 9126 quality standard. The quality characteristics "Ease of deployment", "Usability", "Performance", "Reliability", "Platform independence" and "Security" defined in Capgemini OSMM corresponds to "Installability", "Usability", "Efficiency", "Reliability", "Portability" and "Security" present in ISO/IEC 9126 standard. "Functionality", "Usability", "Security" and "Performance" used in OpenBRR match with "Functionality", "Usability", "Security" and "Efficiency" present in ISO/IEC 9126 standard. OpenBQR, SQO-OSS and EFFORT quality models fully follow ISO/IEC 9126 quality standard as these models have added some OSS specific quality characteristics to the quality characteristics already defined in ISO/IEC 9126 standard. QualiPso-OMM is inspired from CMMI standard. The structure, the assessment process and the content of the QualiPso-OMM model share common elements with CMMI.

**RQ2.**  Does the model involve subjectivity and require expert opinion in its evaluation process?

Except SQO-OSS model, all the remaining 7 models involve subjectivity and require expert opinion of varying degree. SQO-OSS includes only those metrics that can be computed automatically and objectively as the main aim of this model is automation. In Capgemini OSMM, metrics are scored in a range of 1 to 5 according to ambiguous statements. For example the Market Penetration metric is scored with respect to terms: "Unknown", "A viable alternative" or "Market leader". Capgemini asserts that some of the metrics must only be scored by a panel of OSS experts. Similarly, Navicasoft OSMM includes some quality metrics that require expertise on the part of evaluator for appropriately scoring the metrics. In case of OpenBRR, almost half of the metrics (13 out of 28) defined in the model are scored in a range of 1 to 5 using subjective statements which can be interpreted differently by different users making it mandatory certain amount of expertise with OSS on part of the evaluator. QSOS model scores the metrics related to maturity analysis in a range of 1 to 3 on the basis of vague subjective statements, requiring the evaluator to have an in-depth knowledge of open source concept. Following the similar pattern, OpenBQR and EFFORT models have also defined certain quality metrics that can't be evaluated objectively but depend on the expertise and interpretation of the evaluator. QualiPSo OMM is a process quality assessment model that involves filling a questionnaire for quality assessment. The model states that different evaluators can interpret the questions in a different way and hence it requires that evaluators must have some expertise with process quality models like CMMI.

**RQ3.**   Does the model consider the social networking features provided by modern web-based OSS project hosting sites in its evaluation process?

Modern web-based OSS project hosting sites, also called social coding sites, provide standard features of social networking such as starring or subscribing to an OSS project, or follow a project owner or team members etc. These features play an important role in attracting new developers/users to an OSS project as well as encouraging collaboration between existing developers/users thus contributing positively to the project quality. None of the models taken up in this research study take into account any of these social networking features.

**RQ4.**   Does the model provide sufficient tool support and documentation for its application?

Capgemini OSMM, Navicasoft OSMM and EFFORT models are theoretical frameworks and hence do not provide any kind of tool support for model application. These models have been introduced in research publications and apart from their publications no other source of information or documentation, online or offline about the current status of these models is available. For OpenBRR model, the only tool support used to be a spreadsheet template to create the business readiness rating score for an open source project [23]. But as the official website for OpenBRR model is no longer active, this template is also not available. QSOS model provides sufficient tool support for the model application at its official website [24]. Version 2.0 of the model was released on 19 January, 2013. After that no development activity on the model seems to have happened. For OpenBQR model, although in the associated literature, authors claimed that a web based tool was being developed for application of the model, but the tool never came into reality. Initially in 2008, when the SQO-OSS model was first introduced, a quality evaluation tool named as Alitheia was developed to fully automate the evaluation process. Currently the tool is available on GitHub [25], but there is no development activity on this tool since the last 5 years. For QualiPSo-OMM, a website [26] regarding the model documentation is active but no tool support is available on this website. Although the QualiPSo-OMM developers claimed that a tool will be developed for the model application, but till date no such tool is available.

**RQ5.**   Have the model been validated on a wide spectrum of OSS projects?

All the models selected in this research study have been validated on a very narrow range of OSS projects. For Capgemini and Navicasoft OSMM, examples of application of these models on non-existent OSS projects have been given in their respective publications [15, 16]. In the publication introducing the OpenBRR model, it has been stated that users can use the model and provide their evaluation results to the community so that others can also use the readymade evaluations. But no web resource regarding the results of model validation or application on OSS projects currently exists as the official website of OpenBRR project is no longer active. In case of QSOS model, the official website https://www.qsos.org (Last Accessed on 2019-02-10) is active but it contains no data about model application on any real OSS projects. OpenBQR model has been validated only on 3 content management systems: Mambo, Drupal and Web GUI. The results of this validation are present in the research work done by Taibi et al. [19]. The publication introducing the SQO-OSS model claims that in 2008, when the model was first presented, a website was developed that presented the results of the model application on various OSS projects. But currently the mentioned website doesn't exist. Only those model validation results are present that are available in the publications related to SQO-OSS model and that too on very few OSS projects. QualiPSo-OMM model has been validated on a few significantly important open source projects like Mozilla firefox, Google chromium etc. [8]. EFFORT model has been applied on 5 ERP OSS projects to validate the model [27].

**RQ6.**    Does the model consider the impact of commercial involvement on the quality of an OSS project?

Capgemini OSMM, Navicasoft OSMM, OpenBRR and SQO-OSS models don't contain any quality characteristic or metric that takes into account the fact that an OSS project has been owned, maintained or sponsored by a commercial organization. QSOS model takes into consideration the availability of sponsor for OSS project. OpenBQR model states that if a large number of companies are involved in the development of an OSS project, then there is greater probability for a continuative support. It also states that greater the number of developers from a particular company, greater the importance of the OSS project for that company. But it doesn't mention whether these companies are commercial organizations or non-profit open source organizations. Apart from that the model doesn't consider any other impact on the quality of an OSS project owned and maintained by a commercial organization. QualiPSo-OMM includes a metric "Contribution from Software Companies" while evaluating the quality of OSS developing process. The model states that greater the contribution from software companies, higher will be the quality of the development process. In the product attractiveness part of the EFFORT model, for measuring the degree of diffusion, the model uses a metric "Sponsor availability". The model states that presence of a sponsor for the OSS project positively impacts its attractiveness and hence quality. The model doesn't include any metric for checking the commercial involvement.

**RQ7.**    How does the model assign weights to quality characteristics/metrics?

All the 8 quality models selected in this research study allow the evaluator to assign weights to the quality characteristics or metrics but differ in their approach of assignment. Capgemini OSMM provides the facility to weight the quality indicators according to the need of the user or organization. In case of Navicasoft OSMM, the 6 categories defined for quality assessment can be weighted according to what the software is to be used for and by whom. Category wise default weights specified in the model are Product software – 4, Support – 2, Documentation – 1, Training – 1, Integration – 1, Professional services – 1. OpenBRR model has the provision of assigning weighting factors to each metric within each evaluation category to depict the metric's importance within that category. It thus differentiates itself from other quality models in the sense that weights are assigned to metrics rather than quality characteristics. In QSOS model, the OSS project is evaluated along a list of evaluation criteria defined by the model and an absolute score is assigned to each evaluation criterion. The evaluator can adjust the importance of each criterion based on the particular context. This is done by assigning a weight and threshold to each criterion. In OpenBQR, quality characteristics are identified and weighted according to their importance in the evaluation process. OpenBQR is unique in the sense that first the characteristics are weighted and then measured unlike other models where first the characteristics/metrics are measured and then weighted, thus eliminating the need for measuring the metrics with very low importance. The SQO-OSS model allows usage of weights on various metrics, but such practice is not suggested as the authors assume that all the metrics are of equal importance. The SQO-OSS model recommends a profile based quality evaluation process as it uses ordinal scale aggregation method and provides results on ordinal scale (such as good, fair or poor). The QualiPSo-OMM model has the provision of associating a relevance marker with each metric in the form of a numerical value in the range of 1 to 5 depending upon the evaluator's point of view with 1 for metrics of lowest importance and 5 for metrics of highest importance. EFFORT model uses Goal-Question-Metric approach for quality evaluation where a relevance marker in the range of 1 to 5 is associated with a question according to its importance in the evaluation of a goal.

To sum up the discussion, it is observed that existing OSS quality assessment models comprise certain limitations for assessing the quality of OSS projects. Except QSOS and SQO-OSS, the selected models don't provide sufficient tool support for automating the assessment process. Commercial involvement and usage of social networking features by OSS project hosts have not been addressed by existing OSS quality models for quality assessment. Moreover, the models have not been validated on a wider domain of OSS projects so that their reliability can be trusted.

# 6 Conclusion

In the present research work, eight OSS quality models have been compared on the basis of seven research questions. It has been observed that existing OSS quality models don't cover some of the features prevailing in modern OSS development. Web based OSS project hosting sites are integrating various social networking features in project development to catch the attention of a greater number of developers and users. None of the models selected for the present study takes into account the effect of these features. Furthermore, only two models provide tool support for automation of evaluation process. Commercial organizations have also started taking serious interest in OSS development, but none of the models fully consider the impact of commercial involvement on quality of OSS projects. The existing quality assessment models have been validated on a very limited number of OSS projects, rendering their quality assessment process less reliable. Hence it is observed that there is a need for an OSS quality assessment model that addresses the gaps where existing quality models are lagging behind.

# References

1. Grady, R.B.: Practical Software Metrics for Project Management and Process Improvement. Prentice-Hall, Inc., Upper Saddle River (1992)
2. Adewumi, A., Misra, S., Omoregbe, N.: A review of models for evaluating quality in open source software. IERI Procedia **4**, 88–92 (2013)
3. GitHub. https://github.com/. Accessed 10 Feb 2019
4. M. of C. & I. Technology: Policy on Adoption of Open Source Software for Government of India (2014)
5. Li, J., Conradi, R., Bunse, C., Torchiano, M., Slyngstad, O.P.N., Morisio, M.: Development with off-the-shelf components: 10 facts. IEEE Softw. **26**(2), 80–87 (2009)
6. Hauge, Ø., Østerlie, T., Sørensen, C.F., Gerea, M.: An empirical study on selection of open source software - preliminary results. In: ICSE Workshop on Emerging Trends in Free/Libre/Open Source Software Research and Development, pp. 42–47. IEEE (2009)
7. Deprez, J.-C., Alexandre, S.: Comparing assessment methodologies for free/open source software: OpenBRR and QSOS. In: Jedlitschka, A., Salo, O. (eds.) PROFES 2008. LNCS, vol. 5089, pp. 189–203. Springer, Heidelberg (2008). https://doi.org/10.1007/978-3-540-69566-0_17
8. Petrinja, E., Sillitti, A., Succi, G.: Comparing OpenBRR, QSOS, and OMM assessment models. In: Ågerfalk, P., Boldyreff, C., González-Barahona, J.M., Madey, G.R., Noll, J. (eds.) OSS 2010. IFIPAICT, vol. 319, pp. 224–238. Springer, Heidelberg (2010). https://doi.org/10.1007/978-3-642-13244-5_18
9. Stol, K.-J., Ali Babar, M.: A comparison framework for open source software evaluation methods. In: Ågerfalk, P., Boldyreff, C., González-Barahona, J.M., Madey, G.R., Noll, J. (eds.) OSS 2010. IFIPAICT, vol. 319, pp. 389–394. Springer, Heidelberg (2010). https://doi.org/10.1007/978-3-642-13244-5_36
10. Glott, R., Haaland, K., Tannenberg, A.: Quality models for Free/Libre Open Source Software–towards the "Silver Bullet"?. In: 36th EUROMICRO Conference on Software Engineering and Advanced Applications, pp. 439–446. IEEE (2010)
11. Miguel, J.P., Mauricio, D., Rodríguez, G.: A review of software quality models for the evaluation of software products. Int. J. Softw. Eng. Appl. **5**, 31–53 (2014)
12. ISO/IEC: ISO/IEC 9126 Software engineering - Product quality. ISO/IEC (1991)
13. ISO/IEC: ISO/IEC 25010 - Systems and software engineering - Systems and software Quality Requirements and Evaluation (SQuaRE) - System and software quality models. ISO/IEC (2011)
14. CMMI P. T.: CMMI for Software Engineering, Version 1.1, Staged Representation (CMMI-SW, V1.1, Staged). Technical report CMU/SEI-2002-TR-029, Software Engineering Institute, Carnegie Mellon University, Pittsburgh, Pennsylvania (2002)
15. Duijnhouwer, F.W., Widdows, C.: Open Source Maturity Model. Capgemini Expert Letter 18 (2003)
16. Golden, B.: Succeeding with Open Source. Addison-Wesley Information Technology Series. Addison-Wesley Professional, Reading (2004)
17. Wasserman, A.I., Pal, M., Chan, C.: Business readiness rating for open source. In: Proceedings of the EFOSS Workshop, Como, Italy (2006)
18. Origin, A.: Method for Qualification and Selection of Open Source Software (QSOS). http://www.qsos.org/. Accessed 10 Feb 2019
19. Taibi, D., Lavazza, L., Morasca, S.: OpenBQR: a framework for the assessment of OSS. In: Feller, J., Fitzgerald, B., Scacchi, W., Sillitti, A. (eds.) OSS 2007. IFIPAICT, vol. 234, pp. 173–186. Springer, Boston, MA (2007). https://doi.org/10.1007/978-0-387-72486-7_14

20. Samoladas, I., Gousios, G., Spinellis, D., Stamelos, I.: The SQO-OSS quality model: measurement based open source software evaluation. In: Russo, B., Damiani, E., Hissam, S., Lundell, B., Succi, G. (eds.) OSS 2008. IFIPAICT, vol. 275, pp. 237–248. Springer, Boston, MA (2008). https://doi.org/10.1007/978-0-387-09684-1_19
21. Petrinja, E., Succi, G.: Assessing the open source development processes using OMM. In: Advances in Software Engineering, pp. 1–17. Hindawi Publishing Corporation, New York (2012)
22. Aversano, L., Pennino, I., Tortorella, M.: Evaluating the quality of free/open source projects. In: International Conference on Evaluation of Novel Approaches to Software Engineering (ENASE 2010), pp. 186–191 (2010)
23. Leister, W., Christophersen, N.: Open Source, Open Collaboration and Innovation. INF5780 Compendium Autumn 2013, pp. 1–139 (2012)
24. QSOS. https://www.qsos.org. Accessed 10 Feb 2019
25. Alitheia-Core. https://github.com/istlab/Alitheia-Core. Accessed 10 Feb 2019
26. QualiPso-OMM. http://qualipso.icmc.usp.br/OMM. Accessed 10 Feb 2019
27. Aversano, L., Tortorella, M.: Quality evaluation of floss projects: application to ERP systems. Inf. Softw. Technol. 55(7), 1260–1276 (2013)

# VaFLE: Value Flag Length Encoding for Images in a Multithreaded Environment

Bharath A. Kinnal$^{(\boxtimes)}$ [ID], Ujjwal Pasupulety [ID], and V. Geetha

Department of Information Technology, National Institute of Technology
Karnataka Surathkal, Mangaluru 575025, India
1997bhar@gmail.com,
{15it150.ujjwal,geethav}@nitk.edu.in

**Abstract.** The Run Length Encoding (RLE) algorithm substitutes long runs of identical symbols with the value of that symbol followed by the binary representation of the frequency of occurrences of that value. This lossless technique is effective for encoding images where many consecutive pixels have similar intensity values. One of the major problems of RLE for encoding runs of bits is that the encoded runs have their lengths represented as a fixed number of bits in order to simplify decoding. The number of bits assigned is equal to the number required to encode the maximum length run, which results in the addition of padding bits on runs whose lengths do not require as many bits for representation as the maximum length run. Due to this, the encoded output sometimes exceeds the size of the original input, especially for input data where in the runs can have a wide range of sizes. In this paper, we propose VaFLE, a general-purpose lossless data compression algorithm, where the number of bits allocated for representing the length of a given run is a function of the length of the run itself. The total size of an encoded run is independent of the maximum run length of the input data. In order to exploit the inherent data parallelism of RLE, VaFLE was also implemented in a multithreaded OpenMP environment. Our algorithm guarantees better compression rates of upto 3X more than standard RLE. The parallelized algorithm attains a speedup as high as 5X in grayscale and 4X in color images compared to the RLE approach.

**Keywords:** Run length encoding · Data parallelism · OpenMP · Lossless encoding · Compression ratio

## 1 Introduction

Data compression techniques have enabled networked computing systems to send larger amounts of data by effectively utilizing the channel bandwidth. Modern methods make use of complex algorithms to encode the original data bits such that the information they carry is preserved in a smaller number of bits. Data compression can either be lossy or lossless. Lossless compression ensures that the original data can be perfectly reconstructed from the encoded data. By contrast, lossy compression reconstructs only an approximation of the original data, though this usually improves compression rates.

© Springer Nature Singapore Pte Ltd. 2019
A. B. Gani et al. (Eds.): ICICCT 2019, CCIS 1025, pp. 48–60, 2019.
https://doi.org/10.1007/978-981-15-1384-8_5

Run Length Encoding (RLE) is a simple and frequently used lossless encoding method. It involves scanning the input data and storing the input symbol and the length of the symbol run.

RLE is primarily used to compress data from audio, video, and image files. It works exceptionally well with black and white images as well as grayscale images. This is because the data in such files contains large length runs of contiguous 1s or 0s, yielding high compression rates.

One major issue with RLE is that the encoded runs have their lengths stored in a fixed number of bits in order to simplify decoding. This fixed number is equal to the bits required to represent the maximum length resulting in low encoding rates on highly varying input data. The proposed lossless VaFLE algorithm encodes the run lengths such that the number of bits allocated is an increasing function of the run length itself. Contrary to the traditional method of representing runs in the form of (Value, Length) pairs, VaFLE makes use of (Value, Flag, Length Offset) triples. Therefore, the total size of an encoded run independent of the maximum run length of the input data. The end result is an algorithm which offers high compression ratios at lower execution times, which can be seen by analysing the space complexity of VaFLE algorithm's effectiveness, and comparing to that of the RLE approach.

As RLE and RLE-based algorithms possess an inherent form of data parallelism, VaFLE was further optimized to run on multi-core systems by splitting the image file into a number of small chunks, and compressing each chunk in parallel across several processing cores. The OpenMP [4] environment was used to run data parallel operations on different cores, as well as integrating compressed segments in a sequential manner. The algorithm can be used on a variety of multimedia files since it is format independent, but works best on black and white images.

The paper is structured as follows: Sect. 2 provides details on the literature survey. Section 3 explains the novel serial and parallelized VaFLE algorithms along with the space complexity analysis. Section 4 provides details about experimental results. Section 5 demonstrates our results and analysis. The paper concludes in Sect. 6.

## 2 Literature Survey

### 2.1 Previous Work and Applications

The RLE data compression algorithm has been used in a wide area of applications. Vijayvargiya et al. [9] conducted a survey on various image compression techniques which included Run Length encoding as a popular method. Chakraborty and Banerjee [3] used RLE along with special character replacement to develop an efficient lossless colour image compression algorithm. Arif and Anand [2] found another useful application for RLE based compression for speech data. Khan et al. [6] applied RLE in the domain of steganography which has a variety of practical implications in the cybersecurity space.

RLE and RLE-based algorithms inherently have the property of data parallelism which can be exploited at the programming level as well, without the need for dedicated hardware. This can be seen in the publication by Trein et al. [8] who created a hardware implementation for Run Length Encoding where images are transferred as blocks of pixels on operated upon in parallel.

## 2.2  Basics of Run Length Encoding

Run Length Encoding is one of the simplest lossless encoding techniques. RLE compresses a given run of symbols by taking the value of that symbol in the run, and places next to it the number of times the same symbol has occurred in the run consecutively. This significantly reduces the size of the sequence and the new sequence is easy to interpret. For example, consider the following sequence:

$$AABBCCCCDD$$

With RLE, the sequence reduces to:

$$A2B2C4D2$$

The new sequence does not contain as many bits as the previous sequence. However, this works only if there is one bit that occurs much more often in the sequence as the other. In case of equality, RLE proves to be highly inefficient as it would generate a sequence that would be double the size of the original sequence. For example, consider:

$$1010$$

On performing RLE, the following string would be obtained:

$$11011101$$

The output is exactly twice the size of the original sequence, which defeats the purpose of making the smaller string. RLE is useful when it comes to sequences that contain a large quantity of one bit and can be used for characters as well as bits. In the best case, RLE can reduce data to just two numbers if all the values in the original data are exactly the same, regardless of the size of the input. But in the worst case, i.e, if there are no repeating values in the data, RLE can double the size of the original data.

## 2.3  Modified RLE with Bit Stuffing

Performing RLE on smaller sequences actually results in the expansion of the data rather than compression. Amin et al. [1] devised a modified run length encoding scheme which addresses some drawbacks of the original RLE algorithm. This algorithm makes use of bit stuffing and avoids encoding small sequences of one or two bits. The input data is analyzed to highlight if there are any large sequences that will decrease the number of bits to represent the length of each run. The algorithm used in [1] incorporates bit stuffing by inserting non-information bits into the data. The location of stuffed bits is communicated to the receiving end of the data link, where they are removed to recover the original bit streams.

However, the concept of bit stuffing used in this algorithm may be counter-productive, as it could lead to the formation of longer runs. Furthermore, the maximum size for the representing the length of a given bit sequence is still a fixed value. The modified RLE algorithm makes use of bit stuffing to limit the number of consecutive bits of the same value in the data to be transmitted in an optimal manner. The strategy of ignoring small sequences was also incorporated. The combination of bit stuffing and ignoring small sequences results in a much more optimal Run Length Encoding Algorithm.

## 3 Value Flag Length Encoding

Both the original RLE and modified [1] schemes are based on finding the frequency of the sequential bits of a given value. One of the main problems observed is that the maximum size for representing the length of a given run is a fixed value. This constrains the space complexity of a given run of same-valued bits. The proposed VaFLE algorithm ensures the size required for representing length is a function of the original run length itself. This algorithm also ignores the runs of lengths equal to the minimum length specified by a variable called τ.

### 3.1 Serial VaFLE Algorithm

Algorithm 1 shows the basic Value Flag Length Encoding Scheme. The bit stream can be broken up into a sequence of multiple runs of same-valued bits. Only runs having a length greater than the minimum threshold length (specified by Threshold or τ) will be considered for analysis. In the algorithm a given run will be encoded in the form of a triple. The first part indicates the Value of the bits stored (exclusively 1s or exclusively 0s). The length of a given run is represented in the next two parts, namely the Flag and Offset. The Flag which specifies the number of bits required to encode the Offset.

Since Value should store a run of similar bits, its size should be equal to τ, as minimum as possible keeping in mind that it should resemble the sequence of bits of length less than τ. The Flag is indicative of the number of bits required to represent the offset. Consider a given run with a Value segment that contains bit b (exclusively 1 or 0) and the number of bits in the run is n. The Flag is set as k bits of value b, followed by a single bit which is the boolean complementary value of b. The value of k is found using (1).

$$k = \log_2(n - (\tau - 1)) \tag{1}$$

For example, if value = '11', then flag = '111….(k times) 0'. Here, 0 indicates the end of the flag and the start of the length offset. With the same value of k, the offset can be found. Given the original length n, the length offset is calculated using (2) for $n > \tau$.

$$\text{lengthOffset} = (n - 1) - 2^k \tag{2}$$

---

**Algorithm 1:** Serial VaFLE Encoding

---

**Input:** Original file, and threshold length Threshold
**Output:** Compressed file
1 **function** VaFLE_encoding(file, Threshold)
2     $s1 \leftarrow *$ file
3     $s2 \leftarrow *$ compressedFile
4     writeFile(s2, Threshold)
5     size $\leftarrow$ sizeof(file)
6     iRun $\leftarrow 0$
7     while iRun < size do
8        count $\leftarrow$ length of current run
9        if count $\leq$ Threshold then
10           writeFile(s2,bits($s1_{iRun}$, count))
11        else
12           value $\leftarrow$ bits($s1_{iRun}$, Threshold)
13           flag $\leftarrow$ NULL
14           flagSize $\leftarrow \log_2($count $- ($Threshold $-1))$
15           append(flag, intToBits($2^{flagSize \times s1\ iRun} - 1$, flagSize))
16           append(flag, !(bit($s1_{iRun}$)))
17           length $\leftarrow$ NULL
18           lSize $\leftarrow$ flagSize
19           offset $\leftarrow$ count - (Threshold - 1) - $2^{lSize}$
20           append(length,intToBits(offset, lSize))
21           writeFile(s2, value + flag + length)
22        iRun $\leftarrow$ iRun + count
23     return compressedFile

---

**Algorithm 2:** Serial VaFLE Decoding

---

**Input:** Compressed file
**Output:** Original file
1 **function** VaFLE_decoding(compressedFile)
2     $s1 \leftarrow *$ compressedFile
3     $s2 \leftarrow *$ originalFile
4     Threshold $\leftarrow$ (int)extractHeader(s1)
5     iRun $\leftarrow$ sizeof(int)
6     while iRun < sizeof(compressedFile) do
7        runEnd $\leftarrow$ iRun + (length of current run)
8        if runEnd $\leq$ sizeof (compressedFile) then
9           break
10        if runEnd $-$ iRun $\leq$ Threshold then
11           writeFile(s2,bits($s1_{iRun}$, runEnd $-$ iRun))
12           iRun $\leftarrow$ runEnd
13        else
14           offset $\leftarrow$ runEnd $- ($iRun + Threshold + 1$)$
15           tnum $\leftarrow$ bitsToInt($s1_j$, j + offset $- 1$)
16           num $\leftarrow 2^{offset} +$ tnum + Threshold $- 1$
17           writeFile(s2, intToBits($2^{num \times s1\ iRun} - 1$, num))
18           iRun $\leftarrow$ j + offset
19     return originalFile

---

To see how the algorithm runs, consider the following bit stream '0111111111111111' and $\tau = 2$. If the size is not more than $\tau$, it can be ignored. Thus the single 0 is encoded as it is. For the next run, the size, $n = 15$, which is greater than $\tau$. This run can be encoded as a (Value, Flag, Offset) triplet. Being a sequence of 1s and the variable $\tau$ set to 2, the Value will be '11'. The value of k using (1) is equal to 3. Flag is given as k 1s followed by a 0. Next, using Eq. (2), the length offset is found to be equal to 6 whose binary representation is '110' and requires only k bits to store.

$$\text{Flag} = \text{'1110'}$$

$$\text{Length} = \text{'110'}$$

Here, it must be noted that the offset must be of length $= k$, in order to decode the offset as that of a specified length given by the flag. The triplet is given as

$$(11, 1110, 110)$$

Thus the encoded data would be

$$\text{Encoded data}: 0, (11,1110,110)$$
$$\text{Final bit string}: \text{'0111110110'}$$

The length of the encoded bit string is 10. For the given example, the compression ratio is equal to 1.6. Algorithm 1 demonstrates the sequential encoding VaFLE scheme.

Decoding of the encoded bit string, shown in Algorithm 2, can be done by observing the value and finding the length with flag and length offset. To check if a run is encoded into a triplet, the first m consecutive bits having the same bit value should be observed. If $m \leq \tau$, then no change has to be made to run, and it is directly appended to the output. If not, then the length is found by looking at the flag and length offset.

### 3.2 Complexity Analysis of VaFLE

For runs of length $n \leq \tau$, it is assured that they will remain unchanged in the encoded string, and the encoded size equals n. For runs of length $n > \tau$, the total length of the encoded run of bits can be found by calculating the lengths of individual elements in the triplet. The value will have a constant length of $\tau$ bits.

$$v = \tau \tag{3}$$

Deriving the value of k from (1), the flag will have the following length,

$$f = 1 + k \tag{4}$$

The length offset lies in the range of $(n - 1) - 2^k$, where $n \in [2^k, 2^{(k+1)} - 1]$. There can be a total of $2^k$ values. These values can be encoded using k bits.

$$l = k \tag{5}$$

Combining (3), (4), (5), and (6) and substituting k using (1) the compressed size is calculated as follows,

$$
\begin{aligned}
s &= v + f + 1 \\
&= \tau + (k + 1) + k \\
&= \tau + 1 + 2(\log_2(n - (\tau - 1)))
\end{aligned} \tag{6}
$$

Thus, the value of the length s is found to be a function of n, which shows that the length of the encoded run is dependent only on that run length alone, and not on the length of any other run.

As both the encoding and decoding algorithms involve going through all the bits in a sequential manner, the overall time complexity for both algorithms is of the order $O(n)$, where n is the number of bits in the file being encoded or decoded.

### 3.3   Parallelized VaFLE in a Multithreaded Environment

While VaFLE on its own is more efficient than the naive RLE encoding scheme, input data is still scanned in a sequential manner. The algorithm has a time complexity = $O(n)$. A parallel implementation would involve splitting up the input data and compressing each part separately. Data parallelism is a form of parallelization across multiple processing cores in parallel computing environments. It focuses on distributing data across different nodes, which operate on the data in parallel. It can be applied on regular data structures like arrays and matrices by working on each element in parallel.

Algorithm 3 shows the parallel Value Flag Length Encoding Scheme. The inherent data parallelism of RLE algorithms has been exploited in order to create a parallel implementation in C++ through the OpenMP environment. The input bit string is divided into different chunks of similar size, the number of such chunks being equal to the number of threads executing in the parallel environment. The chunks are encoded in parallel, and are then appended sequentially to the output bit string.

One consequence of executing encoding algorithms in a parallel environment is that the formation of chunks may split the runs present. For example, if there are 2 threads and the bit string is '10000001', the threads divide the string at the midpoint, thereby splitting the run of 0s. This may result in a different output. However this does not corrupt the encoded string, and can be decoded to give the original run, as decoding involves observing the value of the initial bits in the encoded run.

This is not the case when running the decoding algorithms in a parallel environment The decoding algorithms would require prior knowledge of the size of the runs before dividing the data into chunks, which is very hard and inefficient. Hence, it is preferable not to execute the decoding algorithms in a parallel environment. Algorithm 2 is used to decode both the parallel and serial encoding outputs.

---

**Algorithm 3:** Parallel VaFLE Encoding

---

**Input:** Original file, and threshold length Threshold, number of threads numT

**Output:** Compressed file

```
1 function VaFLE_encoding(file, Threshold, numT )
2       s1  ← * file
3       s2  ← * compressedFile
4       writeFile(s2, Threshold)
5       size ←sizeof(file)
6       cmprs[numT ] ← NULL
7       #pragma omp parallel num_threads(numT )
8           t ←omp_get_thread_num()
9           beg  ← t * size/numT
10          end  ← (t+ 1) * size/numT
11          if end ≥ size then
12          └   lt ← sz
13          chunk ←bits(s1 beg , lt − beg)
14          chunksize ←sizeof(chunk)
15          iRun ← 0
16          while iRun < chunksize do
17              count ← length of current run
18              if count ≤ Threshold then
19                  │   append(cmprs t ,bits(chunk iRun , count))
20              else
21                  value ←bits(chunk iRun , Threshold)
22                  flag ← NULL
23                  flagSize ←log(count − (Threshold - 1))
24                  append(flag,intToBits(2 flagSize×chunk iRun − 1, flagSize))
25                  append(flag, !(bit(chunk iRun )))
26                  length ← NULL
27                  lSize ← flagSize
28                  offset ← count − (Threshold − 1) - 2 lSize
29                  append(length, intToBits(offset, lSize))
30              └   append(cmprs t , value + f lag + length)
31          └ └   iRun ← iRun + count
32      for j ∈ 1, ...numT do
33      └   writeFile(s2, cmprs j )
34  └ return compressedFile
```

---

# 4  Experimental Methodology

The performance of the parallelized VaFLE algorithm was evaluated and compared with a basic implementation of the RLE algorithm using various lossy and lossless formats and sizes of the standard Lenna image (shown in Table 1). The hardware configuration consisted of a 56-core server grade Intel Xeon CPU with 125 Gigabytes of RAM. The Relative Speedup (RS) and Relative Compression Ratios (RCR) were calculated using formulae (7) and (8) respectively. For JPEG color images, multiple $512 \times 512$ samples of varying quality were considered. The file sizes range from 5 KB to 30 KB. The threshold value $\tau$ was set to 2, in order to ignore the encoding of lone and pair bits. Dividing the original file size by the compressed file size gives the compression ratio of an algorithm for a given input. We also compare VaFLE's compression ratio against RLE, LZW [10] and Huffman [5] encoding, three well known algorithms.

**Table 1.** Image formats used for the performance evaluation of VaFLE

| Format | Scheme (Color and dimension in pixels) |
|--------|----------------------------------------|
| PNG | Gray (512 × 512), Color (480 × 480, 512 × 512) |
| BMP | B/W (512 × 512), Gray (256 × 256, 512 × 512), Color (512 × 512) |
| JPG | Gray (128 × 128, 256 × 256), Color (512 × 512) |
| TIF | Gray (64 × 64, 256 × 256, 512 × 512, 1028 × 1028), Color (512 × 512) |

## 5  Results and Analysis

Figure 1 shows the difference in relative execution times achieved while comparing the serial RLE against the parallel VaFLE scheme using Eq. (7). In terms of latency, parallel VaFLE shows a speedup in most of the image formats over serial VaFLE except in JPGs (which are lossy by design, Fig. 1(d)). The lossless formats such as PNG (Fig. 1(a)) and BMP (Fig. 1(b)) take the least time to encode. Each thread gets to work on a small portion of the data and quickly completes their encoding task. Parallel VaFLE on most image formats gives a speedup roughly proportional to the image size due to reduced overhead during communication between threads.

$$RS = \frac{\text{Serial RLE Execution Time}}{\text{Parallel VaFLE Execution Time}} \tag{7}$$

$$RCR = \frac{\text{VaFLE compression ratio}}{\text{RLE compression ratio}} \tag{8}$$

It can be inferred from Fig. 2 that the VaFLE algorithm gives performs similarly in serial and parallel environments, both being relatively better than RLE, although there are minute differences in the compression ratio upon parallelizing VaFLE. Although the same data gets compressed in a manner similar to the serial algorithm, splitting up image data will result in the breaking of long sequences of consecutive bits.

Figure 3 shows the compression ratio of VaFLE, LZW, Huffman and RLE encoding. LZW is a very effective, general purpose algorithm and achieves more the 2X compression ratio for most image formats (30X in the case of the black and white image). While Huffman encoding surpasses the compression ratio performance of VaFLE, it must be noted that Huffman coding spends some time mapping image intensity values to an 8 bit representation. VaFLE does not have such a computational overhead and will have a better runtime compared to Huffman encoding. VaFLE does perform better (10X compression ratio) than Huffman encoding (5X compression ratio) on the black and white image which shows the effectiveness of a Run Length Encoding based scheme when the input contains a large number of similar bits in a sequence.

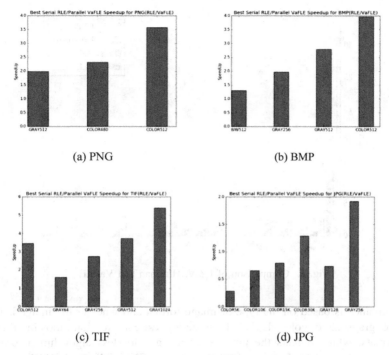

**Fig. 1.** Speedup achieved by parallel VaFLE against serial RLE in various image formats.

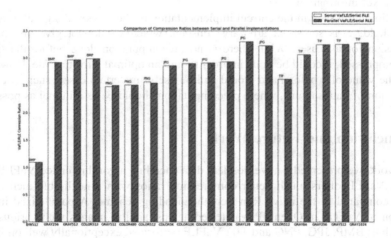

**Fig. 2.** Comparison of serial and parallel compression ratios (VaFLE/RLE)

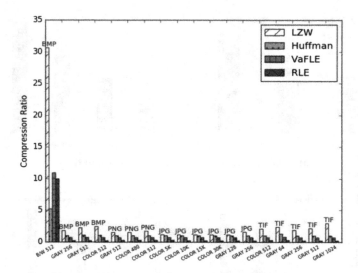

**Fig. 3.** Comparison of LZW, Huffman and VaFLE

Performance is based on the type of image format and whether the image is black and white, grayscale or colored. JPG attain the highest compression ratios but they are lossy formats, which defeats the purpose of using a lossless algorithm in order to preserve raw image data. Lossless formats, such as PNG, TIF, and BMP, are more favourable media to demonstrate high compression ratios while still retaining the original pixel information.

One limitation seen in the current implementation of the proposed algorithm is that the threshold, $\tau$, is set as an arbitrary constant, which would not give an optimal compression for all cases. While there is no precomputation done before the file is being compressed, it could be done in order to find an optimal $\tau$ for that file. However, setting the value of $\tau$ beforehand would still be useful in certain cases, such as online compression of data streams, where precomputation should be as minimal as possible.

## 6   Conclusion and Future Work

In this work, we presented a novel lossless data encoding algorithm called VaFLE. On its own, VaFLE offers much better compression ratios with a negligible increase in runtime compared to the naive Run Length Encoding scheme. A parallelized implementation of VaFLE in OpenMP was tested against similar images with various formats such as BMP, JPG, PNG and TIF. VaFLE performed exceptionally well on Black and White images. Our algorithm guarantees better compression rates of upto 3X greater than standard RLE. Parallelized VaFLE attains a speedup as high as 5X in grayscale and 4X in color images compared to the naive RLE approach. Our implementation of this algorithm can be particularly useful for online compression of data streams. Being a Run Length Encoding based algorithm, VaFLE does not have better compression rates compared to other well known encoding schemes such as LZW and Huffman encoding. However, VaFLE is faster compared to Huffman encoding as there are no pre-computation steps involved.

In the future, we plan to implement an optimized version of this algorithm, in which the optimal value of the threshold length $\tau$ can be calculated for a given file before compression takes place, so as to improve the compression rate. Also, since it performs well with binary images or with images where there is a low number of varying intensity values, it is possible for grayscale image intensities to be mapped to a lower number of intensity values and appended as a header so that the mapped image has a high compression ratio but the entire grayscale image can still be reconstructed at the receiver end using the header information. Furthermore, we plan to implement VaFLE using CUDA [7] and further improve the runtime of the algorithm using a large number of GPU threads.

**Acknowledgements.** We sincerely thank Dr. Sidney Rosario, Department of Information Technology, NITK Surathkal, who helped give us the opportunity to work on this project, and providing valuable inputs during the development of the algorithm itself. We also thank Dr. Basavaraj Talwar, Department of Computer Science, NITK Surathkal, and Mr. Swaroop Ranganath for providing valuable inputs on how the paper could be improved.

Finally, we thank the Department of Information Technology for providing to us the necessary computer facilities and infrastructure needed to implement our project.

# References

1. Amin, A., Qureshi, H.A., Junaid, M., Habib, M.Y., Anjum, W.: Modified run length encoding scheme with introduction of bit stuffing for efficient data compression. In: 2011 International Conference for Internet Technology and Secured Transactions, pp. 668–672, December 2011
2. Arif, M., Anand, R.S.: Run length encoding for speech data compression. In: 2012 IEEE International Conference on Computational Intelligence and Computing Research, pp. 1–5, December 2012. https://doi.org/10.1109/ICCIC.2012.6510185
3. Chakraborty, D., Banerjee, S.: Efficient lossless colour image compression using run length encoding and special character replacement (2011)
4. Dagum, L., Menon, R.: OpenMP: an industry-standard API for shared-memory programming. IEEE Comput. Sci. Eng. **5**(1), 46–55 (1998)
5. Huffman, D.A.: A method for the construction of minimum-redundancy codes. Proc. IRE **40**(9), 1098–1101 (1952). https://doi.org/10.1109/JRPROC.1952.273898
6. Khan, S., Khan, T., Naem, M., Ahmad, N.: Run-length encoding based lossless compressed image steganography. SURJ **47**, 541–544 (2015)
7. Nickolls, J., Buck, I., Garland, M., Skadron, K.: Scalable parallel programming with cuda. Queue **6**(2), 40–53 (2008). https://doi.org/10.1145/1365490.1365500
8. Trein, J., Schwarzbacher, A.T., Hoppe, B., Noff, K.: A hardware implementation of a run length encoding compression algorithm with parallel inputs. In: IET Irish Signals and Systems Conference (ISSC 2008). pp. 337–342, June 2008. https://doi.org/10.1049/cp:20080685

9. Vijayvargiya, G., Silakari, S., Pandey, R.: A survey: various techniques of image compression. CoRR abs/1311.6877 (2013). http://arxiv.org/abs/1311.6877
10. Welch, T.A.: A technique for high-performance data compression. Computer 17(6), 8–19 (1984). https://doi.org/10.1109/MC.1984.1659158

# Impact of Sink Location in the Routing of Wireless Sensor Networks

Mohit Sajwan[1(✉)] ⓘ, Karan Verma[1], and Ajay K. Sharma[2]

[1] National Institute of Technology, Delhi, New Delhi, India
{mohitsajwan, karanverma}@nitdelhi.ac.in
[2] I K J Gujral Technical University, Jalandhar, India
sharmaajayk@rediffmail.com

**Abstract.** Wireless sensor networks is a collection of many low power sensing devices having local processing and wireless communication capabilities. In these wireless networks, energy constraint is significant. To provide communication between sensor devices, many routing protocols have been proposed. In each routing protocol, their main task is to transmit data (via single hop or multi-hop) towards the sink. Hence, location of the sink has a vital impact on the various factors like performance of the routing protocol. As sink can be static and mobile, but in many applications, sink mobility is not possible. In this paper, we examine the performance of extanttree and clustered based routing WSN protocol, when the different location of a static sink (i.e., at centre, corner and far) is deployed. We perform analytical simulation in terms of networks lifetime. We test the existing network based routing algorithm (i.e. HEEMP, CAMP, GSTEB, TBC, PEGASIS, LEACH) extensively. For each routing algorithm we have evaluated the impact of sink locality. The experimental results show that the HEEMP, CAMP algorithm outperforms other existing algorithms.

**Keywords:** Protocol · Sink location · WSNs · Network lifetime

## 1 Introduction

In this Digital evolution, the contribution of Wireless sensor networks (WSNs) plays a pivotal role, as it senses the environment and accordingly beneficially adapts their behavior. It has a lot of applications in the domain of health, traffic monitoring, security surveillance, military, environmental, etc. [1]. WSNs consist of plenty sensing nodes (called as motes) which are placed in a terrain (i.e. Region of interest). Every node will participate in constructing a wireless network. Each sensor node consists of four components: (1) sensing component (2) power supply component (3) computing component (4) communication component. In most of the applications, batteries (i.e., comes under power supply component) are irreplaceable or not rechargeable. Hence our main task is to make the effective utilization of battery. While experimentally observing it, communication component (i.e. transceiver) consumes maximum energy [2]. Therefore our aim in WSNs is to utilize the communication component efficiently. Figure 1 depicts the protocol stack of WSN, where network layer, consist of routing protocol, helps in the construction of a virtual path between the sink and sensor. It has a

© Springer Nature Singapore Pte Ltd. 2019
A. B. Gani et al. (Eds.): ICICCT 2019, CCIS 1025, pp. 61–71, 2019.
https://doi.org/10.1007/978-981-15-1384-8_6

significant impact in the communication component, many different routing protocols (i.e., classified into at, hierarchical and geographical) have been proposed for increasing the network lifetime (mentioned below in Sect. 3) [3].

The base station (Sink) is a centralized system, either mobile or fixed, that gets all the data collected by the WSN. After which it analyzes the data and show the results to the end user. In real, the location of a base station (sink) is as essential as routing protocol. In a similar work, [4] author proposed the impact of sink location only on LEACH routing protocol and observe that when sink is located at the center of the network, its overall performance is better than other scenarios. In this paper, we examine the repercussions of sink location placed at center, corner and far with well-established routing protocols like cluster based (i.e. LEACH [5]), chain based (i.e. PEGASIS [6]), tree based (i.e., GSTEB [7], TBC [8]), and hybrid based (i.e.] HEEMP [9], CAMP [10]) protocols against network lifetime. Various simulations are performed by using the different sink location of the network. It was ascertained that HEEMP outperforms all the existing protocol when a sink is deployed at center and corner but at far CAMP outperformed others. Similarly PEGASIS as well as CAMP has a less impact on sink locality whereas LEACH and TBC has the maximum. The remainder of the paper is organized as follows: initially, related research work is explained in Sect. 2. Later, the system model and problem statement for our simulation model are discussed in Sects. 3 and 4, respectively. After which experimental results consisting of the aforementioned routing protocol comparison is conferred in Sect. 5. Lastly, Sect. 6 discussed about the conclusion.

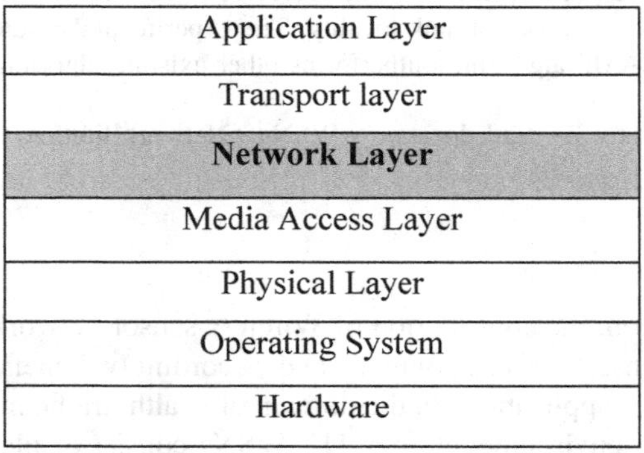

| Application Layer |
| :---: |
| Transport layer |
| **Network Layer** |
| Media Access Layer |
| Physical Layer |
| Operating System |
| Hardware |

**Fig. 1.** Protocol stack of wireless sensor networks

## 2   Related Work

In Wireless Sensor Networks routing in one of the foremost component, because sensor nodes consume maximum energy during data transmission, hence the role of routing protocol is to transmit data is such a manner so that it can perform the task (i.e., sending data to the sink) by considering the energy minimization factor. In literature, routing protocols of WSN has been broadly classified into two classes: flat and hierarchical class [11]. In the first class (i.e., flat routing), during the transmission phase, each sensor nodes of the monitored network perform the same role. Where each sensor node follows the multi-hop communication and transmits (i.e., forwarded) its data to the base station (i.e., sink), this approach is in-effective as it costs huge node overhead leads to massive energy consumption, still it is used because its implementation is easy. To solve the issues of flat routing, second class (i.e., hierarchical routing) is introduced, where a tree-like structure is formed. The main outline of this approach is to make some nodes as special nodes (i.e., sometimes called as a leader, cluster head, etc.). Again this class is divided into subclasses: cluster, tree, and hybrid. In the first subclass entire region of interest is split into multiple clusters, where each cluster is designated by a cluster head. The main reason of introducing a cluster head; it acts differently. When sensor nodes sense its data after which they forward their data to their respective cluster head, then each CH aggregate the data and forward its data towards the sink, hence using this approach the number of transmissions becomes less due to CH data aggregation and leads to minimum energy consumption.

Therefore with the clustering concept, many researchers have proposed many clustering based routing protocol. Where LEACH [5] is one of the most well-established cluster-based routing protocol. It starts with the cluster head election, where each node participated in CH election, but only p% nodes are randomly elected as cluster head. Later using Euclidean distance, each node latches with the minimum distance CH nodes and form p clusters. During transmission, each cluster member node transmits its data packets to its allocated CHs, after which each Cluster head process the data aggregation and transmitted towards the base station (i.e., sink). There are many shortcomings of LEACH routing protocol, hence in order to rectify all the problems, researchers have proposed many routing protocol HEED [12], HEER [15], Q-LEACH [16], CL-LEACH [13], where their main aim is to increase the lifetime of wireless sensor networks. Later, PEGASIS [6] a chain based hierarchical routing protocol is proposed. In this protocol, initially, a node farthest from the sink is selected, which is also known starting node. It starts the chain construction using a greedy approach (latch with the nearest and unvisited neighbour). After by considering all nodes, the entire chain is constructed, after which one leader is elected. During data transmission each node sends its data to its chained neighbour node, in such a way it should reach to the leader node, once it reaches the leader node, it aggregates the data and forward it towards the sink. In the second subclass, tree-based routing protocol is proposed. In Tree-Based Clustering (TBC) [8] routing protocol, its cluster head election and clustering formation is similar to LEACH, but to provide the nodes load balancing within a cluster, each cluster is divided into levels, such that CH lies in the 0th level, and for transmission outer level node latch with the inner level node (i.e., 1st level node latch with the 0th level node). At each level, aggregation takes place; once data reached the CH, it will forward (via transmission) the data in the direction of the sink node. In GSTEB [7], a new tree-based routing is proposed, where based on the highest residual energy a node is elected as leader by the sink, after that each node tries to latch with the leader node, in such a manner if a node rely in between leader node and itself, it latches with the neighbouring node, and it forms a tree-like structure. Hence, each node transmits its data to its parent node, where the parent node aggregates the data, similarly once leader receive the data, it transmits directly towards the sink. In the third sub division hybrid routing protocols are proposed which is a combination of flat and hierarchical based routing scheme. In cluster based multipath routing protocol (CBMR [14]), initially by incorporating the factors of residual energy and node degree, cluster heads are elected, and cluster are formed. After which multi-hop path got constructed between sink and CHs. During data transmission data start transmitting from cluster member node to CH and CH node to the sink using a multi-hop path. In clustered aided multipath routing protocol (CAMP [10]), it begins with the division of terrain into equal-sized virtual zones. Each zone nodes participates in the CHs elections using chances of election and nodes which comes under the communication range of the CHs latches with the CHs, each zone node then using intelligent routing process transmits its data towards the sink. In HEEMP [9], a combination of cluster, tree and flat scheme are incorporated in a single scheme, where initially cluster is formed by the CH election process, such that no two CHs overlap with each other (i.e., using dist_threshold factor), later cluster are formed, where within a cluster, instead of transmitting data directly to the CH node, cluster member nodes latches with a parent node. At last CH doesn't directly send its data to the sink, while it forms the route, using route selection mechanism and forwards to the sink.

## 3   System Model

In this entire paper, the following system model is used:

– Network Model: For network model, (x) sensor nodes are deployed over rectangular terrain having terrain area (m × n square meter). Later each node creates a neighbor table G(V, E) consisting of it neighbouring node (V) and distance among nodes (E).
– Network Lifetime Model [9]: This model is used for the critical application where the significance of all nodes are required. Hence lifetime of a network is defined as the number of round where sink receiving all the data till first node dies.
– Energy Model [10]: In this model, energy consumption by a node during transmission and reception is discussed. As mentioned earlier transmission consumes maximum amount of energy. In this paper first order radio model is adopted, where node transmit (p) bits of data to its neighbouring node which is (dist) distance apart.

The following equations of receiving and transmission energy consumption by a node (i.e. mote) are mentioned beneath:-

Where a receiving node consuming the energy is equated as:

$$E_{Rec}(p, dist) = E_{elcs} * p \tag{1}$$

$$E_{Trans}(p, dist) = \begin{cases} E_{elec} * p + E_{fs} * p * dist^2 & if\ (dist < d0) \\ E_{elec} * k + E_{mp} * p * dist^4 & if\ (dist \geq d0) \end{cases} \tag{2}$$

## 4  Problem Statement

Our primary focus for this paper, is to test the efficacy and compare the performance of well-established routing protocol HEEMP, CAMP, LEACH, PEGASIS, GSTEB, TBC by implementing the concept of Sink locality. In the monitored region single static sink is deployed for the data collection. Sink locality is defined as the process of changing the location of the static sink for a given routing protocol.

In this paper, we have considered three static locations of the sink such that i = 3, at center, corner and at far as depicted in Fig. 2. The sink is categorized into static sink and mobile sink. In static sink as the name suggest once the sink is deployed its location cannot be changed. Similarly, in the case of the mobile sink, it can change its location and can move randomly or in a controlled manner. But in many applications, the mobile sink is not possible. Therefore for these application static sink is one of the foremost option. Hence in this article our main aim is to analyze the impact of sink location in the well-established routing algorithm. Figure 3 shows the location of sink at center in a similar fashion Figs. 4 and 5 shows the location of sink at center and far respectively.

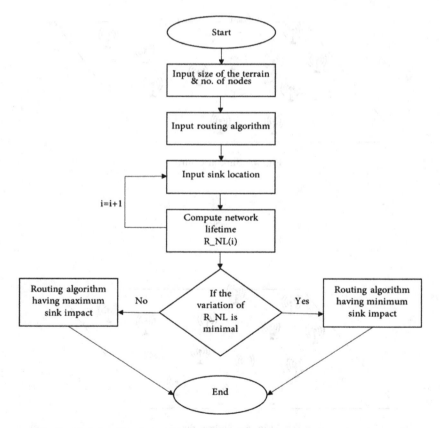

**Fig. 2.**  Sink locality factor (flow chart) of a network based routing protocol.

Therefore this paper helps us to find which routing protocol gives a better result and at which location. Also help us to find, whether designated routing protocol has a uniform or non-uniform impact on the sink location.

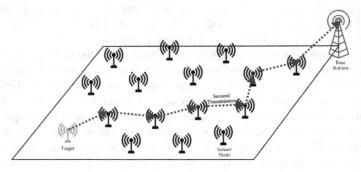

**Fig. 3.** Sink at corner.

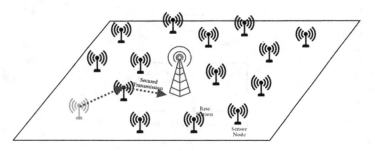

**Fig. 4.** Sink at centre.

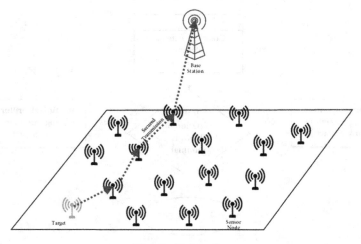

**Fig. 5.** Sink at far.

# 5 Comparative Analysis and Simulation Results

In this segment, we demonstrate the experimental results of the existing hierarchical based routing algorithm (HEEMP, CAMP, GSTEB, TBC, PEGASIS, LEACH) with an impact of sink location. All the simulations are performed using parameters mentioned in Table 1 over MATLAB.

**Table 1.** Simulation parameters

| Parameter | Values |
|---|---|
| Size of a data packet | 2k bits |
| Nodes number deployed over terrain | 150 |
| Terrain Size | $200 \times 200$ (m$^2$) |
| Initial energy of each node | 0.5 J |
| Each nodes maximum communication range | 30 m |
| Size of Control packet | 200 bits |
| Data packet aggregation $(E_{DA})$ energy consumption | $5 \times 10^{-9}$ J/bit/signal |
| Free space $(E_{fs})$ energy consumption | $10 \times 10^{-12}$ J/bit/m$^2$ |
| Multipath $(E_{mp})$ energy consumption | $0.0013 \times 10^{-12}$ J/bit/m$^4$ |
| Transceiver $(E_{Trans\_elec})$ energy consumption | $50 \times 10^{-9}$ nJ/bit |
| Receiver $(E_{Rec\_elec})$ energy consumption | $50 \times 10^{-9}$ nJ/bit |

We have chosen the sink location at (center, corner and far) of the terrain. To evaluate the aforementioned algorithm we have considered network lifetime (FND statistics) as a measure. Sink locality is defined as, the process of changing location of the sink by following the constant constraints (i.e., terrain size, deployed number of nodes, the sink location, routing protocol, & other parameter mentioned in Table 1). Here each above mentioned routing protocol is simulated against varying location of the static sink by placing it, at center, corner of the terrain or far from the terrain (i.e., sink must be d0 distance away from any node of the terrain) in the terrain size of $200 \times 200$ m$^2$. Figure 6 depicts the impact of sink location on network lifetime (i.e., FND statistics), it clearly shows that LEACH protocol works best when sink placed at center but it goes on decreasing when the location is at corner and far because of the load on the cluster head. Hence network lifetime keeps on degrading from corner to far. Figure 7 shows PEGASIS has a less impact on sink locality. Because in PEGASIS, initially chain construction method got invoke in a greedy manner and maximum residual energy node elected as a leader in each round.

In Fig. 8 similar to LEACH routing it is not consistent under sink locality factor because of the load in CHs.

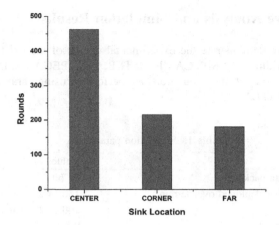

**Fig. 6.** FND statistics of LEACH sink locality.

While GSTEB sink locality factor is somehow better than LEACH and TBC, from Fig. 9 it can be depicted that GSTEB algorithm is nearly consistent for center and corner but for far it is not consistent. Hence within the network terrain, GSTEB network lifetime has less impact on sink locality. But at far it has more impact because of their tree construction and load on leader node.

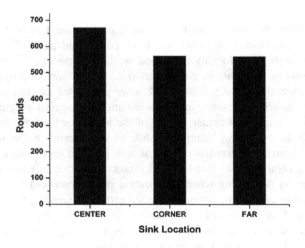

**Fig. 7.** FND statistics of PEGASIS sink locality.

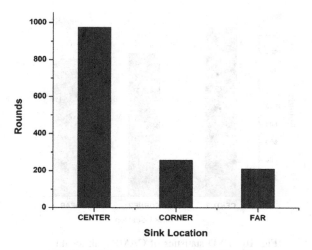

**Fig. 8.** FND statistics of TBC sink locality.

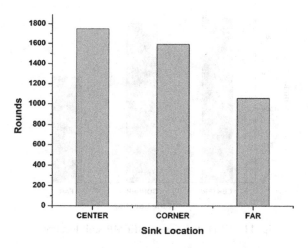

**Fig. 9.** FND statistics of GSTEB sink locality.

Figure 10 depicts that CAMP has a less variation on sink locality factor, as a sink at center and corner are nearly the same. Because in CAMP intelligent routing process works, which helps to make entire network uniformly (wrt energy consumption). Figure 11 clearly shows that HEEMP has the highest FND statistics among others when the sink is placed at center and corner. While at far CAMP outperforms HEEMP. Hence, sink locality has a minimal impact on HEEMP performance.

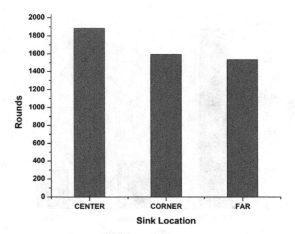

**Fig. 10.** FND statistics of CAMP sink locality.

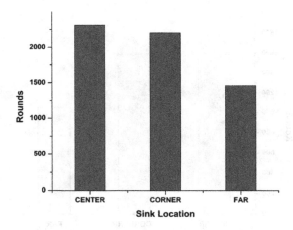

**Fig. 11.** FND statistics of HEEMP sink locality.

## 6   Conclusion

In wireless sensor networks network lifetime and sink locality are the demanding issues. In this paper well established routing protocols have been gone through under these issues. Therefore, various hierarchical routing scheme evaluates the sink locality of above mentioned scheme. At first, we simulate the various existing hierarchical network layer based routing protocol for maximizing the lifetime of a network. Routing protocols like HEEMP, CAMP, GSTEB, TBC, PEGASIS and LEACH have been simulated in this paper against the sink locality factor. It is concluded that PEGASIS and CAMP have the best SINK LOCALITY FACTOR because they have more consistent FND statistics towards sink location throughout. In a similar fashion, all the aforementioned algorithms performed well when the sink is placed at center. Similarly, performed worst when the sink is placed at far. We also confirm that TBC and LEACH have the worst sink locality factor. From our simulation, we also observe that HEEMP outperforms against the above mentioned network layer routing scheme when the sink is placed at center and corner. While CAMP outperforms when it is placed at far.

# References

1. Yick, J., Mukherjee, B., Ghosal, D.: Wireless sensor network survey. Comput. Netw. **52**(12), 2292–2330 (2008)
2. Sarkar, A., Murugan, T.S.: Cluster head selection for energy efficient and delay-less routing in wireless sensor network. Wirel. Netw. **25**(1), 303–320 (2019)
3. Abbasi, A.A., Younis, M.: A survey on clustering algorithms for wireless sensor networks. Comput. Commun. **30**(14–15), 2826–2841 (2007)
4. Tamandani, Y.K., Bokhari, M.U.: The impact of sink location on the performance, throughput and energy efficiency of the WSNs. In: 2015 4th International Conference on Reliability, Infocom Technologies and Optimization (ICRITO) (Trends and Future Directions), pp. 1–5. IEEE (2015)
5. Heinzelman, W.R, Chandrakasan, A., Balakrishnan, H.: Energy-efficient communication protocol for wireless micro sensor networks. In: Proceedings of the 33rd Annual Hawaii International Conference on IEEE System sciences, pp 1–10 (2000)
6. Lindsey, S., Raghavendra, C.S.: PEGASIS: power-efficient gathering in sensor information systems. In Aerospace Conference Proceedings, pp 1125–1130. IEEE (2002)
7. Han, Z., Wu, J., Zhang, J., Liu, L., Tian, K.: A general self-organized tree-based energy-balance routing protocol for wireless sensor network. IEEE Trans. Nucl. Sci. **61**(2), 732–740 (2014)
8. Kim, K.T, Lyu, C.H, Moon, S.S., Youn, H.Y.: Tree-based clustering (TBC) for energy efficient wireless sensor networks. In 2010 IEEE 24th International Conference on Advanced Information Networking and Applications Workshops (WAINA), pp. 680–685. IEEE (2010)
9. Sajwan, M., Gosain, D., Sharma, A.K.: Hybrid energy-efficient multi-path routing for wireless sensor networks. Comput. Electr. Eng. **67**, 96–113 (2018)
10. Sajwan, M., Gosain, D., Sharma, A.K.: CAMP: cluster aided multi-path routing protocol for wireless sensor networks. Wirel. Netw. **25**, 2603–2620 (2018)
11. Kumar, D., Aseri, T.C., Patel, R.B.: EEHC: Energy efficient heterogeneous clustered scheme for wireless sensor networks. Comput. Commun. **32**(4), 662–667 (2009)
12. Younis, O., Fahmy, S.: HEED: a hybrid, energy-efficient, distributed clustering approach for ad hoc sensor networks. IEEE Trans. Mob. Comput. **3**(4), 366–379 (2004)
13. Marappan, P., Rodrigues, P.: An energy efficient routing protocol for correlated data using CL-LEACH in WSN. Wirel. Netw. **22**(4), 1415–1423 (2016)
14. Sharma, S., Jena, S.K.: Cluster based multipath routing protocol for wireless sensor networks. ACM SIGCOMM Comput. Commun. Rev. **45**(2), 14–20 (2015)
15. Yi, D., Yang, H.: HEERA delay-aware and energy-efficient routing protocol for wireless sensor networks. Comput. Netw. **104**, 155–173 (2016)
16. Manzoor, B., et al.: Q-LEACH: a new routing protocol for WSNs. Procedia Comput. Sci. **19**, 926–931 (2013)

# Development of a WSN Based Data Logger for Electrical Power System

Sanjay Chaturvedi$^{(\boxtimes)}$ , Arun Parakh, and H. K. Verma

Shri G.S. Institute of Technology & Science, Indore, India
sanjaychaturvedi36@gmail.com,
arunparakh.eed@gmail.com, vermaharishgs@gmail.com

**Abstract.** Proper operation of any process is ensured by its continuous monitoring and monitoring of a system depends solely on the availability of data about the process variables. Traditionally wired sensor networks have been used to collect the data of the process variables as it enables the remote monitoring of the entire process from a place. The major problem with the wired sensor network includes the time needed to realize these networks as well as the cabling needed for these networks. Wireless sensor networks appear to be the best alternative to wired sensor networks in terms of ease of implementation and the time and space needed for implementation. But a wireless sensor network requires an ad-hoc network for communication and data transfer. Also, monitoring systems collect a large amount of data that itself presents a problem of handling bulk data and logging this data for future studies. For that, we need to develop a database to save this data and a mechanism for logging this data into the database automatically.

This paper presents development of two vital building blocks needed in the making of a wireless sensor network, using open source tools and open standards. First is a wireless serial link for connecting nodes in a wireless sensor network. Second is the development of a database and a data logging system to log the data received from the sensor nodes into the database. The wireless serial link is realized using XBee radio modules which work on ZigBee protocol. Open source database management system, MySQL, is used to make database and to automate the task of logging the data into the database, we have used Python scripts. Both the developed systems have been used in our work of "wireless sensor network based power monitoring system". And the results obtained there validate our developed systems.

**Keywords:** Wireless sensor network · Ad-hoc network · ZigBee · Python · MySQL · Open source

## 1 Introduction

A wireless sensor network is defined as a network of sensors, spatially distributed which continuously measures the value of the parameters related to the system being monitored and communicates the sensed data to a central unit through wireless links. Wireless sensor networks are being used in a variety of applications which includes monitoring of health [1], power in smart grids [2], flame detection [3], water content in

© Springer Nature Singapore Pte Ltd. 2019
A. B. Gani et al. (Eds.): ICICCT 2019, CCIS 1025, pp. 72–83, 2019.
https://doi.org/10.1007/978-981-15-1384-8_7

civil engineering works [4], factory automation systems [5] and for energy management systems [6] to name few.

The three main components of a wireless sensor network include sensor nodes, wireless communication link, and a central unit. Sensor nodes are based on some processing element like a micro-controller which when integrated with sensors and a wireless transceiver forms the sensor node. Wireless communication is done either by the use of available wireless network services or using an ad-hoc network. And the central unit is a generally a computer that is responsible for coordinating with all the nodes in the sensor network and managing the data received from the sensor nodes. This work is focused on the development of two of the three main components of a wireless sensor network viz. the wireless network and the central unit.

As mentioned above, for the wireless network we have the option of using the already available communication networks like the GSM and GPRS and secondly, we have the option to develop ad-hoc network using available wireless technologies like Bluetooth, WiFi, and ZigBee. The problem with the existing networks such as GSM and GPRS is of their availability and quality of service (QoS) at the desired location. Also using these networks makes us dependent on a third party for our services and eventually they add up to the operational cost of our monitoring systems. One more issue with the use of these networks is that they offer less or no choice to customize the network to our needs. Under aforementioned conditions, ad-hoc wireless networks appear to be a better choice over the existing wireless networks. These ad-hoc wireless networks are customizable and can be deployed anywhere and are independent of any third party service hence makes possible the use wireless sensor networks as standalone monitoring systems [8].

After deciding to go with the ad-hoc networks, we come to question about the criteria for selecting a wireless technology to develop our network. The selection of wireless technology is based on the analysis of the three aforementioned wireless technologies viz. Bluetooth, Wi-Fi, and ZigBee on certain important criteria. These important criteria are the payload capacity, scalability of the network, security of data transmission, the ability of the network to operate in noisy environments, cost of implementation, power consumption and time to deploy the network [9, 10]. One more important feature to look for is the feasibility of its integration with different wireless technologies, for example, interfacing a Bluetooth network to a Wi-Fi network. After comparing the three technologies on the before mentioned criteria and a comparative study as shown in Table 1, ZigBee technology presents itself as the suitable choice for the development of a wireless network.

The other building block of a WSN, the central unit, is responsible for coordinating with all the nodes in the sensor network and handling the data received from the sensor nodes. Apart from that, we are also interested in saving the data for analyzing the operation of the system being monitored. As we are continuously receiving the data from the sensor network, it presents a challenging problem to manage the data manually; where by "managing the data" we mean the task of requesting data from the sensor nodes, receiving that data and loading it into a database. The best and efficient solution to this problem is to automate the task of data management. The automation should be such that the central unit sequentially requests data from each sensor node in the network, receives that data and loads it automatically into the database.

This automation can be done using a script. Considering the easiness of handling the serial communication and of accessing the database in python we decided to use python for writing our automation script. After done with above setup, we now create a database using open source MySQL database management system. Here we preferred MySQL over other options such as MariaDB, SQLite, MongoDB because being an industry standard it is compatible with nearly all the operating systems.

**Table 1.** Comparison of Wi-Fi, Bluetooth and ZigBee [7]

| Features | Wi-Fi IEEE 802.11 | Bluetooth IEEE 802.15.1 | ZigBee IEEE 802.15.4 |
|---|---|---|---|
| Application | Wireless LAN | Cable Replacement | Control and Monitor |
| Frequency Bands | 2.4 GHz | 2.4 GHz | 2.4 GHz 868 MHz 815 MHz |
| Nodes Per Network | 30 | 7 | 65000 |
| Bandwidth | 2–100 Mbps | 1 Mbps | 20–250 Kbps |
| Range (Meters) | 1–100 | 1–10 | 1–75 and more |
| Topology | Tree | Tree | Star, Tree, Cluster Tree, Mesh |

In this paper, we present the process of creating a wireless link using an open standard, ZigBee. Then we continue with the configuration of the central unit. Here central unit is a computer with free and open source Linux operating system. We have used Ubuntu 16.04 in our system. In that we create a database using open source database management system MySQL and automate the task of data logging using Python, which is again open source. Use of open source tools reduces significantly the development cost of a WSN based system hence promoting their use for the development of monitoring and control systems.

The paper proceeds with the introduction and development of a ZigBee based wireless network in the next section. The following section deals with the development of the data logging system. In section four, we test this network in a wireless sensor network prototype developed to measure electrical parameters at a power distribution point. The last section concludes our work.

## 2    Development of Wireless Network

This section present the steps involved in the creation of the wireless network. It starts with a brief introduction to ZigBee technology and then proceeds to development of the wireless link and its testing.

## 2.1  ZigBee

ZigBee is a communication standard built on top of IEEE 802.15.4 standard, specially built for wireless personal area network. It is developed by ZigBee Alliance and is an open standard hence it can incorporate in its network, items from multiple manufacturers [14]. ZigBee network is a low power consuming network is being used widely in control and monitoring applications [11]. Also, the low cost of ZigBee as compared to other proprietary wireless technology like Wi-Fi make it a preferable option for developing an ad-hoc network. One important feature of ZigBee Network is that all the communication are encrypted by default which assures a bare minimum level of security in data transmission. Also, ZigBee supports a variety of network topology which gives users multiple options for customization.

As Mentioned above, a ZigBee network supports multiple network topologies. But all the ZigBee networks have a maximum three types of devices which are a coordinator, router and end device. These devices are identical radio modules being configured to work in different roles. And a ZigBee network must have one and only one coordinator. Depending on the topology of the network there can be routers otherwise end devices along with a coordinator are a must for creating a ZigBee network. A ZigBee network depending upon the application can work either in AT or in API mode.

## 2.2  Setting Up the Network

In this work, we are developing a wireless link that will have a coordinator and an end device and since we are focused on creating a wireless serial link hence our network will work in AT mode. For our network, we are using XBee radio modules which are produced by Digi-Key specifically for developing ZigBee networks. By default the XBee modules come configured as "Router AT" therefore to be used in a network, we need to update their firmware as per their assigned roles. We update the firmware using XCTU software tool provided by Digi-Key free of cost for configuration and setup of XBee radio modules. In standard practice, the firmware of all the XBee modules is updated before use. Note that the XBee modules to be used as end devices are configured as "Router" in their respective modes of operation.

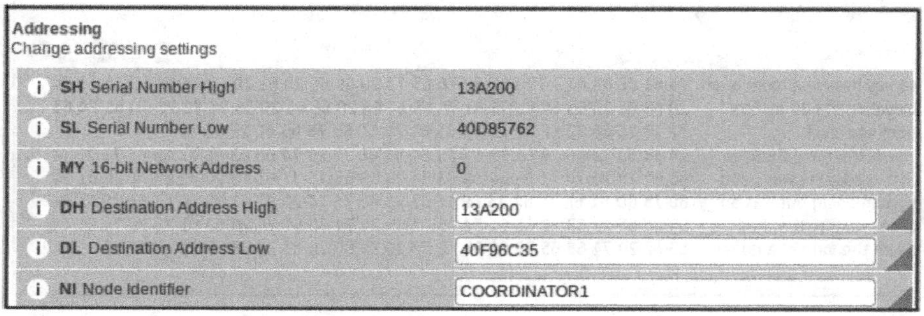

| Addressing | |
|---|---|
| Change addressing settings | |
| ⓘ SH Serial Number High | 13A200 |
| ⓘ SL Serial Number Low | 40D85762 |
| ⓘ MY 16-bit Network Address | 0 |
| ⓘ DH Destination Address High | 13A200 |
| ⓘ DL Destination Address Low | 40F96C35 |
| ⓘ NI Node Identifier | COORDINATOR1 |

**Fig. 1.** Coordinator setup in XCTU

Once the XBee radio module is configured as a coordinator and end device, we now have to set up certain minimum parameters in each module as shown in Figs. 1 and 2. The list of parameters to be set up is as follows:

- PAN ID
- NODE ID
- Source Address High(SH)
- Source Address Low(SL)
- Destination Address High(DH)
- Destination Address Low(DL)
- Module Address(MY)

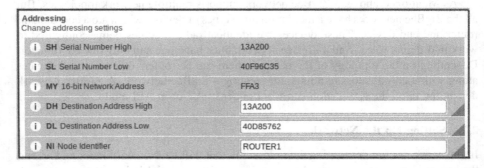

**Fig. 2.** Router setup in XCTU

The modules configured as Router will have the Source address of the Coordinator in their destination address. The "MY" address of the Coordinator is set to a default value i.e. "0". The Router can have any value of MY address from 1 to FFFF, but it should be unique for each Router or End device.

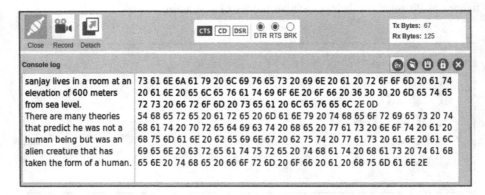

**Fig. 3.** Data sent from Router in Blue colour and received at coordinator in Red colour (Color figure online)

## 2.3  Testing the Network

Once both the modules have been configured and set up for their determined role, we now test the wireless link between the coordinator and the node using the console in XCTU. The connection is activated by closing the connection in the console. A message is sent from the console of the router and the coordinator and we check whether the same message is received at the coordinator and router console respectively or not, as shown in Figs. 3 and 4.

**Fig. 4.** Data received by Router in Red colour and sent from Coordinator in Blue colour (Color figure online)

# 3  Development of Data Logging System

This section deals with the development of the central unit where we will have our database. After an introduction to the software's used for the development and management of the database and for the automation, we proceed to the steps involved the development of data logging system.

## 3.1  MySQL and Python

A database stores data and to manage this data we need a database management system. Nowadays, relational database management system are used for management of bulk data. In terms of RDBMS, a database is a collection of tables where tables are defined as matrix with data.

Among the available options of RDBMS we selected MySQL for our purpose. MySQL is an open source relational database management system [12]. It is compatible with many platforms and works easily with many programming languages. MySQL is customizable and its GPL license allows its modification to fit to users requirements. Also MySQL supports large databases.

Use of Python in our work is motivated by the fact that Python is open source and is compatible with all the major platforms [13]. Python is also easy to use and supports object-oriented programming, structured programming and many other programming paradigms. It also eases the task of serial data transfer and accessing database. Ultimately the selection a programming language for any work is done on the basis of its compatibility with the platform and with the database application being used. And in both the cases Python presents itself as a suitable option.

### 3.2 Setting Up the Central Unit

We open the MySQL database management system in Linux terminal using the log-in password that is set up at the time of installation as shown in Fig. 5.

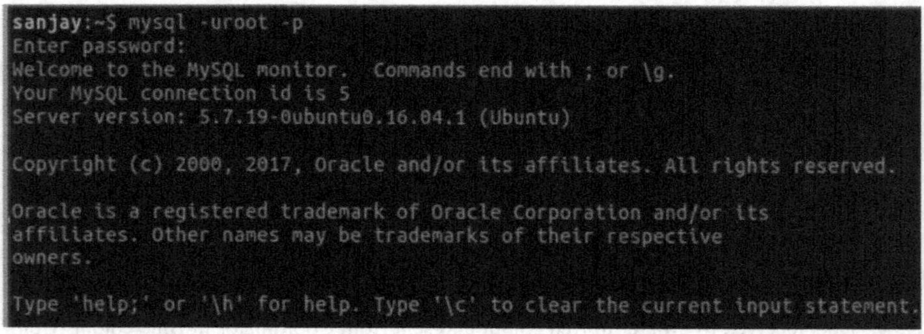

```
sanjay:~$ mysql -uroot -p
Enter password:
Welcome to the MySQL monitor.  Commands end with ; or \g.
Your MySQL connection id is 5
Server version: 5.7.19-0ubuntu0.16.04.1 (Ubuntu)

Copyright (c) 2000, 2017, Oracle and/or its affiliates. All rights reserved.

Oracle is a registered trademark of Oracle Corporation and/or its
affiliates. Other names may be trademarks of their respective
owners.

Type 'help;' or '\h' for help. Type '\c' to clear the current input statement.
```

**Fig. 5.** MySQL opened in Ubuntu terminal

Now we create a database using the command "create database <name of the database>" and within that database we make a table using the command "create table <table name> (table attributes)" as shown in Fig. 6.

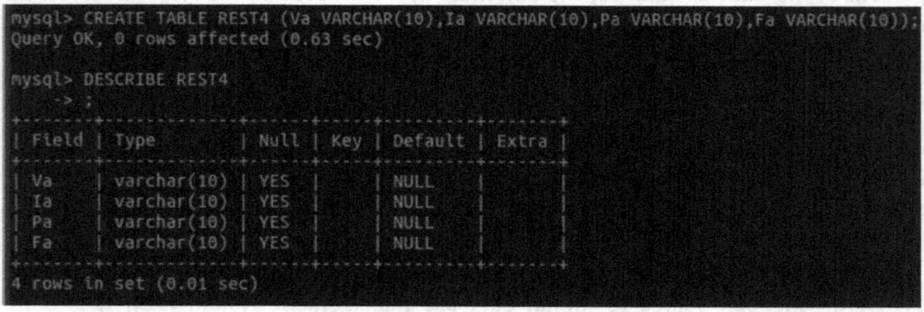

```
mysql> CREATE TABLE REST4 (Va VARCHAR(10),Ia VARCHAR(10),Pa VARCHAR(10),Fa VARCHAR(10));
Query OK, 0 rows affected (0.63 sec)

mysql> DESCRIBE REST4
    -> ;
+-------+-------------+------+-----+---------+-------+
| Field | Type        | Null | Key | Default | Extra |
+-------+-------------+------+-----+---------+-------+
| Va    | varchar(10) | YES  |     | NULL    |       |
| Ia    | varchar(10) | YES  |     | NULL    |       |
| Pa    | varchar(10) | YES  |     | NULL    |       |
| Fa    | varchar(10) | YES  |     | NULL    |       |
+-------+-------------+------+-----+---------+-------+
4 rows in set (0.01 sec)
```

**Fig. 6.** Creating a table in a database in MySQL

Once done with the database, we now move to write the Python script in Python IDLE. Libraries used are "serial" for serial port communication and

"MySQLdb" to access the database. After importing the libraries we now make an object to read and write using the serial port as shown in Fig. 7.

```
#------Establishing serial communication-------------
ser = serial.Serial('/dev/ttyACM0', baudrate = 9600)# port name and baud rate.
time.sleep(3)
```

**Fig. 7.** Connection with the serial port

Then we create a connection with the database in order to access and make changes in it. For that, we have to declare the 'host' where MySQL is present and the 'port' assigned for MySQL use. Along with that, we have to declare the "user" and "password" used for accessing the database and finally the name of the database being used, as shown in Fig. 8.

```
#---------------Libraries imported------------------

import serial  # Serial library
import MySQLdb # MySQL connector
import time

#------Establishing connection to MySQL databse------

db = MySQLdb.connect(host="127.0.0.1", # "localhost"
                     port=3306, # port for MySQL
                     user="root", # username
                     passwd="shivoham", # password
                     db="test") # name of the database
cur = db.cursor()
```

**Fig. 8.** Connection with the database

Once the serial connection is made and the database is accessed, we now request the data from the sensor node. The idea is to include a simple token of acknowledgment for initiating any kind of data transfer. Through coordinator, the script broadcasts a letter as a token and the node receives that token. Node verifies the received token and transmits back the data that is associated with that token, as we will see in the next section. This simple method of using a token of acknowledgment synchronizes the communication between the central unit and the node.

The data sent by the nodes to the coordinator are then accessed through the serial port and inserted into the database using "INSERT INTO" command, as shown in Fig. 12. And in this manner, the above process keeps repeating in a loop without any need for human intervention.

## 4    Implementation of Developed Systems

We test the above developed wireless communication link and the data logging system in one of our work where we developed a Wireless Sensor Network based power monitoring system for electrical power distribution networks, as shown in Fig. 9.

**Fig. 9.** Setup of WSN based power monitoring system

In that work, we built a wireless power monitoring system to monitor the electrical parameters at any distribution point in an electric power distribution network. The system has sensor nodes which are built using open source microcontroller development board, Arduino UNO. Sensor node measures the value of voltage, current, frequency and power at a distribution point and on request from the central unit it transmits the requested data back to it, as shown in Figs. 10 and 11.

```
while 1:
    ser.write('a')
    voltageA = getValues()
    ser.write('b')
    currentA = getValues()
    ser.write('c')
    powerA = getValues()
    ser.write('d')
    frequencyA = getValues()
    time.sleep(15)
```

**Fig. 10.** Data request by central unit to the node

```
/---------Serial Transmission of codes |--------------------------
if (Serial.available() > 0) {
   int inByte = Serial.read();

   switch (inByte) {
     case 'a':
       Serial.println(voltage);
       break;
     case 'b':
       Serial.println(current);
       break;
     case 'c':
       Serial.println(power);
       break;
     case 'd':
       Serial.println(frequency);
       break;
     default:
     Serial.println("00");

   }
```

**Fig. 11.** Sensor node built using Arduino UNO

The data received at the central unit is then parsed to obtain useful information and then is loaded into the database, as shown in Fig. 12.

```
--------Inserting data into MySQL databse----------------

sql = "INSERT INTO REST4(`Va`, `Ia`, `Pa`, `Fa`) VALUES (%s, %s, %s, %s)"
data = (voltageA , currentA, powerA, frequencyA)
cur.execute(sql,data)
db.commit()
```

**Fig. 12.** Loading data into database

We then verify the data logging system by the data received at the serial port and the data loaded into the table in the database, as shown in Fig. 13. We find that the data logging system works perfectly and does its job of loading the serial data into the database.

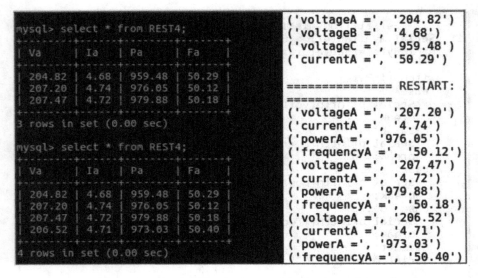

**Fig. 13.** Data loaded into database and received at serial port

## 5  Conclusion and Future Scope

A ZigBee based wireless serial link has been created along with a data logging system. The task of requesting the data from nodes and loading the received data into the database has been automated using Python. We have presented that we can develop the components of a wireless sensor network (WSN) using open source tools like MySQL, Python and open standards like ZigBee, which ultimately reduces the development cost of a WSN based monitoring and control system.

In future we can implement a mesh network using the same XBee modules. Use of a mesh network will enable us to expand our network. Also, it allows us to have non line of sight communication between nodes and the coordinator unit by routing through the intermediate nodes.

## References

1. Lee, D.S., Lee, Y.D., Chung, W.Y., Myllyla, R.: Vital sign monitoring system with life emergency event detection using wireless sensor network. In: 2006 IEEE SENSORS, pp. 518–521. IEEE, October 2006
2. Yerra, R.V.P., Bharathi, A.K., Rajalakshmi, P., Desai, U.B.: WSN based power monitoring in smart grids. In: 2011 Seventh International Conference on Intelligent Sensors, Sensor Networks and Information Processing, pp. 401–406. IEEE, December 2011
3. Cheong, P., Chang, K.F., Lai, Y.H., Ho, S.K., Sou, I.K., Tam, K.W.: A ZigBee-based wireless sensor network node for ultraviolet detection of flame. IEEE Trans. Ind. Electron. **58**(11), 5271–5277 (2011)

4. Ong, J.B., You, Z., Mills-Beale, J., Tan, E.L., Pereles, B.D., Ong, K.G.: A wireless, passive embedded sensor for real-time monitoring of water content in civil engineering materials. IEEE Sens. J. **8**(12), 2053–2058 (2008)
5. Korber, H.J., Wattar, H., Scholl, G.: Modular wireless real-time sensor/actuator network for factory automation applications. IEEE Trans. Ind. Inform. **3**(2), 111–119 (2007)
6. Han, D.M., Lim, J.H.: Smart home energy management system using IEEE 802.15. 4 and ZigBee. IEEE Trans. Consum. Electron. **56**(3), 1403–1410 (2010)
7. Aju, O.G.: A survey of ZigBee wireless sensor network technology: topology, applications and challenges. Int. J. Comput. Appl. **130**(9), 47–55 (2015)
8. Buratti, C., Conti, A., Dardari, D., Verdone, R.: An overview on wireless sensor networks technology and evolution. Sensors **9**(9), 6869–6896 (2009)
9. Gungor, V.C., Lu, B., Hancke, G.P.: Opportunities and challenges of wireless sensor networks in smart grid. IEEE Trans. Ind. Electron. **57**(10), 3557–3564 (2010)
10. Dzung, D., Apneseth, C., Endresen, J., Frey, J.E.: Design and implementation of a real-time wireless sensor/actuator communication system. In: 2005 IEEE Conference on Emerging Technologies and Factory Automation, vol. 2, p. 10. IEEE, September 2005
11. de Almeida Oliveira, T., Godoy, E.P.: ZigBee wireless dynamic sensor networks: feasibility analysis and implementation guide. IEEE Sens. J. **16**(11), 4614–4621 (2016)
12. MySQL Server. https://dev.mysql.com/doc/refman/5.7/en/what-is-mysql.html
13. About Python. https://www.python.org/doc/essays/blurb/
14. ZigBee Technology Overview. https://www.digikey.in/en/ptm/f/freescale-semiconductor/zigbee-technology-overview

# Embedded Subscriber Identity Module
# with Context Switching

Kaushik Sarker⊙ and K. M. Muzahidul Islam(⊠) ⊙

Department of Software Engineering, Daffodil International University,
Dhaka, Bangladesh
kaushik.swe@daffodilvarsity.edu.bd,
muzahidul670@diu.edu.bd

**Abstract.** Telecommunications technology user wants quality channels including security, customization, personalization and autonomy. These requirements are encouraging customers to use multiple identity modules. Use of multiple modules brings the necessity of caring multiple cellular phones or multiple cellular units in a mobile device. Moreover, the switching between the identity module is physical and less secure with the possibility of an identity module cloning or loss of module in case of theft. Several research works have been found in the area of embedded Subscriber Identity Module (eSIM) and Virtual Subscriber Identity module (VSIM). However, the limitations of both eSIM and VSIM have given the authors of this paper a scope to study and come up with a solution by proposing a new model. In the proposed model authors have considered the benefits of eSIM and VSIM together and reducing the limitations such as switching between the modules, parallel activation of the modules, module cloning etc. At the end of the paper, authors have compared ten such limitations, known as features with the existing models and presented a graphical simulation of the proposed model with the proposed methodology of context switching between embedded subscriber identity modules.

**Keywords:** eSIM · VSIM · Multiple SIM activation · SIM Context Switching · Subscriber Profile Manager

## 1 Introduction

Gradually consumers are shifting from the bulky device to tiny device. Moreover, their needs have grown up invariably. On the contrary, mobile companies are trying to connect customers' needs into a small space [7]. In this circumstance, the conventional removable SIM device has occupied a large amount of space and created some limitation as SIM can be removed and cloned without user's permission which has created vulnerability by producing immoral activities.

Each device holds IMEI which is stored unsafely into the device. Moreover, IMEI Tracking Technology is not accessible to the public and many of them do not conserve the IMEI number in a safe place. Besides, IMEI number can be changed or masked.

GSMA introduces Embedded Subscriber Identity Module (eSIM) technology to overwhelm some vulnerabilities of removal SIM and this technology stores the

A. B. Gani et al. (Eds.): ICICCT 2019, CCIS 1025, pp. 84–97, 2019.
https://doi.org/10.1007/978-981-15-1384-8_8

customer identity profiles and installs it to the device through a central server [3]. In contrast, the VSIM concept is the revolution of ensuring user autonomy. However, unsafe IMEI problems are still present. After the introduction the limitation of eSIM and VSIM have been discussed in the literature review. Next, in the research methodology section a conceptual architecture has been proposed. After that, a simulation is provided in the result and discussion section along with the comparison between existing and the proposed model and finally concluded in the next section.

## 2  Literature Review

Inadequate network coverage led to the interrupting of communication. Besides, different network operators offer special tariffs with different bundles. On the other hand, customers do not want to detach the old number. As a result, consumers are motivated to use multiple Subscriber Identity Modules one for personal and one for official use, etc. [13].

### 2.1  Architectures with Their Limitations

**GSMA eUICC Architecture.** Embedded UICC (eUICC) requirements, definitions, roles and procedures are standardized by European Telecommunications Standards Institute in the year 2013 [15]. Besides, GSMA develops a standard specification based on ETSI which covered OTA installations, enablement, disablement and profile deletion process to the eUICC [4].

Proprietary five interfaces (1) Subscription Manager-Data Preparation (SM–DP), (2) Subscription Manager–Secure Routing (SM–SR), (3) Mobile Network Operator (MNO), (4) Certificate Issuer, and (5) eUICC Vendor have been connected to eight technical connection of OTA profile management which helped enough to well-define the eSIM technology [15]. Moreover, it is a rewritable built-in hardware component that allows the SIM profile installation remotely over the air by a Mobile Network Operator (MNO) through a universal server [1] which has three categories namely (1) machine to machine (M2M), (2) Machine to Person (M2P) and (3) Hybrid. Besides, four groups of players are involved, they are (1) eSIM Vendor, (2) Original Equipment Manufacturer (OEM), (3) Mobile Virtual Network Operator, and (4) Independent Profile Manager. In contrast, to serve this technology to the consumer, three network configurations have been maintained namely (1) OEM-Centered, (2) MNO-Centered, and (3) Independent Party. However, Mobile Network Operator (MNO) sells M2P categories eSIM. Although, this technology has been designed to reduce the space and cost of the device including multiple SIM Profiles installation without considering simultaneous activations. Nonetheless, in this technology to change the SIM Profile, it does not require exchange of physical components [3–6]. Besides, eSIM technology reduced integration, testing and handling costs for M2M SIM products which have done a little change and used existing SIM factors including (1) MFF1, MFF2 for embedded, (2) 2FF, 3FF for removable and same hardware component [5].

**Limitations of GSMA eUICC Architecture.** Performance and security issues including lacking in requirements [9, 11] have been found out through analyzing all documents and thirteen eUICC procedures including Registration, Profile Verification, Ordering, Download and Installation, Enabling, Disabling, etc. [4, 13, 14]. Many of the security issues already solved by author A. Vesselkov [15]. However, some limitations still exist and they are related to problem with multiple SIM activation, sharing of contact from inactive SIM, longer response time, managing profile remotely by subscriber.

### 2.2 Virtual SIM (VSIM) Architecture with Limitations

**Virtual SIM (VSIM) Architecture.** Virtual Subscriber Identity Module (VSIM) which is introduced in the year 2012 provides different framework and technique epitomes for holding and exchanging individual information contained inside the memory of portable handset gadgets. A few with versatile handset with the capacity to download individual information to a server after validation and Verification, whereas, a few use alphanumeric passwords for client confirmation and check purposes [12].

**Limitations of Virtual SIM (VSIM).** VSIM is the first concept where user autonomy has been ensured by providing some legal management power. However, some lack of requirements, performance and security issues still present [12], which are related to issues of SIM cloning, problem with multiple SIM activation and longer response time.

## 3   Research Methodology

eSIM Context Switching model is a combination of eSIM [3–5] and VSIM [12] model. Although, some modification has been made to reduce cost, space and increases usability. Moreover, this modification also reduced the eSIM and VSIM limitations which are mentioned before. The following four primary features have been considered along with six other features while proposing this model.

1. Multiple SIM activation: Multiple SIM activation is done through artificial switching which activate the SIM according to the user activity by integrating M2M [3] with artificial methodology.
2. Sharing: Closed or offline SIM can share necessary data with the device such as a contact list and others.
3. Reducing response time: If SIM is closed, busy, waiting or already has an established voice channel then it can send feedback without requesting the end device which can reduce time than the traditional system.
4. Preventing sim cloning: This model does not allow IMEI modification through the user rather IMEI is updated continuously.

## 3.1    Proposed System Architecture

**Context Switching Model** [10, 12].

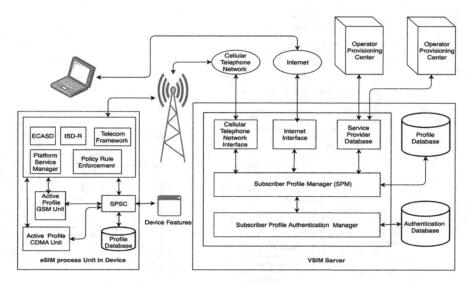

**Fig. 1.**  Context switching model diagram.

**Profile Activation/Deletion By network Request.**

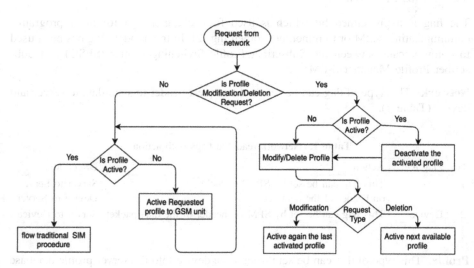

**Fig. 2.**  Profile activation/deletion by network request (in device)

**Request routing through Subscriber Profile Manager.**

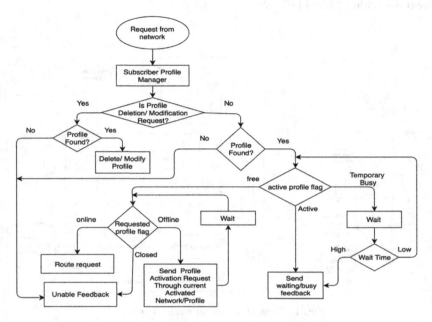

**Fig. 3.** Request routing through subscriber profile manager (Server)

### 3.2   Status Flag Details

The flag is a predefined bit which is used for leaving a sign for other programs, communicating M2M or to remember something [2]. In this model, flag has been used to communicate between the Subscriber Profile Switching Center (SPSC) and Subscriber Profile Manager (SPM).

**Network.** This type of flag can be set on the request header to send data to server and device (Table 1).

**Table 1.**  Network readable flags with actions

| No | Flag name | Action | Directions |
|---|---|---|---|
| 1 |  | This flag can be set by SPSC or SPM on header of the data packet | Server to Server Device to Server |
| 2 | Device | This flag can be set by SPM on header of the data packet | Server to Device |

**Profile.** This type of flag can be set through the device into the server profile database (Table 2).

**Table 2.** SPSC or SPM readable flags with actions

| No | Flag name | Action |
|---|---|---|
| 1 | Active | When any communication channel has to be established between two end device then active profile can be set |
| 2 | Online | It does indicate that Profile is active and it has established network |
| 3 | Offline | This flag denoted that currently profile not active or has not established network but it can be active by SPM request |
| 4 | Closed | It does indicate unable to activate this profile |
| 5 | Temporary Busy | If activate profile receive any data or signal for a while then this flag can be set to the profile |
| 6 | Busy | This flag denoted that it can handle only message request |

### 3.3   Profile Database

Profile database [12] must be at least four attributes, namely (1) serial no, (2) profile raw data, (3) flag, (4) IMEI. This table's name may have depended on the user's primary key. Profile flag and IMEI must have updated continuously through the device.

### 3.4   Steps of Context Switching Model

1. ECASD, ISD-R, Telecom Framework, Platform Service Manager, Policy Rule Enforcement will work like eSIM procedure [10] however has to use Subscriber Profile Switching Center (SPSC) as media to communicate with Subscriber Identity Profile, as shown in "Fig. 1".
2. SPSC will activate profile into CDMA activation unit only through own device request. Steps 3 to 14 are only for the GSM network.
3. When procedure 1 handovers the task to SPSC then it will check the request type. Two types of request can be sent by the GSM network, one of them devoted to profile management, As shown in "Fig. 2". Another, will follow procedures of traditional SIM [8].
4. If SPSC gets any other requests except profile related request from SPM then it will follow the procedure of the eSIM, as shown in "Fig. 2".
5. SPSC must have to send all individual Subscriber Profiles flag (online, offline, etc.) and all information of the device (IMEI, etc.) to the SPM continuously.
6. If there is any data header that contains the SERVER or DEVICE flag then MNO must have to be proceed.
7. Cellular Telephone Network Interface, Internet Interface, Service Provider Database and Subscriber Profile Authentication Manager will follow the VSIM procedure [12]. But it has to use SPM as a media to handle SERVER flagged request or any database related request.
8. Traditional network procedure will be followed when the SERVER or DEVICE flag is missing [8].
9. In the end device, the eSIM procedure will be followed to SIM installation and deletion [10] but it has to be handed over to the SPSC before the final stage of the

procedure. In the server part, the VSIM procedure will be followed to SIM deletion and installation [12] but it has to hand over to the SPM before the final stage.

10. Subscriber Profile Manager (SPM) can get two types of network requests, one of them is Profile related and another is routing related. Both requests first check profile existence from Profile Database, if the profile does not exist then sends acknowledgment which is shown in "Fig. 3". On the other hand, if the profile exists then it follows step 11 or 12.

11. If SPM gets profile deletion or modification request through remote source then it routes that request to SPSC and waits for feedback before action. When it gets feedback from end device then it triggers the action.

12. If SPM gets routing request then it considers profile status, there are three types of profile status available. Profile flags with actions (Table 3).

**Table 3.** Actions for profile flag of active profile

| Profile flag name | Condition check | Action |
|---|---|---|
| Temporary busy | Wait time high | Wait few moments and check again its profile |
| | Wait time low | Send busy feedback |
| Active | Null | Send waiting/Busy feedback |
| Free | Null | Flow procedure "12" |

13. To route the request, it checks three types of flag for the requested profile to take further actions. Profile flag with actions (Table 4).

**Table 4.** Actions of Profile Flag of Requested Profile

| Profile flag name | Action |
|---|---|
| Online | Route the current request |
| Offline | First send profile activation request to the SPSC and wait and check again. |
| Closed | Send unable feedback |

14. To establish a channel between two devices, communicating with the server is required.

## 4  Result and Discussion

Proposed eSIM Context Switching model through merging two systems [3, 12] with the mentioned steps is compared in the following table with eSIM and VSIM. From the table it is observed that the proposed model solves all mentioned issues related to previously mentioned features (Table 5).

**Table 5.** Features comparison with proposed model

| No | Features | eSIM | VSIM | Proposed model |
|---|---|---|---|---|
| 01 | Multi profile active | No [3–5, 15] | No [12] | Yes |
| 02 | Offline profile can share contacts | No [4] | No [12] | Yes |
| 03 | Profile backup from device | Yes [4] | Yes [12] | No |
| 04 | Unable or waiting feedback | From Device [4] | From Device [12] | From SPM |
| 05 | Unable or waiting feedback processing time | Long [4] | Long [12] | Short |
| 06 | User Autonomy | No [4] | Yes [12] | Yes |
| 07 | Device owner authentication Required To add profile | Not Specified | Not Specified | Yes |
| 08 | Profile owner authentication required to add profile | Yes [4] | Yes [12] | Yes |
| 09 | Master password for modification | Not Specified | Yes [12] | Yes |
| 10 | CDMA and GSM Active at a time | No [4] | No [12] | Yes |

## 4.1 Simulation of the Proposed Model

Since this a conceptual model, therefore has been proved by developing a virtual simulator where only context switching part has been considered. To test this model in the simulator, three Cellular Network Towers [16], two Dialer Devices [17] and one receiver device [17] have been used (Fig. 4).

**Fig. 4.** Simulator color specifications.

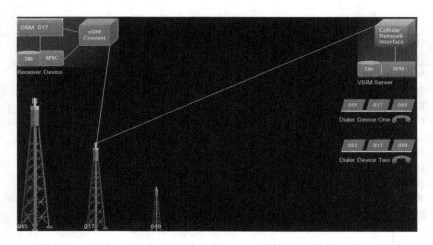

**Fig. 5.** Simulator initial state.

In "Fig. 5", GSM activation unit, SPSC and eSIM constant have established a regular connection within the receiver device. A regular connection is also established between Cellular Telephone Network Interface and SPM in VSIM server. eSIM constant of Receiver Device has established a regular connection with Cellular Telephone Network Interface of VSIM server through GSM network tower 017.

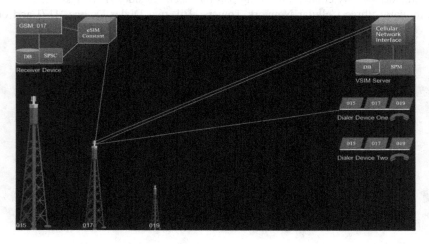

**Fig. 6.** Dialer one requests through same connection.

In "Fig. 6", Dialer Device One has established a channel with receiver device SIM 017 by following 1 to 4 steps.

1. The request was sent by Dialer Device One to GSM network tower 017.
2. GSM network 017 had to route that request to the Cellular Telephone Network Interface.
3. Cellular Telephone Network Interface had to handover the request to the SPM.
4. SPM had to analyze flags of the requested SIM and had gotten ONLINE flag [Fig. 5]. Therefore, SPM directly has routed the request to the Receiver Device to establish a voice channel.

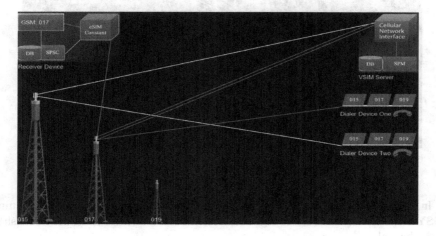

**Fig. 7.** Dialer two requests for another network.

In "Fig. 7", Dialer Device Two wanted to establish a channel with Receiver Device SIM 015 while having an established channel between Dialer One and Receiver Device SIM 017.

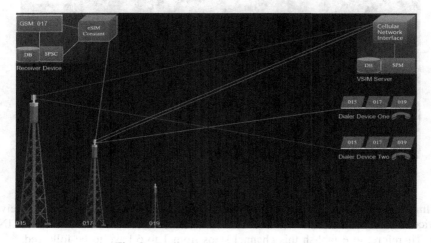

**Fig. 8.** Server sends activation feedback.

Since Dialer Device One has already an established channel (Active flag), therefore SPM has responded a BUSY flag to the dialer device which is shown in "Fig. 8".

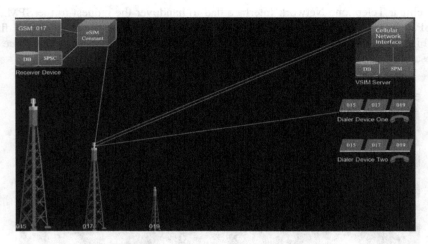

**Fig. 9.** Request of dialer two rejected.

In "Fig. 9", Dialer Device Two has rejected the request automatically by getting BUSY response from SPM. In "Fig. 10", Dialer Device One has freed the established voice channel.

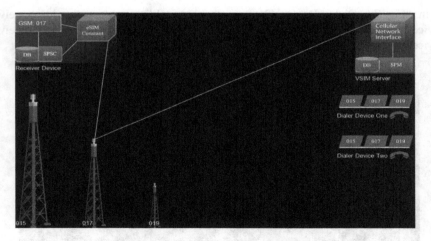

**Fig. 10.** Dialer one has freed the channel again.

In "Fig. 11", Dialer Device Two wants to establish a channel with Receiver Device SIM 015 but Receiver Device has an active different SIM with the ONLINE flag. Therefore, to establish this channel steps from 1 to 6 have to be followed.

1. Dialer Device Two has sent a request to the Cellular Telephone Network Interface of VSIM server through GSM network tower 015.
2. Cellular Telephone Network Interface has handovered this request to the SPM.
3. SPM has analyzed flags of the requested SIM and has realized an OFFLINE flag. Besides, it also got a different ONLINE flag SIM which is located in same Receiver Device that is wanted by Dialer Device Two. Therefore, SPM has sent SIM Switching request to SPSC through Cellular Telephone Network Interface, GSM network 017 and eSIM constant. Moreover, SPM rechecks again after waiting for a while.

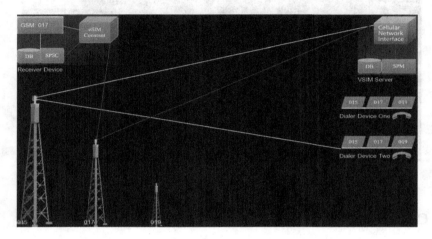

**Fig. 11.** Dialer two requests again and server sends request to SPSC.

4. eSIM constant has checked SIM flag again. After checking, if ACTIVE flag is not found then it does the route to the SPSC.

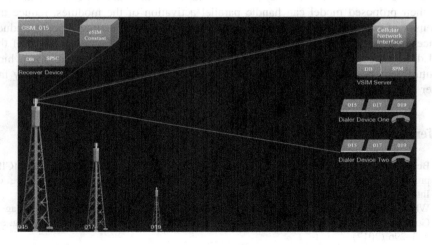

**Fig. 12.** SPSC has activated requested profile to the GSM Unit.

5. In "Fig. 12", SPSC has activated the requested SIM after removing the activated SIM from GSM SIM activation unit.
6. In "Fig. 13", finally, a voice channel has been established between Receiver Device and Dialer Device Two.

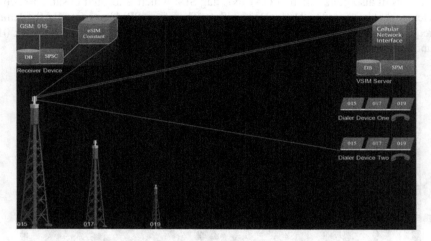

**Fig. 13.** After profile switching, channel has been established.

## 5    Conclusion

The main focus of the study was to find out a technique for proper switching between the embedded subscriber identity modules by eliminating the limitations of the existing eSIM and VSIM techniques. By providing the required modification in the existing model and combining the two existing techniques with required modification by introducing the proper methodology of working, the authors have been able to show that their proposed model can handle parallel activation of the modules, reduce the expenses of physical switching, increase the security by eliminating cloning, reduce device theft etc. However, the system did not consider the roaming capability of the SIM during this switching. Moreover, the proposed model is a conceptual one, which in future may be researched and implemented in a real-life device and those may fall under the scope of the future research.

## References

1. Bender, H., Lehmann, G.: Evolution of SIM provisioning towards a flexible MCIM provisioning in M2M vertical industries. In: 2012 16th International Conference on Intelligence in Next Generation Networks. IEEE (2012)
2. What is flag? - Definition from WhatIs.com. https://whatis.techtarget.com/definition/flag
3. Gerpott, T.J., May, S.: Embedded subscriber identity module eSIM. Bus. Inf. Syst. Eng. **59**, 293–296 (2017)

4. GSMA, Embedded SIM Remote Provisioning Architecture. https://www.gsma.com/iot/wp-content/uploads/2014/01/1.-GSMA-Embedded-SIM-Remote-Provisioning-Architecture-Version-1.1.pdf

5. GSMA, GSMA Embedded SIM Specification Remote SIM Provisioning for M2M. https://www.gsma.com/iot/wp-content/uploads/2014/10/Embedded-SIM-Toolkit-Oct-14-updated1.pdf

6. GSMA, Leading M2M alliance back the GSMA embedded SIM specification to accelerate the internet of things. https://www.gsma.com/newsroom/press-release/leading-m2m-alliances-back-the-gsma-embedded-sim/

7. Mobile network operator on-demand subscription management study. https://www.ey.com/Publication/vwLUAssets/EY-mobile-network-operator-on-demand-subscription-management/$FILE/EY-mobile-network-operator-on-demand-subscription-management.pdf

8. Mouly, M., Pautet, M.: The GSM system for mobile communications. In: Cell&Sys, Palaiseau, France (1992)

9. Park, J., Baek, K., Kang, C.: Secure profile provisioning architecture for embedded UICC. In: 2013 International Conference on Availability, Reliability and Security. IEEE (2013)

10. GSMA, Remote Provisioning Architecture for Embedded UICC Technical Specification. https://www.gsma.com/newsroom/wp-content/uploads/SGP.02_v3.2_updated.pdf

11. Richarme, M.: The virtual SIM - a feasibility study. Technical University of Denmark, Department of Applied Mathematics and Computer Science, Lyngby, Denmark (2008)

12. Shi, G., Tangirala, V., Durand, J., Dudani, A.: Virtual SIM card for mobile handsets. USA Patent US11963918 (2007)

13. Sutherland, E.: Counting mobile phones, sim cards & customers. SSRN, 10 (2009)

14. Vahidian, E.: Evolution of the SIM to eSIM. NTNU Open (2013)

15. Vesselkov, A., Hammainen, H., Ikalainen, P.: Value networks of embedded SIM-based remote subscription management. In: 2015 Conference of Telecommunication, Media and Internet Techno-Economics (CTTE). IEEE (2015)

16. Erran, L., Morley, Z., Rexford, J.: CellSDN: software-defined cellular networks. In: ACM SIGCOMM, Hong Kong, China (2013)

17. Andrus, J., Dall, C., Hof, A.V., Laadan, O., Nieh, J.: Cells: a virtual mobile smartphone architecture. In: Proceedings of the 23rd SOSP (2011)

# Statistical Analysis of Cloud Based Scheduling Heuristics

Sudha Narang[1]([⊠]) ⓘ, Puneet Goswami[1] ⓘ, and Anurag Jain[2] ⓘ

[1] Department of Computer Science and Engineering,
SRM University, Sonipat, Haryana, India
sudhanarang@mait.ac.in, goswamipuneet@gmail.com
[2] Virtualization Department, School of Computer Science,
Energy Acres Building, University of Petroleum and Energy Studies (UPES),
Dehradun, India
er.anuragjain14@gmail.com

**Abstract.** Scheduling of cloudlets (tasks) on virtual machines in cloud has always been of prime concern. Various heuristics have already been proposed in this area of research and are well documented. In this work, authors have proposed a unique method of statistically evaluating the results of simulation of these heuristics for cloud-based model. The results are evaluated for a standard set of performance metrics. The statistical method applied proves the reliability of simulation results obtained and can be applied to evaluation of all heuristics. In addition to this a recent and more advanced CloudSim Plus simulation tool is used as there is paucity of work that demonstrates using this tool for this research problem. The simulations use a standard model of task and machine heterogeneity that is pertinent to cloud computing. To make the simulation environment more realistic, Poisson distribution is used for the arrival of cloudlets, and exponential distribution for length (size) of cloudlets (tasks).

**Keywords:** Cloud computing · Virtual Machine (VM) · Makespan · CloudSim Plus · Cloudlet · Max-Min · Min-Min · Sufferage · Throughput

## 1 Introduction

Cloud Computing is conceptually aligned to the canonical definition of grid computing wherein a grid is a cluster of networked machines processing a big set of tasks [1]. Cloud computing additionally includes the ability to scale and "pay as you go" economic model amongst other additional characteristics [2]. A significant part of the research work in cloud computing is related to scheduling of tasks and inherits from similar problem addressed in grid computing. The difference lies in using cloud-based models and cloud-based simulation tools to arrive at optimal scheduling heuristics. While scheduling tasks on machines, it is ensured that the task load is properly balanced among various machines in a distributed system. Thus load balancing improves the response time and utilization of resources and avoids the problem of overloading or underloading of tasks on machines [3].

© Springer Nature Singapore Pte Ltd. 2019
A. B. Gani et al. (Eds.): ICICCT 2019, CCIS 1025, pp. 98–112, 2019.
https://doi.org/10.1007/978-981-15-1384-8_9

The task scheduling problem can be considered on the basis of the arrival of different tasks. If the tasks are scheduled as soon as they arrive, then it is referred to as dynamic or online mode, or if the tasks can be aggregated into groups and scheduled together, then this is referred to as static or batch mode. However different authors classified the load balancing Algorithms on different basis as seen in [4–9].

This work compares standard load balancing heuristics as applied to cloud computing using cloud-based simulation models implemented using CloudSim Plus, a recent and advanced tool for simulating cloud computing scenarios. It employs a novel way for researchers in the field to apply statistics to interpretation of results and improve reliability.

Section 2 illustrates different pseudocodes for implementing the standard heuristics Sect. 3 briefs the important performance metrics that are used to compare the heuristics. The implementation of different online and batch mode heuristics using CloudSim plus simulation tool are given in Sect. 4. A typical set of performance metrics is used to compare a set of standard heuristics. Section 5 ends with the scope for future work.

## 2 Design of Different Heuristics Used for Comparison

For the immediate or online mode following heuristics were chosen for simulation:

- MCT (Minimum Completion Time),
- MET (Minimum Execution Time),
- SA (Switching Algorithm),
- KPB (K Percent Best) and
- OLB (Opportunistic Load Balancing).

For the batch mode, following heuristics were chosen for simulation:

- Min-Min,
- Max-Min and
- Sufferage

## 2.1   Pseudocode of Online Heuristics

The Pseudocode of different approaches simulated in CloudSim Plus is given below:

- **Minimum Completion Time (MCT)**

```
For each Cloudlet (Task) to be scheduled:

    Calculate execution time on each VM for the cloudlet
    (Cloudlet length/VM MIPS)
    For each VM:
        Update ready time =
            Current ready time for VM -
        Submission Delay of current cloudlet

    For each VM :
        Calculate Total Completion Time =
            Execution Time + Ready Time


Find VM_min with Minimum Completion Time
Assign VM_min to this Cloudlet
Update Ready Time for VM_min with current Cloudlet comple-
tion time
```

- **Minimum Execution Time (MET)**

```
    For each Cloudlet (Task) to be scheduled:

        Calculate execution time on each VM for the
            cloudlet (Cloudlet length/VM  MIPS)


        Find VM_min with Minimum Execution Time
        Assign VM_min to this Cloudlet
```

- **Switching Algorithm (SA)**

Determine empirically the best performing LOWER_THRESHOLD and UPPER_-THRESHOLD for below pseudo code.

```
For each Cloudlet (Task) to be scheduled:

        Calculate execution time on each VM for the cloud-
        let (Cloudlet length/VM MIPS)

        For each VM:
             Update ready time =
                     Current ready time for VM -
                     Submission Delay of current cloudlet

        From VM ready times find minimum ready time R_min
        and maximum ready time R_max
        Find the ratio r=R_min/R_max
        If ( r >= LOWER_THRESHOLD && r <= UPPER_THRESHOLD)
             Switch to MCT  -- refer MCT pseudo code
        else :
             Switch to MET  -- refer MET pseudo code
```

- **K percent best**

```
For each Cloudlet (Task) to be scheduled:

      Calculate execution time on each VM for the cloud-
      let (Cloudlet length/VM MIPS)

      For each VM:
           Update ready time =
                 Current ready time for VM -
                 Submission Delay of current cloudlet


      Find best K percent VMs as:
                 Sort VMs in decreasing order of MIPS
                 Find the subset of top K percent of sorted
                 list
      For each VM in K percent VMs as above:
                 Calculate Total Completion Time = Execution
                 Time + Ready Time


      Find VM_min with Minimum Completion Time
      Assign VM_min to this Cloudlet
      Update Ready Time for VM_min with current Cloudlet com-
      pletion time
```

- **Opportunistic Load Balancing (OLB)**

```
For each Cloudlet (Task) to be scheduled:

      Calculate execution time on each VM for the cloud-
      let (Cloudlet length/VM MIPS)
      For each VM:
           Update ready time =
                 Current ready time for VM -
                 Submission Delay of current cloudlet

      Find minimum ready time of all VMs
      Assign VM with minimum ready time to this Cloudlet
```

## 2.2   Pseudocode of Batch Mode Heuristics

- **MaxMin**

```
Create 2D matrix for Cloudlet Execution Time on each VM
for every cloudlet using (Cloudlet length/VM MIPS)

do:
        For each element in 2D Matrix:
                Update Element as Total Completion Times
                =   Execution Time + Ready Times of VMs

        For each Cloudlet find minimum completion time and
        associated VM

        Assign VM to the Cloudlet with maximum of these
        minimum completion times

        Update Ready time of the VM assigned.

        Remove Assigned Cloudlet from the 2D Matrix

  while (2DMatrix.size != 0)
```

- **MinMin**

```
Create 2D matrix for Cloudlet Execution Time on each VM
for every cloudlet using (Cloudlet length/VM MIPS)

do:
        For each element in 2D Matrix:
        Update Element as Total Completion Times =
            Execution Time + Ready Times of VMs
        For each Cloudlet find minimum completion time and
        associated VM

        Assign VM to the Cloudlet with minimum of these
        minimum completion times

        Update Ready time of the VM assigned.

        Remove Assigned Cloudlet from the 2D Matrix

  while (2DMatrix.size != 0)
```

- **Sufferage**

```
Create 2D matrix for Cloudlet Execution Time on each VM
for every cloudlet using (Cloudlet length/VM MIPS)

do:
          For each element in 2D Matrix :
                  Update Element as Total Completion Times =
                          Execution Time + Ready Times of VMs

          For each Cloudlet find top two minimum completion
          times and associated VMs

          Find Sufferage Value as the difference of these
          minimum times

          If no Cloudlets competing for the VM with minimum
          completion time:

              Assign VM to the Cloudlet with minimum of
              these minimum completion times

          else:

              Compare the competing cloudlets Sufferage
          times and assign the VM to the highest Sufferage
          value cloudlet

          Update Ready time of the VM assigned.

          Remove Assigned Cloudlet from the 2D Matrix

   while (2DMatrix.size != 0)
```

## 3   Performance Metrics

The following performance metrics have been used in this work for comparing
heuristics for task scheduling in the cloud.

- **Makespan**
  This metric is most often used for comparison of task scheduling heuristic. It is the finish time of the last cloudlet, which implies that all submitted tasks (cloudlets) have been processed by the pool of virtual machines.

$$\text{Makespan} = \max(\text{Cloudlet Finish Times})$$

- **Resource Utilization**
  This metric provides a view on the level of utilisation of virtual machines pool by averaging the utilisation of each machine. Utilisation is the difference of busy and idle times of machine through the length of the execution (i.e. makespan). The value is calculated as

$$\text{RU} = \sum \forall j, Rij = 1 \ \frac{R_{rt} - R_{it}}{makespan}$$

where Rrt and Rit are resource running time and resource idle time.

$$\text{Avg RU} = \sum_{i=1}^{n} ru$$

where n = number of resources
It ranges between −1 (completely idle) and 1 (fully busy) [10]. Generally, higher utilisation of the virtual machines farm is the objective which would imply a busier set of virtual machines.

- **Throughput**
  This metric captures the number of tasks processed over a standard unit of time (typically seconds). In the case of cloud computing, this is calculated as the number of cloudlets processed over the makespan of the system and is thus a derived metric.

- **Average Response Time**
  Another derived metric is the average of execution times for a set of cloudlets [11].

$$\text{Expected Response Time} = F - A + T_{delay}$$

where:
F: time to complete the task.
A: arrival time of the task.
$T_{delay}$: transfer time of the task. (for this simulation is 0)

## 4  Comparison of Standard Heuristics

We performed simulations to study the performance of some standard heuristics as defined in [12] in typical cloud computing scenarios. From the point of view of simulation, the heuristics in immediate mode and batch mode require different experimental setups and were therefore simulated and analysed separately. The immediate mode, also referred to as the online mode, requires a cloudlet to be assigned to a Virtual Machine as soon as it arrives. Whereas in the batch mode, the cloudlets are aggregated together to form a meta-task and the heuristic is used to determine the assignment of all the cloudlets to Virtual Machines available in one batch instead of as and when they arrive.

All simulations in this study were performed using Cloudsim Plus, a fork of Cloudsim3 that resolves the issues of having to write duplicate code for heuristic implementation in Cloudsim [13]. Cloudsim Plus is Java 8 compatible, which enables writing more concise code using lambda expressions [14]. It is an open source project available at http://cloudsimplus.org. Additionally, it has some exclusive features that enable implementation of heuristics requiring dynamic creation of Virtual machines and Cloudlets. Its structure consists of 4 modules whose description can be found at http://cloudsimplus.org. In the present study new modules were developed in Cloudsim Plus to simulate the immediate and batch heuristics.

### 4.1    Immediate Mode

For the immediate or online mode five heuristics were chosen for comparison: MCT (Minimum Completion Time), MET (Minimum Execution Time), SA (Switching Algorithm), KPB (K Percent Best) and OLB (Opportunistic Load Balancing). The typical benchmark of these heuristics is the MCT, which takes into account both the ready time and completion times of the Cloudlet. The simulations were performed in CloudSim plus, which allows dynamic creation of cloudlets, with cloudlets arriving with a delay modelled by Poisson distribution. This aligns more closely with the real time expected behavior. The cloudlet lengths were modeled as having an exponential distribution and the heterogeneity of the Virtual machines was modeled as having processing capacity in progressively higher step increments. While the processing power of virtual machines is heterogenous, homogeneity is maintained in terms of other machine characteristics like Ram, Bandwidth, and Storage for the purpose of this simulation.

In our simulations, cloudlets were considered independent of each other and of the same priority. We assumed a direct correlation between virtual machine capacity (MIPS) and cloudlet completion time as a higher capacity (MIPS) virtual machine is expected to process the cloudlets faster than a machine with lower capacity (MIPS). This assumption results in a consistent ETC (Expected Time to Compute) matrix which is typical to cloud scenarios. This aspect has a significant bearing on the results of simulation and relative comparison of heuristics. Since our simulations had high task heterogeneity given the exponential distribution (Fig. 1) of cloudlet length whereas the Virtual Machine variations are relatively minimal (Table 1), they more closely resembled an overall HiLo heterogeneity scenario as described by Maheswaran et al. [12]. In this context, the experiment results may be interpreted as follows.

In case of MET the above assumptions invariably lead to higher capacity Virtual Machines having better execution times for all cloudlets and all the cloudlet traffic being directed to the highest specification machine (Table 2, Fig. 2), which is consistent with the known drawback of this heuristic in terms of load balancing. This translates into specific performance metrics getting impacted as seen in the simulation results where MET has the highest the average response time and lowest throughput. In case of KPB, the best k-percent Virtual Machines are always selected, which is essentially a subset of the most performant machines in the initial set of Virtual Machines, where the performance of a machine is assumed to be determined entirely by the processing capacity of the machine, measured in MIPS. KPB appears to overcome the limitations of MET in the scenarios simulated in this study.

MCT And KPB (K = 0.67) outperformed other heuristics in terms of average response times as in Fig. 2. Whereas on makespan and throughput metrics all heuristics performed similarly except for MCT and KBP outperforming the others on makespan.

In terms of utilization, other than MET, all other heuristics have more idle time for Virtual Machines than busy times, as indicated by the negative values of total utilization, with MCT having the most aggregated idle time across the infrastructure thus indicating more in line with cloud principles of horizontal scalability (more low end machines) as opposed to vertical scalability (higher processing capacity machines). MET superficially seems to have the most optimal utilization in this set up, however given this heuristic results in all traffic directed to a single, most performant machine, this is effectively poor utilization of the broader cloud infrastructure. It demonstrates MET being more suitable for infrastructure that is vertically scalable and a poor choice for cloud.

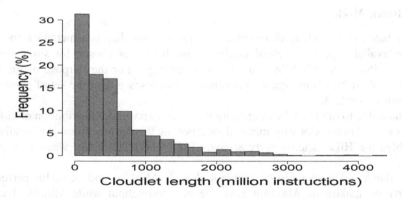

**Fig. 1.** Distribution of cloudlet lengths in simulation of immediate mode heuristics.

**Table 1.** Parameters used for the simulation of load balancing heuristics in immediate mode.

| | |
|---|---|
| DATACENTERS | 1 |
| HOSTS | 2 |
| HOST_PES | 8 |
| VMS | 6 |
| VM_MIPS | {150, 200, 250, 300, 350, 400} |
| CLOUDLETS | 1000 |
| MEAN_CLOUDLET_ARRIVAL_SECOND (Poisson) | 2 |
| MEAN_CLOUDLET_LENGTH (Exponential) | 600 |
| KPB K value | K = 67% |
| Switching Algorithm Lower Threshold | 0 |
| Switching Algorithm Upper Threshold | 0.2 |

**Table 2.** Performance of heuristics in immediate mode. Mean and standard deviation from five Cloudsim Plus iterations is provided for each heuristic.

| | Makespan | Throughput | Avg response time (secs.) | Utilization |
|---|---|---|---|---|
| MET | 2005 ± 23 | 0.5 ± 0.01 | 7.37 ± 1.9 | 0.23 ± 0.32 |
| MCT | 1972 ± 23 | 0.51 ± 0 | 2.11 ± 0.08 | −0.64 ± 0.01 |
| SA | 2007 ± 53 | 0.5 ± 0.02 | 2.33 ± 0.11 | −0.61 ± 0.03 |
| OLB | 1995 ± 45 | 0.5 ± 0.01 | 2.66 ± 0.17 | −0.56 ± 0.03 |
| KPB | 1973 ± 31 | 0.51 ± 0.01 | 2.01 ± 0.05 | −0.24 ± 0.02 |

## 4.2 Batch Mode

For the batch mode, where cloudlets are aggregated together as a meta-task for mapping to available pool of virtual machines, the heuristics chosen for simulation on Cloudsim Plus were MinMin, MaxMin and Sufferage. For the purpose of the comparison of these heuristics against performance metrics, we used the set of parameters as shown in Table 3.

Again, due to high task heterogeneity given the exponential distribution of cloudlet lengths (Fig. 3) and relatively minimal variation in Virtual Machines, the simulations resembled the HiLo heterogeneity scenario [12]. The ETC matrix was consistent in nature.

As illustrated in the results (Table 4, Fig. 4) MinMin and MaxMin performed similarly on makespan, MaxMin gave the best throughput while MinMin had the lowest response time. Sufferage lagged on these three parameters and proved to be a poor choice on this cloud-based model. In terms of utilization, MaxMin kept the VMs busiest, almost 100%, whereas Sufferage had the least load on machines, and a poor choice given the sub-optimal results on other parameters.

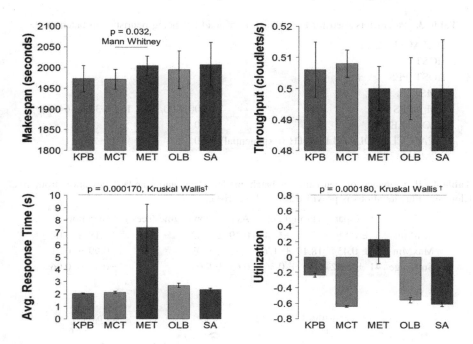

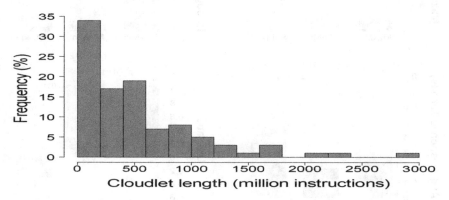

A. Fig. 2. and Fig. 3

<p>† All pairwise comparisons are significant (p < 0.05) by Mann Whitney U test</p>

**Fig. 2.** Performance of heuristics in immediate mode. The bar charts show the mean and standard deviation of performance metrics for the heuristics. Mann Whitney and Kruksal Wallis tests were conducted to compare the means.

**Fig. 3.** Distribution of cloudlet lengths in simulation of batch mode heuristics.

**Table 3.** Parameters used for the simulation of load balancing heuristics in batch mode.

| | |
|---|---|
| DATACENTERS | 1 |
| HOSTS | 2 |
| HOST_PES | 8 |
| VMS | 6 |
| VM_MIPS | {200, 400, 800, 1600, 3200, 6400} |
| CLOUDLETS | 100 |
| MEAN_CLOUDLET_LENGTH (Exponential) | 600 |

**Table 4.** Performance of heuristics in batch mode. Mean and standard deviation from five Cloudsim Plus iterations is provided for each heuristic.

| | Makespan | Throughput | Avg response time (secs.) | Utilization |
|---|---|---|---|---|
| MinMin | 5 ± 0.55 | 17.17 ± 1.42 | 0.19 ± 0.02 | 0.18 ± 0.04 |
| MaxMin | 5 ± 0.45 | 18.19 ± 1.74 | 0.31 ± 0.03 | 0.99 ± 0.01 |
| Sufferage | 41 ± 5.42 | 2.45 ± 0.36 | 0.62 ± 0.04 | −0.44 ± 0.06 |

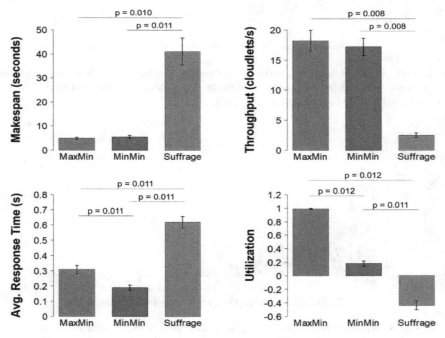

All p-values were computed by Mann Whitney U test

**Fig. 4.** Performance of heuristics in batch mode. The bar charts show the mean and standard deviation of performance metrics for the heuristics. Mann Whitney and Kruksal Wallis tests were conducted to compare the means.

# 5   Conclusion and Future Research

The problem of scheduling tasks onto virtual machines for achieving load balancing in Cloud computing is a well-researched problem having its roots in grid computing. In this study the performance of typical task scheduling heuristics was evaluated in a statistical framework where the cloudlet length and arrival times followed a statistical distribution and the mean performance of heuristics was compared for statistical significance. Experiments were run on novel cloud simulator Cloudsim Plus using five online and three batch heuristics. The key performance metrics were compared for differences in their means using nonparametric hypothesis tests. Evaluation of results using statistical tools presents a novel way for researchers in the field to improve the reliability of the interpretation of results.

In the simulated scenarios, MCT and KPB came out to be better performing heuristics for online mode Cloud models, by having lower makespan, better throughput and lower response times. MET was found to be the worst performing and clearly not suitable for cloud-based models. For Batch mode, Min-Min came out be the better heuristic than the other two, on an overall basis. While Max-Min performed better on throughput, Sufferage gave better results on resource utilization.

In terms of future work, this proposed model of evaluating heuristics can be extended to many other heuristics found in the literature as well as for any new proposed ones. A variety of statistical distributions can be considered for cloudlet length and arrival times emulating real scenarios. Simulations can be implemented in Cloudsim Plus which supports multiple scenarios relevant to Cloud Computing and is extensible.

In summary, the use of statistical methods mentioned herein provides a more robust and reliable interpretation of the results of simulations comparing load balancing heuristics in Cloud computing.

# References

1. Foster, I., Kesselman, C. (eds.): The Grid 2: Blueprint for a New Computing Infrastructure. Elsevier, San Francisco (2003)
2. Mell, P., Grance, T.: The NIST definition of cloud computing (2011)
3. Chaczko, Z., et al.: Availability and load balancing in cloud computing. In: International Conference on Computer and Software Modeling, Singapore, vol. 14 (2011)
4. Thakur, A., Goraya, M.S.: A taxonomic survey on load balancing in cloud. J. Netw. Comput. Appl. **98**, 43–57 (2017)
5. Kaur, R., Luthra, P.: Load balancing in cloud computing. In: Proceedings of International Conference on Recent Trends in Information, Telecommunication and Computing, ITC 2012 (2012)
6. Kunwar, V., Agarwal, N., Rana, A., Pandey, J.P.: Load balancing in cloud—a systematic review. In: Aggarwal, V.B., Bhatnagar, V., Mishra, D.K. (eds.) Big Data Analytics. AISC, vol. 654, pp. 583–593. Springer, Singapore (2018). https://doi.org/10.1007/978-981-10-6620-7_56
7. Khiyaita, A., et al.: Load balancing cloud computing: state of art. In: 2012 National Days of Network Security and Systems (JNS2). IEEE (2012)

8. Mayanka, K., Mishra, A.: A comparative study of load balancing algorithms in cloud computing environment. arXiv preprint: arXiv:1403.6918 (2014)
9. Mishra, S.K., Sahoo, B., Parida, P.P.: Load balancing in cloud computing: a big picture. J. King Saud Univ. Comput. Inf. Sci. (2018)
10. Phi, N., et al.: Proposed load balancing algorithm to reduce response time and processing time on cloud computing. Int. J. Comput. Netw. Commun. (IJCNC) **10**(3), 87–98 (2018)
11. Maipan-uku, J.Y., Rabiu, I., Mishra, A.: Immediate/batch mode scheduling algorithms for grid computing: a review
12. Maheswaran, M., et al.: Dynamic matching and scheduling of a class of independent tasks onto heterogeneous computing systems. In: Proceedings of the Eighth Heterogeneous Computing Workshop (HCW 1999). IEEE (1999)
13. Calheiros, R.N., et al.: CloudSim: a toolkit for modeling and simulation of cloud computing environments and evaluation of resource provisioning algorithms. Softw. Pract. Exp. **41**(1), 23–50 (2011)
14. Silva Filho, M.C., et al.: CloudSim plus: a cloud computing simulation framework pursuing software engineering principles for improved modularity, extensibility and correctness. In: 2017 IFIP/IEEE Symposium on Integrated Network and Service Management (IM). IEEE (2017)

# A Risk Transfer Based DDoS Mitigation Framework for Cloud Environment

B. B. Gupta(✉) and S. A. Harish

Department of Computer Engineering, National Institute of Technology,
Kurukshetra, Thanesar 136119, India
gupta.brij@gmail.com, harishsa85@gmail.com

**Abstract.** The impact of Cloud computing on the current information technology infrastructure has undeniably lead to a paradigm shift. The software, Platform and Infrastructure services offered by Cloud computing has been widely adopted by industries and academia alike. Protecting the core architecture of Cloud computing environment against the wake of Distributed Denial of Service attacks is necessary. Any disruptions in Cloud services reduce availability causing losses to the organizations involved. Firms lose revenue and customers loose trust on Cloud providers. This paper discusses a risk transfer based approach to handle such attacks in Cloud environment employing Fog nodes. Fog nodes work in tandem with Autonomous systems possessing unused bandwidth which can be leveraged by the Cloud during an attack. The burden of protection is partially transferred to willing third parties. Such a proactive conceptual defensive framework has been proposed in this paper.

**Keywords:** Internet of Things (IoT) · Mobile Cloud Computing (MCC) ·
Fog computing · Autonomous Systems (AS) · Distributed Denial of Service
(DDoS) attacks

## 1 Introduction

Cloud computing has become the staple of commercial and enterprise IT network service model for many users. It's scalability and robust nature has allowed many IT industries to consider Cloud as their backbone. The National Institute of Standards and Technology (NIST) attributes certain fundamental characteristics to Cloud computing [1]. Rapid scalability, On-demand self-service, resource pooling and measured service have been mentioned as its most distinct characters. Cloud has enabled industries to shift their focus to product development rather than the implementation. In essence, valuable manpower and resources can be directed towards improving the application rather than managing the essentials required by the application. These low level tasks can be offset to a third party provider which in this case is a Cloud service provider. Providers may charge incentives accordingly. Infrastructure and their maintenance is no longer a concern for the Cloud consumer due to scalability and fluidity. It is a win-win situation for both the consumer and the service provider. Recent additions to this paradigm enables a pay-as-you-go model which allows the customer to append resources during demanding and peak service periods. The ease with which IT facilities

© Springer Nature Singapore Pte Ltd. 2019
A. B. Gani et al. (Eds.): ICICCT 2019, CCIS 1025, pp. 113–127, 2019.
https://doi.org/10.1007/978-981-15-1384-8_10

can scale their products or services without outrageous overheads has truly been the trump card for Cloud computing. Predictive planning for enterprises has been simplified in the wake of Cloud services. The rapid boom of Cloud computing stands as a self-evident reason for its success. Cloud infrastructure can take three distinct service models namely Software-as-a-service (SaaS), Platform-as-a-Service (PaaS) and Infrastructure-as-a-Service (IaaS). It can be deployed as either a public, private, community or hybrid Cloud [2]. In spite of the benefits associated with Cloud computing, some inherent security and privacy risks exist. Resource pooling, multi-tenancy and share ability features can be exploited by cyber offenders and anyone with a malicious intent [3]. Similar issues can compound to create a Denial of Service (DoS) risk for the Cloud user and the provider alike.

Uninterrupted service and zero down time are major factors in determining the commercial viability of many industries and organizations. Any deviations lead to lack of user trust and revenue loss for the industry. Some countries mandate that Cloud providers in the telecommunications sector be allowed only five minutes of down time per year [4]. Thus, it can be stated that Cloud providers are under considerable pressure to provide continuous and reliable service to clients. These interruptions can either be due to inherent architectural failures or due to external attacks like Distributed Denial of Service (DDoS). Such targeted attacks on Cloud providers can prove to be devastating for IT facilities. A DoS or DDoS attack works by exhausting the resources of a server by sending multitudes of requests. The effect can be caused by either depletion of Cloud service bandwidth or the exhaustion of Cloud server resources. A DDoS attack employs the use of compromised machines called botnets to generate traffic that bombards the server. These attacks can be performed with relative ease. It may not require specialized knowledge from the attacker's point of view which is an alarming state of affair. For example, a cybercriminal group called 'Vickingdom2015' had brought down Cloud services in the year 2015 causing considerable disruption and loss [3]. Traditional DDoS mitigation strategies differ from Cloud DDoS defense strategies on the basis of the entity which is in control of the resources. The key difference is that Cloud providers are the ones carrying network and computational overheads. As a consequence, these pooled resources are shared by users of the Cloud.

The term Fog computing was introduced by Cisco [5]. The two computing frameworks namely Fog and Cloud are intertwined. Fog computing essentially brings the computational capabilities of a Cloud closer to the user resulting in faster access. Integration of the Cloud with Fog has given Clouds the option to scale down their central architecture. Resource intensive tasks can be offset to Fog nodes which are closer to the user. Internet of Things (IoT) employs Fog computing techniques to reduce latency in communication which is essential in real-time deployments. Although Cloud and Fog are individually utilized for a myriad of applications, a holistic and integrated approach can give rise to a more secure, robust and flexible infrastructure. From a security perspective, Fog nodes can be leveraged to provide protection against DDoS attacks. Real-time detection and prevention of the attack can be achieved with relative ease in the Fog computing era. Much of the detection overhead can be drawn away from existing central Cloud architecture. Figure 1 shows a simplistic view of Fog computing architecture.

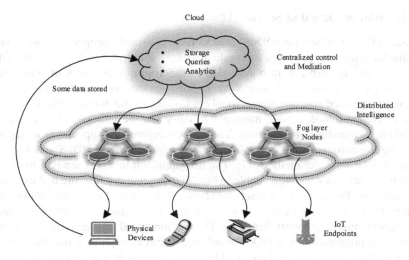

**Fig. 1.** Fog computing architecture

In essence, a mitigatory framework to address the particular issue of DDoS attacks in Clouds is the focus of this paper. Existing solutions to defend against DDoS attacks have had unfavorable overheads. We have proposed a framework leveraging the power of Fog nodes employing risk transfer mechanics to tackle DDoS attacks. The framework has been simulated using OMNeT++ from which results have been drawn. The rest of the paper is summarized as follows. Section 2 describes technical details of the concepts involved. Section 3 contains related works. Section 4 discusses the proposed Fog based risk transfer framework. Section 5 records the simulation results and its analysis. Finally, Sect. 6 concludes the paper with future work.

## 2 Background

Cloud computing as a paradigm that has carved a niche in today's internet landscape. According to statistics mentioned by Forbes, 83% of all enterprise workloads will be managed with the help of Cloud services by the year 2020 [6]. Public Clouds are estimated to grow considerably by 2020. It is estimated that 41% of all Cloud services shall turn to public Clouds compared to 31% as of today. It is undeniable that Cloud services have gained a multifold increase in consumption over the past ten-year period. However, the number one reason against adoption of Cloud services is security concern. The same concern can be estimated to grow based on the projections for 2020. Shift to public Clouds expose a larger surface area to attackers who aim to try and disrupt services. In this section, we discuss the fundamentals of DDoS attacks, Cloud computing and Fog computing. This is done with a view of grasping the essential aspects required to build an integrated mitigatory framework.

## 2.1   Distributed Denial of Service Attacks

Distributed Denial of Service (DDoS) attacks have constantly been an internet security issue that have not yet received a full scale migratory solution. More specifically DDoS flooding attacks are difficult to handle due to the voluminous influx of traffic at the network edge. Uninterrupted access to online services ensure that organizations maintain undisturbed revenue flow. Attackers usually recruit botnets which are a collection of compromised zombie machines to carry out the attack. Vulnerabilities are exploited to take control of a machine. The attack army once set up, can be engaged on a target by the command and control center that manages them. DDoS attacks can be performed in two distinct methods [7]. The first approach involves sending malformed packets to the servers which are unable to handle them, thereby freezing and exhausting their resources. The second approach involves either exhausting bandwidth or server resources by bombarding large volumes of traffic. Conventional defense systems are either Source-based, Network-based or Destination-based in terms of deployment location. The propagation of a DDoS attack vector can be separated into three distinct phases namely scan, recruit and attack. The most common attack strategy is to try and exhaust the bandwidth of the victim with high volume attacks. This targets the network's capabilities. The resources of the server are targeted in the subsequent attack where legitimate packets are constructed in order to make the server overuse its resources. An example would be the TCP SYN attack where the victim is flooded with TCP SYN packets. The server's connection queue is soon filled with open connections expecting an ACK. The ACK is never provided. Victim is overloaded and does not accept any new connections. IP spoofing can also be utilized to perform Smurf like attacks.

## 2.2   Cloud Computing

Previous sections have already introduced the word Cloud. At its bare bones, Cloud is a network of connected devices and computers that provide a collective service. From the perspective of an onlooker outside the network, these nodes appear as a cohesive unit possessing the same goal. The definition of Cloud computing has been reported by National Institute of Standards and Technologies (NIST) as: "Cloud computing is a model for enabling ubiquitous, convenient, on-demand network access to a shared pool of configurable computing resources (e.g., networks, servers, storage, applications, and services) that can be rapidly provisioned and released with minimal management effort or service provider interaction" [1]. Over the years, Cloud computing has changed the landscape of the IT industry. Industry leaders like Google, Microsoft and Amazon have comprehensively shifted their focus to a Cloud centric approach. Cloud computing can be differentiated primarily based on the service it provides [8]. These classifications are Infrastructure-as-a-service (IaaS), Platform-as-a-service (PaaS) and Software-as-a-service (SaaS). Clouds may also be classified according to their delivery models [8]. They can either be private, public or Hybrid in nature. Private Clouds are closed for use within an organization. They are deployed within the intranet and outsiders have no access to these Cloud systems. Public Clouds are open and placed on the internet, accessible to anyone from around the world. They are the most prevalent implementation in the current scenario. Hybrid Clouds are primarily private Clouds that can be extended to users beyond the intranet perimeter. All the three delivery models may provide any of the three services. Public Clouds are the most vulnerable due to their exposure to the internet. Figure 2 shows the growth projections of public Clouds until the year 2021.

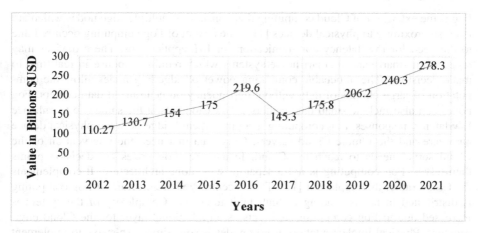

**Fig. 2.** Growth of public Clouds services market from 2012 to 2021 (estimated), Source: Statista

Cloud computing services are shared. Necessary demarcation between clients may not be properly implemented. Thus threats could originate from other clients within the Cloud service. Employee negligence is in general a threat to all systems. Particularly in Cloud, it can prove devastating due to the ease of access that Clouds provide. One may log into their Cloud space keeping the connection open for exploits from other applications or human factors. Non-compliance with regulatory mandates like PCI DSS, HIPAA/HITECH, GLBA, FISMA, FERPA, EU data protection, etc. by the providers of Cloud services transfer the liability on to the Cloud user. The threat of lawsuits and penalties as a result of the non-compliance can prove to be costly. The fear of loss of control inhibits companies from transferring their data to a Cloud provider. The subscriber may not have a say in how the data is stored and processed. 38% of the 200 IT firms that were included in a survey by Cloud Security Alliance (CSA) echoed the same sentiment [9]. Thirty percent of the CSA survey respondents languished about insider threats within the Cloud service provider. A high number of companies have employees who possess compromised credentials for sale on the dark web. The threat to availability of data tops the list of concerns echoed. DDoS attacks target this aspect of data security which leads to irreparable losses especially in terms of trust [10]. The attacker can understand the working of the Cloud through reconnaissance and launch a scathing DDoS attack to bring the system to a grinding halt for legitimate users. If not a halt, significant disruptions are caused. For most organizations that are dependent on these Cloud services, unavailability of their own data is not acceptable. Thus, DDoS attacks are to be taken seriously and significant resources are to be deployed in defending against such an attack. If enough traffic is initiated towards a Cloud computing system, it might go down completely. In summary, High latency, downtime, security and privacy risks are major concerns associated with Cloud computing.

## 2.3 Fog Computing

Fog is the extension of Cloud computing that consists of multiple edge nodes which are in close proximity to physical devices [11]. The advent of Fog computing occurred due to the need for low latency communication for IoT applications. These devices may range from smart cars to smart home systems which require response in real-time to make decisions. The noticeable computing power of edge/Fog nodes allow the computation of large chunks of data without dependency on centralized data centers. Fog nodes can also include small data centers called Cloudlets at the same site to enhance low latency responses. Fog computing is an intelligent bridge between physical device hardware and the remote Cloud servers. Only data that need the intervention of the Cloud server needs to reach the Cloud. In a sense, Fog nodes are decision points themselves. Fog computing is not a separate standalone architecture. It complements the Cloud rather than replace it [12]. This decentralized architecture of Fog computing is distributed in nature leading to high fault tolerance. Complexity of the system is increased due to Fog computing as it acts as an additional layer for the Cloud environment. Physical implementation of Fog nodes is sometimes expensive to implement at the network edge. These nodes do not share the same scalability as Cloud does. Future technologies that require agile and seamless connections shall be complemented by the existence of an evolved Fog computing architecture. In this paper, Fog nodes are leveraged for a protective framework against DDoS attacks.

## 3    Related Work

Various challenges for DDoS detection and mitigation have been outlined in the work done by Gupta et al. [13]. It was noticed that a considerable number of solutions employing Software Defined Networking (SDN) for DDoS protection have been proposed. Work presented by Yan et al. [14] explores the possibilities of using SDN frameworks to secure Clouds from DDoS attacks. Cloud DDoS defense deployment can occur in key locations in the architecture. Source-end deployment can ensure that IP addresses cannot be spoofed effortlessly. High rate attacks from the source can be detected and rate limited to protect the intermediate and the target networks. Machine learning algorithms are used in the work done by He et al. [15] to extract features from attack streams. This machine learning approach is claimed to be effective against flooding, spoofing and brute-force DDoS attacks. Intermediate network deployment is done on the network nodes between the source and target Autonomous systems. Issues with such systems is the requirement for modification of existing network devices. Also, intermediate networks cannot perform deep packet inspections owing to their inability to access data beyond the network layer. In the Cloud computing environment, these defense mechanisms are not very effective due to their deployment in a different Autonomous system. IP traceback mechanism using deterministic packet marking developed by Shui et al. [16] could identify the source Autonomous system which is the origin of the attack. Usage of this information for effective DDoS defense is once again a gray area. Distributed defense against DDoS attacks are seen as the best prospects for effective mitigation. Jakaria et al. [17] proposed a distributed denial of service defense module using Network Function Virtualization (NFV) called VFence. It concentrated on SYN flood attacks that target the victim network leading to bandwidth and resource depletion. It is safe to conclude that distributed hybrid defense mechanisms are relatively more effective solutions against DDoS attacks.

DDoS attack detection mechanisms can be classified into signature based detection and anomaly based detection techniques. Signature based detection schemes compare a set of rules and known attack signatures. The detection module can extract only previously known attacks. Lo et al. [18] propose an Intrusion Detection System (IDS) based distributed mechanism within the Cloud environment. Lonea et al. have deployed a virtual machine based IDS that detects deviations in signatures. Anomaly based detection schemes check for deviations in normal traffic behavior. Vissers et al. [19] use a gaussian model to defend against application layer attacks on Cloud services using the simple object access protocol (SOAP). Girma et al. [20] introduce a hybrid statistical model which employs entropy based detection techniques to categorize DDoS attack vectors. Deepali et al. [21] proposed the use of Fog layer for protection of Cloud against DDoS intrusions. The positioning of the Fog layer is convenient for effective use against DDoS attacks.

## 4 Proposed Framework

### 4.1 Framework Entities

In this paper, we propose a dynamic risk-transfer approach using Fog nodes to protect against the threat of DDoS attack. It has already been established that Fog nodes are connections points for various services that the Cloud provides. Genuine users and attackers alike connect to the Cloud service through a Fog node. Numerous works have already been carried out in detecting DDoS attacks and classifying the connections as genuine or malicious. Our proposed framework employs existing DDoS detection and classification methods discussed in the previous section. Classification may be achieved through intelligent machine learning algorithms or through static classifiers. It is assumed that Fog nodes are provisioned with these detection capabilities beforehand. Moreover, the Fog layer has decisive power in handling the traffic entering the Cloud.

The framework performs defense in two distinct stages: identification and response. Identification starts with the detection of an attack by the Fog nodes. The nodes are in constant communication with the central Cloud architecture regarding their load levels. Once a threshold level of traffic $(T_c)$ is exceeded, the Cloud is notified of a possible oncoming attack. Along with the notification, attack information data set (AD_S) is also propagated to the central Cloud. The AD_S consists of two distinct classifications of the existing connections: trusted and suspicious. Three metrics are attributed to each live connection namely uptime $(U_t)$, bandwidth used $(B_w)$ and change ratio $(C_r)$ as shown in Fig. 3. Change ratio is calculated using the previous two attributes as:

$$C_r = U_t/B_w \qquad (1)$$

The algorithm which is executed at the Fog nodes is shown in Fig. 4. A low change ratio implies that the connection is new but is soliciting a huge bandwidth. Depending upon the value, existing connections are categorized into trusted or suspicious. It is not claimed that the proposed metric is highly accurate in identifying malicious traffic. Detection accuracy is heightened when the metric is used in tandem with inherent detection mechanisms at the Fog layer. The unit of measurements do not matter since the change ratios are proportional.

**Fig. 3.** Attack information dataset (AD_S)

When the threshold $T_c$ is exceeded, AD_S packets are sent to the central Cloud along with other information such as protocol used, packet size and address information. The algorithm shown in table is executed at the edge of the Fog node. The Cloud infrastructure contains a pool of external Autonomous systems that volunteer their unused bandwidth ($U_{bw}$).

| Algorithm 1 | |
|---|---|
| 1. | **if** traffic $> T_c$ |
| 2. | create AD_S |
| 3. | calculate $C_r$ |
| 4. | encode $C_r$ in AD_S |
| 5. | send AD_S information to Cloud |
| 6. | **end if** |

**Fig. 4.** Algorithm executed at the Fog site during threshold violation

In the response stage, the Cloud conducts the decision making process using the algorithm shown in Fig. 5. The existing Autonomous systems are measured for their bandwidth availability $U_{bw}$. The suitable Autonomous system is chosen based on proximity and latency to route traffic through them. An update message is sent to the trusted clients which are identified using the parameters received from Fog nodes. The message contains the route information indicating the availability of an alternate route through the chosen Autonomous system to reach the destination sever. Thus, if the original route starts getting congested due to an actual attack, an alternative route has already been created through bandwidth provisioning. The risk of DDoS attack has been shared by the willing Autonomous systems.

| Algorithm 2 | |
| --- | --- |
| 1. | for every AD_S received |
| 2. | calculate trusted connections ($C_t$) |
| 3. | calculate suspicious connections ($C_s$) |
| 4. | if ( $C_t$ / $C_s$ ) < response threshold |
| 5. | provision == true |
| 6. | end if |
| 7. | end for |
| 8. | if provision == true |
| 9. | Get $U_{bw}$ from all AS |
| 10. | select the AS with maximum $U_{bw}$ |
| 11. | for every trusted connection ($C_t$) |
| 12. | send updated route packet |
| 13. | end for |
| 14. | end if |

**Fig. 5.** Algorithm executed at the central Cloud site during threshold violation

## 4.2  Insights About the Proposed Framework

The framework has been visualized in Fig. 6. The main advantage of the proposed mechanism is its proactive nature. If the threshold $T_c$ is set low enough, an alternative route is established to share the traffic whether the attack takes place or not. The clients are also proactively informed about the existence of a possible attack scenario. Thus, there are possibilities of leveraging this aspect to activate source based filtering mechanisms for the particular Fog outlet. Conformance of the proposed system with existing internet protocols and standards ensure ease of implementation. No new network devices are introduced in this design. The risk of DDoS attacks are collectively shared by the community. Most Autonomous systems procure extra bandwidth to be used for peak time usage. During non-peak hours, majority of the bandwidth is lying dormant. The proposed system exploits the potential of this bandwidth.

The design focusses on general DDoS attacks on the Cloud. These include connection-oriented and connection-less attacks. However, the framework excels when most legitimate connections are connection-oriented in nature. These connections are automatically preserved. For example, connection-oriented SYN attacks tend to attain a very low change ratio ($C_r$). Connection-less attacks like UDP floods can be identified at the Fog nodes. The design can be extended to other types of DDoS attacks by changing the monitored parameters appropriately.

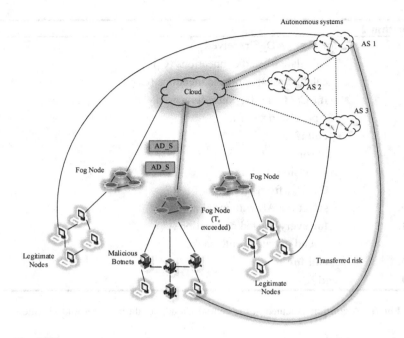

**Fig. 6.** Proposed Fog based risk-transfer defense approach for DDoS attacks

## 5   Simulation and Results

In essence, a mitigatory framework to address the particular issue of DDoS attacks in Clouds is the prime objective of this paper. More specifically, this work aims (i) to reduce the bandwidth load on fog nodes under attack and (ii) to ensure that legitimate connections are kept alive during a DDoS attack event. The simulation also highlights the use of external Autonomous systems for routing traffic. Fulfillment of the objectives shall effectively ensure that the attack has been diffused. Ultimately, availability of the Cloud service is to be maintained. Discrete event simulator OMNeT++ has been used to simulate the defensive framework. The average packet size has been assumed to be 600 Bytes based on the work done by Piotr et al. [22]. Consequently, the capacity of an average internet connection infrastructure has been assumed to have a value between 20 to 50 thousand packets per second. Figures 7 and 8 show the framework in execution under the discrete event simulation environment. Each end node represents a collection of Autonomous systems. For the purpose of simulation, the nodes are named as either Genuine or Malicious. A malicious node does not signify that all its constituents are compromised machines. It is taken only as the propagator of DDoS attack. Two fog nodes are introduced in the simulation. FogNode1 faces the DDoS attack vector. FogNode2 has no malicious nodes and is a participant in the simulation in order to stress the connection capacity. AsynchSys denotes all the volunteering Autonomous systems which have agreed to lease their bandwidth during an attack scenario. The inbuilt cMessage class has been extended to build custom messages carrying critical information between the nodes. AD_S has been represented as a special message packet carrying the change ratio $C_r$ of all live connections when the threshold is violated. The estimation is not meant to be highly accurate but merely serve as a pointer for the Cloud to divert trusted connections to the newly sanctioned route.

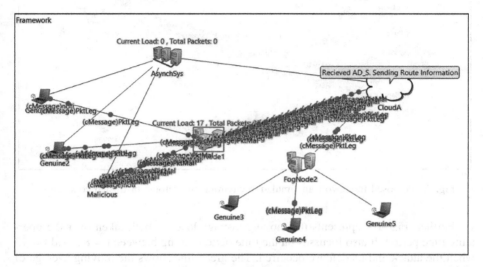

**Fig. 7.** Proposed framework in simulation environment: reception of AD_S

Figure 7 shows the point in time just after the reception of an AD_S packet by the Cloud. Prior to this event, the malicious node had commenced its attack on FogNode1 which is evident from the same figure. The attack stressed the fog node leading to dropping of packets and eventually violate the threshold. The AD_S packet was generated by FogNode1 once threshold ($T_c$) was breached. The threshold was set at 20 thousand packets per time unit. Figure 8 denotes the scenario after the new route was introduced to selected trusted connections. The load on the fog node as well as the Cloud drastically reduced once the defenses activated. Figure 9 shows a graph depicting the events from beginning to the end of simulation along with the total mean load on both the nodes. Consequently, the load on the connections between both the Fog nodes and CloudA have been recorded against simulation time taken as a parameter in the x-axis. Between the intervals of 0 and 4.5, normal traffic is observed. At t = 5, the malicious node starts bombardment of traffic on FogNode1. Between t = 5.5 and t = 10.5, denial of service for legitimate users begin, leading to packet drops. Peak load occurs at t = 9.2. The first AD_S packet is sent at t = 9.2. The threshold $T_c$ was deliberately configured high to test the response time. The experiment was parameterized and multiple simulations were conducted. The reported response time over these runs averaged between 1.1 to 1.3 time units. The green and yellow lines show total mean load on FogNode1 and AsynchSys respectively which is calculated cumulatively since the beginning of the attack.

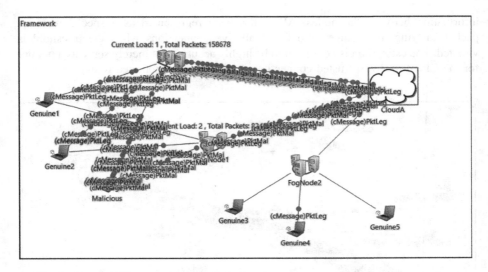

**Fig. 8.** Proposed framework in simulation environment: reroute of trusted connections

Further, Fig. 10 represents the moving average load on both taken over the operating time period. It also focusses on the time period during between t = 8.5 and t = 13 when the attack and its defense occurred. The green line shows the moving average of the load on FogNode1 and the yellow line depicts the moving average load on the newly utilized Autonomous system. The mean is projected to reach lower values eventually as seen from the graph. The Autonomous system is chosen based on its current available bandwidth. Once a new route is sanctioned with the dispatch of route packets, legitimate connections are directed to use it. Malicious nodes which are wrongly classified as trusted are also directed to utilize the new route. This does not impact the overall efficiency due to surplus provisioned bandwidth. Border Gateway Protocol (BGP) peering contracts have to be established beforehand to enable the use of available Autonomous systems. Establishment of said contracts may be hindered by policies and heterogeneous protocols but the benefits of doing so are easily perceivable. Our proposed framework makes use of such a scenario where defense against DDoS attack is achieved by a collective cooperative mechanism. The cloud core is flanked by fog nodes which handles the majority overhead in the defense process. The risk of attack is shared between all the volunteering entities. The impact on a single entity is minimized. Small and medium Cloud based organizations in need of defense benefit massively from this setup. The key to a successful defense is the presence of well-established and large Autonomous systems as participants. These self-sufficient networks can decisively create the shift in the DDoS defense paradigm if all their unused bandwidth are put into collective use.

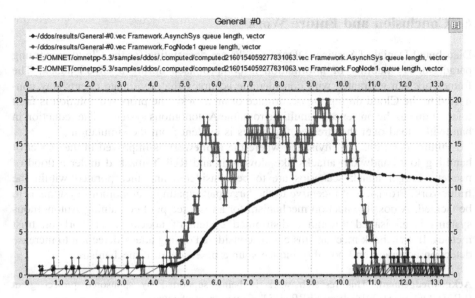

**Fig. 9.** Graph record of the entire DDoS attack event (Color figure online)

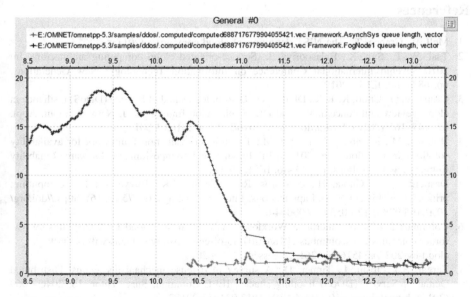

**Fig. 10.** Moving time averaged load towards the end of the attack (Color figure online)

# 6 Conclusion and Future Work

Distributed Denial of Service (DDoS) attacks are devastating for the Cloud computing paradigm. Availability of information is compromised during a DDoS attack. The framework can guarantee with considerable accuracy that legitimate clients stay connected to the Cloud during the occurrence of an attack. The proposed solution is fault tolerant due to the presence of multiple routable Autonomous systems. The reduction in bandwidth load over the stressed fog nodes is evident from the simulation.

Future directions are envisaged where the framework is improved in the aspect of handling low bandwidth attacks like slow loris and RUDY masked under a flood of packets. Larger attack scenarios are to be considered and incorporated within the framework. Provision of incentive to the large participating Autonomous systems is to be tackled. A possible auction mechanism for a resource pool containing Autonomous systems is envisaged through a distributed system of nodes which work on trust metrics. IP traceback mechanisms can be considered during attack detection to increase detection accuracy and possibly induce source based defense mechanisms.

**Acknowledgement.** This research work is being supported by sponsored project grant (SB/FTP/ETA-131/2014) from SERB, DST, Government of India.

# References

1. Mell, P., Grance, T.: The NIST definition of cloud computing (2011)
2. Tsai, W.-T., Sun, X., Balasooriya, J.: Service-oriented cloud computing architecture. In: 2010 Seventh International Conference on Information Technology: New Generations, pp. 684–689. IEEE (2010)
3. Osanaiye, O., Choo, K.-K.R., Dlodlo, M.: Distributed Denial of Service (DDoS) resilience in cloud: review and conceptual cloud DDoS mitigation framework. J. Netw. Comput. Appl. **67**, 147–165 (2016). https://doi.org/10.1016/J.JNCA.2016.01.001
4. Hormati, M., Khendek, F., Toeroe, M.: Towards an evaluation framework for availability solutions in the cloud. In: 2014 IEEE International Symposium on Software Reliability Engineering Workshops, pp. 43–46. IEEE (2014)
5. Dastjerdi, A.V., Gupta, H., Calheiros, R.N., Ghosh, S.K., Buyya, R.: Fog computing: principles, architectures, and applications. Internet of Things, 61–75 (2016). https://doi.org/10.1016/b978-0-12-805395-9.00004-6
6. Columbus L 83% of Enterprise Workloads Will Be in the Cloud by 2020. https://www.forbes.com/sites/louiscolumbus/2018/01/07/83-of-enterprise-workloads-will-be-in-the-cloud-by-2020/#3451605e6261. Accessed 31 Jan 2019
7. Zargar, S.T., Joshi, J., Tipper, D.: A survey of defense mechanisms against Distributed Denial of Service (DDoS) flooding attacks. IEEE Commun. Surv. Tutor. **15**, 2046–2069 (2013). https://doi.org/10.1109/SURV.2013.031413.00127
8. Botta, A., de Donato, W., Persico, V., Pescapé, A.: Integration of cloud computing and Internet of Things: a survey. Futur. Gener. Comput. Syst. **56**, 684–700 (2016). https://doi.org/10.1016/J.FUTURE.2015.09.021
9. Coles C Top 6 Cloud Security Issues in Cloud Computing. https://www.skyhighnetworks.com/cloud-security-blog/6-cloud-security-issues-that-businesses-experience/. Accessed 31 Jan 2019

10. Bhushan, K., Gupta, B.B.: Security challenges in cloud computing: state-of-art. Int. J. Big Data Intell. **4**, 81 (2017). https://doi.org/10.1504/IJBDI.2017.083116

11. Global State of the Internet Security &amp; DDoS Attack Reports, Akamai. https://www.akamai.com/us/en/resources/our-thinking/state-of-the-internet-report/global-state-of-the-internet-security-ddos-attack-reports.jsp. Accessed 18 Mar 2019

12. Iorga, M., Feldman, L., Barton, R., Martin, M.J., Goren, N., Mahmoudi, C.: Fog computing conceptual model, Gaithersburg, MD (2018)

13. Gupta, B.B., Badve, O.P.: Taxonomy of DoS and DDoS attacks and desirable defense mechanism in a Cloud computing environment. Neural Comput. Appl. **28**, 3655–3682 (2017). https://doi.org/10.1007/s00521-016-2317-5

14. Yan, Q., Yu, F.R., Gong, Q., Li, J.: Software-Defined Networking (SDN) and Distributed Denial of Service (DDoS) attacks in cloud computing environments: a survey, some research issues, and challenges. IEEE Commun. Surv. Tutor. **18**, 602–622 (2016). https://doi.org/10.1109/COMST.2015.2487361

15. He, Z., Zhang, T., Lee, R.B.: Machine learning based DDoS attack detection from source side in cloud. In: 2017 IEEE 4th International Conference on Cyber Security and Cloud Computing (CSCloud), pp. 114–120. IEEE (2017)

16. Yu, S., Zhou, W., Guo, S., Guo, M.: A feasible IP traceback framework through dynamic deterministic packet marking. IEEE Trans. Comput. **65**, 1418–1427 (2016). https://doi.org/10.1109/TC.2015.2439287

17. Jakaria, A.H.M., Yang, W., Rashidi, B., Fung, C., Rahman, M.A.: VFence: a defense against Distributed Denial of Service attacks using network function virtualization. In: 2016 IEEE 40th Annual Computer Software and Applications Conference (COMPSAC), pp. 431–436. IEEE (2016)

18. Lo, C.-C., Huang, C.-C., Ku, J.: A cooperative intrusion detection system framework for cloud computing networks. In: 2010 39th International Conference on Parallel Processing Workshops, pp. 280–284. IEEE (2010)

19. Vissers, T., Somasundaram, T.S., Pieters, L., Govindarajan, K., Hellinckx, P.: DDoS defense system for web services in a cloud environment. Futur. Gener. Comput. Syst. **37**, 37–45 (2014). https://doi.org/10.1016/J.FUTURE.2014.03.003

20. Girma, A., Garuba, M., Li, J., Liu, C.: Analysis of DDoS attacks and an introduction of a hybrid statistical model to detect DDoS attacks on cloud computing environment. In: 2015 12th International Conference on Information Technology - New Generations, pp. 212–217. IEEE (2015)

21. Deepali, B.K.: DDoS attack mitigation and resource provisioning in cloud using fog computing. In: 2017 International Conference on Smart Technologies for Smart Nation (SmartTechCon), pp. 308–313. IEEE (2017)

22. Jurkiewicz, P., Rzym, G., Boryło, P.: Flow length and size distributions in campus internet traffic, September 2018. https://arxiv.org/abs/1809.03486. Accessed 1 July 2019

# A Comparative Analysis of Mobility Models for Network of UAVs

Ashima Adya$^{(\boxtimes)}$, Krishna Pal Sharma, and Nonita

Dr. B. R. Ambedkar National Institute of Technology, Jalandhar, Punjab, India
ashima.adya@gmail.com, kpsharma17vce@gmail.com,
nonita@nitj.com

**Abstract.** Flying Adhoc Network (FANET) is an emerging research area gaining lot of attention of researchers nowadays. FANET as the name suggests, is an ad hoc network of Unarmed Aerial Vehicles (UAVs) flying in the space and forming a connected network to accomplish a common task with cooperation of each other. Mobility of nodes in such networks has always been a challenging task and thus researchers have proposed many solutions over the time for directing nodes mobility within the region of interest. Since, FANET nodes tend to move at much greater speed as compared to nodes in other networks like mobile ad hoc networks (MANETs) and drain more energy pertaining to its self-organizing nature, various new mobility models have been suggested and traditional models for MANET have been modified in accordance with the need of FANETs. In this paper, we present comparative study of both old as well as the new mobility models. A systematic comparative analysis is done based on certain parameters like their ability to cover the area, maintenance of connectivity, collision avoidance and energy consumption. Also, the paper explores some future directions and research problems related to the FANETs.

**Keywords:** FANET · Mobility models · Adhoc network · UAV · Virtual forces · Area coverage · Connectivity · Drones · Mobile networks

## 1 Introduction

Unarmed Aerial Vehicles (UAVs) also known as drones, are autonomous aircrafts without human pilot on board, and has become a thriving area in which many researchers want to contribute [1]. The collection of UAVs working together to achieve a single goal is known as UAV Fleet, they all connect and communicate to form an Adhoc network called Flying Adhoc Network (FANET) as depicted in Fig. 1. Usage of multiple nodes has much greater benefits as compared to single node as it allows coverage of large area (multiple nodes can coordinate with each other to cover different coordinates), reliability (if any node fails other can take its place), cost effectiveness (cost for setting single node structure is same as multiple and in the same cost we can have advantage of multiple nodes), unlimited mission duration [2] (if some UAVs are refilling their energy others can continue with the mission), load balancing (collection of data about a mission can be balanced among multiple nodes to maintain energy levels). Flying Networks are very useful in many real world scenarios such as search

© Springer Nature Singapore Pte Ltd. 2019
A. B. Gani et al. (Eds.): ICICCT 2019, CCIS 1025, pp. 128–143, 2019.
https://doi.org/10.1007/978-981-15-1384-8_11

and rescue operations, battlefield monitoring, architecture planning, and firefighting [3]. In many hostile environments, where human intervention is not possible, UAV based networks can be utilize to accomplish applications efficiently. For example a network can be formed for searching a specific target on ground in search and rescues context. In traffic monitoring context, they are able to monitor the road's traffic and can inform appropriate authorities when there is any accident or miss happening. For battlefield monitoring, they can monitor enemy's base area for any unwanted disturbances. It can also be used for monitoring environment using multiple sensors such as pressure, humidity, temperature sensors, etc. Precision farming has also gained a lot of hype in terms of flying network. Similarly, there are various areas where network of flying nodes can be used to achieve great benefits.

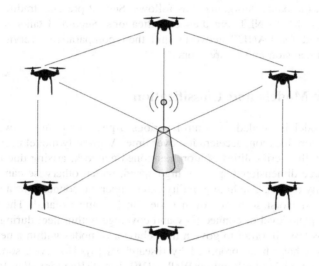

**Fig. 1.** Illustration of a typical Flying Adhoc Network

The issues pertaining to flying network which differentiate it from traditional MANET networks are high mobility, aerodynamics constraints, and limited bandwidth. High mobility leads to a lot of problems in such networks like breakage of communication and connectivity among moving nodes, sudden topological changes in the network and consumption of lot of energy. A mobility model that can govern the realistic UAV movements and can handle unique attributes of such nodes is a critical requirement for flying network. The importance of a mobility model lies in the fact that field testing is lot costly for evaluation of network performance which can be done in more efficient and cost effective manner using mobility models [4]. A lot of mobility models have been designed by the researchers, but most of these models are remodeled from MANET networks. Not much of research has been done on the mobility models for FANET keeping unique attributes of flying nodes in consideration. A mobility model should consider problems faced in real time scenarios such as sudden change in trajectory, obstacle detection, coverage of large area while maintaining the connectivity, maintenance of required energy levels among UAVs so as to complete mission oriented tasks.

Various mobility models have been adapted from traditional mobile networks in to FANET network. Random Walk (RW) [3, 5], Random Way Point (RWP) [6], Gauss Markov (GM) [7], Column mobility model (CLMN) [3] have been proposed for traditional networks and attuned for flying networks. But the mobility of flying nodes is different and traditional models can not accurately reproduce their behavior. Some of the mobility models like Alpha Based Model (ABM) [8], Multiple Pheromone UAV Mobility Model (MPU) [9], Coverage Connect model (CCM) [10] and Paparazzi mobility model (PPRZM) [11] are designed especially for FANET networks attributing to its distinctive constraints.

The rest of the paper is organized as follows: Sect. 2 presents traditional and new mobility models for FANET and their classifications. Section 3 talks about mobility models designed for FANET networks and their comparative overview. Section 4 presents the conclusion and future scope.

## 2   Mobility Models and Classification

A mobility model is needed to control various aspects of flying networks such as velocity, direction, location, acceleration over time. A mobility model needs to be good enough for handling criticalities of aforementioned network, arising due to movement of nodes in three-dimensional space at high speed, which otherwise can lead to inaccurate and faulty results. The high mobility can hamper the connections amongst nodes and sometime in worst scenarios even node can become isolated. Therefore, maintaining an adequate level of connectivity and coverage within fleet during operation is really a tedious task. In order to govern the behavior of nodes within a network, many mobility models have been proposed by researchers [3]. However, some traditional MANET based models like Random Walk (RW), RWP (Random Way Point) [4] can be utilized in FANETs also but in some applications scenarios where a coordinated fleet moving is required, they are not able to mock mobility pattern of nodes well and can give misleading results. Thus, over the time, many models for flying network have been proposed like Semi Random Circular Movement (SRCM) [18], Alpha Based Model (ABM), which takes mobility and frequent change in arrangement of flying nodes in consideration. Attempts have been made to fit traditional models in FANETs and new models have been designed specifically for flying network. Figure 2 represents a Venn diagram depicting relationship between existing models adapted from MANET and new models designed specifically for FANET. This section is more focused about the classifications of mobility models as per our study of related work. We consider certain important properties needed for mobile environment and possible application scenarios and based on these properties classify models in different categories.

A mobility model can belong to one or more groups based on its characteristics for example Group Force Mobility Model [22] (GFMM) can be classified as group model as well as connectivity model because it possess characteristics of both the models. Following are the possible classifications.

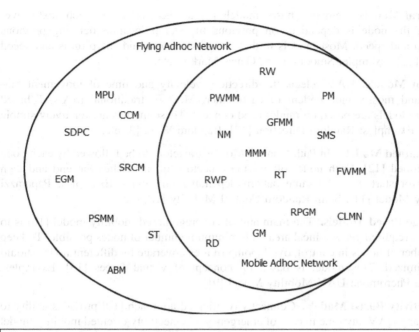

RW-Random Walk, RWP-Random Waypoint, NM-Nomadic Community, CLMN-Column Mobility Model, RD-Random Direction, RT-Random Trip, GFMM-Group Force Mobility Model, MMM-Manhattan Mobility Model, FWMM-Free Way Mobility Model, RPGM-Reference Point Group Mobility Model, SMS-Semi-Markov Group Mobility Model, PM-Pursue Mobility Model, GM-Gauss-Markov Mobility Model, SDPC-Self-Deployable Point Coverage, ABM-Alpha based Mobility Model, SRCM-Semi-Random Circular Movement, EGM-Enhanced Gauss-Markov Mobility Model, CCM-Connected Coverage Model, MPU-Multiple Pheromone UAV, PSMM-Particle Swarm Mobility Model, PPRZM-Paparazzi Mobility Model

**Fig. 2.** Venn diagram representing new mobility models and traditional MANET models applied on FANET.

**Group/Spatial Models:** In group based mobility models, multiple nodes move together following a common point [3]. Such mobility models are best suited for flying networks as nodes need to move in collaboration and take decisions on their own without human intervention. In such models there is a spatial dependency among flying nodes because position and movement of node is dependent on other nodes [4]. Example: Pheromone mobility model, Nomadic Community [3], Reference Point Group Mobility Model [5] etc.

**Entity Models.** In entity based mobility model [16] each UAV is independent of each other. In this model a single node is given responsibility of covering a predefined area without any interaction with other nodes. In this model mostly decision making is done by a centralized system located on ground. Example: SRCM [18], Random Model [3] etc.

**Temporal Models.** Temporal based models [5] are the models in which next movement of the node is dependent on previous timestamp location defining previous direction and speed. Most models in this category try to avoid sharp turns and speed change [12]. Example: Smooth Turn, Gauss Markov etc.

**Random Models.** UAV selects the direction, velocity and time of movement randomly and mostly independent of each other. Most of traditional MANET based mobility models are based on randomized concept. These models are relatively simple models. Examples: Random Direction [12], Random Walk [3] etc.

**Path Planned Models.** In Path Planned Models trajectory to be followed by each node is predefined [12]. Each node follows the trajectory till it reaches the end and then changes or start over the same pattern depending on model. Examples: Paparazzi Mobility Model [11], Semi Random Control Mobility [18] etc.

**Coverage Based Models.** The main aim of coverage based mobility model [10] is to cover the required geographical area with minimum number of nodes possible. To keep the number of nodes in control, overlapping of area coverage by different nodes should be minimized. This model is based on concept of virtual forces [13]. Examples: Multiple Pheromone UAV Mobility Model [9].

**Connectivity Based Models.** Connectivity oriented approach [10] provides ability to contact any UAV any time in case of emergency. Connectivity oriented mobility model is generally based on spatial mobility models in MANETS i.e. their decision is generally influenced based on local information or neighborhood. In such models main focus is on maintenance of connection among nodes all the time. Example: Group Force Mobility Model [23].

## 3    Mobility Models for Flying Ad Hoc Networks

MANET mobility models do not fit well for flying environment, due to some unique challenges and high mobility in 3D region of interest. Therefore, new models which can imitate the movement of flying nodes are required. In Flying Adhoc Network, for most of the applications, network needs to be self-organizing and it should cover the area of interest. While covering the area, connection between nodes should not break. Naïve mobility models have been designed keeping above constraints in consideration and some of these models use controlled mobility [2, 17] using virtual forces [8, 13] to maintain both coverage and connectivity. In this section, we explore some of the new FANET mobility models in depth and presented a comparative analysis (Table 1).

**Table 1.** Comparative analysis of mobility models.

| Mobility model | Category | Description | Contributions and limitations | Others |
|---|---|---|---|---|
| RW [3, 5] | Random, Entity | Memory less model, nodes choose random speed and direction to move for a constant time or distance and change direction randomly | Very simple and doesn't consider sudden stop, change and direction of speed | |
| RWP [6] | Random, Entity | Similar to RW, considers a pause time and chooses a random destination and start travelling towards the chosen destination | The inclusion of pause time helps to smooth out sudden change in direction but acceleration and deceleration nodes is not taken in consideration | RD [3], RT [5] |
| MMM [22] | Random, partially Path Planned, Spatial | Nodes move in an urban street like grid, horizontal and vertical lines intersecting each other. On an intersection point UAV can choose any random probabilistic direction for movement | Suits well only in a grid like structure not for other scenarios Sharp turns and change of speed is not considered | FWMM [3], PWMM [5] |
| GM [7] | Temporal | Temporal correlative movement which is based on Gaussian equations where the change of speed and direction for next move is dependent on a parameter $\alpha$ | Well suited for sudden movements but not perform well for FANET due to constraints in movement | SMS [5], EGM [25] |
| CLMN [3] | Spatiotemporal, Group | Nodes lying on a straight line but each node move around a reference point randomly. New reference point is created based on previous point | Well suited for FANET network and prevent collisions but sudden change of speed and direction is not considered | NM [3], PM [3], RPGM [5] |

(*continued*)

**Table 1.**  (*continued*)

| Mobility model | Category | Description | Contributions and limitations | Others |
|---|---|---|---|---|
| GFMM [23] | Group, Connectivity | Nodes are divided in to groups, nodes belonging to same group apply both attractive and repulsive forces on each other to maintain connectivity and avoid collision respectively and nodes of different groups repel each other | Collision and obstacle avoidance is considered in this model but coverage of the area which is must for FANET network is not taken in to account and also change of a node's group is not considered | |
| MPUMM [9] | Group, Coverage | Attractive and repulsive pheromone is used to attract and repel nodes to an area of interest | Not suited for real time scenarios as it doesn't consider connectivity and collision of nodes in consideration | DPR [15] |
| PSMM [16] | Group, Synthetic, Spatiotemporal | The velocity of a node is calculated based on velocity and location of center in previous time slot considering collision free adjustments | Calculation of velocity and waypoints can sometimes converge prematurely arising unrealistic situations. Also it doesn't consider coverage much | |
| PPRZM [11] | Path Planned, Entity | Paparazzi nodes move based on predefined shapes. Each UAV chooses a movement and altitude which remain fixed during entire simulation | Not suited for most of the FANET applications as it follows a fixed trajectory | |
| CCM [2] | Group, Spatial, Coverage and Connectivity hybrid, Temporal | Used Ant Colony Optimization and pheromone based approach to attract the UAVs to less covered areas<br>One hop information is used to determine the next position of UAV and best among the viable solution is chosen based on pheromone approach | Good in terms of coverage and connectivity but maintenance of connectivity with small number of nodes is a challenge | SDPC [3, 24] |

(*continued*)

**Table 1.** (*continued*)

| Mobility model | Category | Description | Contributions and limitations | Others |
|---|---|---|---|---|
| ABM [8] | Group, Spatial, Coverage and Connectivity hybrid | The node to be followed is decided based on followship weight which is calculated based on different parameters | Suitable for FANET networks but to determine the value of α is tedious. No emphasis has been given on collision avoidance | |
| SRCM [18] | Random, Entity and Path Planned | Nodes move in circular motion. On reaching a destination point in circle node chooses another circle with same center randomly and start moving. Clearly ending point of previous period is starting for next period | Avoids collision as multiple UAVS can chooses different circles with different center and cover the area but change in sudden radius while on move is not feasible for FANET | ST [20] |

## 3.1 Pheromone Based Mobility Model

Pheromone based approach is inspired with natural phenomenon of stigmergy used in ant colony [9]. Just as ants deposits pheromones to indicate the already explored area and this pheromone tend to disappear with time. Similar concept is used in Multiple Pheromone UAV Mobility Model [9] (MPU), an attractive chemical called attractive pheromone can be used to attract UAVs to the area where density of nodes is less and a repulsive pheromone can be used to indicate already covered area to repel UAVs from that location. As depicted in Fig. 3, a node gets attracted towards region 2 where number of nodes is less and they spread an attractive chemical to attract more nodes. Repulsive pheromone repels the incoming UAV in region 1.

Another variation of pheromone based model is Distributed Pheromone Repel Mobility [15] (DPR). In this model each node maintains a pheromone map, each cell of which is marked with the timestamp of last time UAV scanned that area. Once an area is scanned local pheromone maps are broadcasted and each UAV merge these map with their own maps. A node tend to move to the area with low pheromone smell as these are the areas which are not recently visited and area outside the search areas are marked with high pheromone smell to avoid those areas.

## 3.2 Particle Swarm Mobility Model

Particle Swarm Mobility Model [16] (PSMM) is based on temporal as well as spatial relation between the nodes, keeping in account the safe distance between nodes to avoid collision. It is based on PSO [26] (Particle Swarm Optimization) and has two phases (i) Generation of velocities and waypoints (ii) Collision-free adjustments. In first

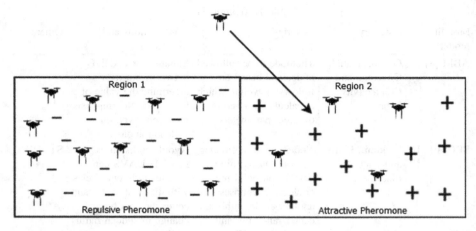

**Fig. 3.** A UAV getting attracted towards Region 2 because of the attractive pheromone.

step, the path or trajectory of node is assumed to be sequence of waypoints at discrete times. The next waypoint to be followed at time t is dependent on velocity at time t − 1 (temporal dependency) and the location of center at time t − 1 (spatial dependency) as depicted in Fig. 4(a).

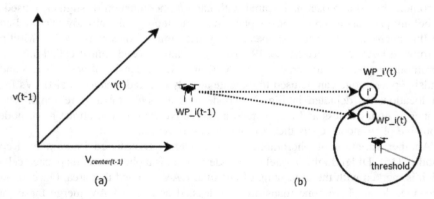

**Fig. 4.** (a) Velocity at time t is resultant of velocity at t − 1 and guidance of center location (b) A collision free adjustment for node i.

In second phase nodes which are not at a safe distance are identified and necessary adjustments are made to avoid collision keeping change in spatiotemporal properties as minimal as possible. These adjustments are made assuming nodes are sharing information among group, initial distribution of nodes is safe and collision if happen in between time interval t and t + 1 is not considered. In Fig. 4(b), a node i can move to position WP_i(t) in next time stamp and can collide with node j, a collision free adjustment is made and node can now move to WP_i'(t).

### 3.3    Paparazzi Model

Paparazzi model is kind of path planning mobility model. PPRZM [11], paparazzi UAV can make five movements Stay-At, Eight, Waypoint, Oval and Scan as depicted in Fig. 5. Each node is given a particular shape and altitude for movement which will remain same for whole simulation. Probability of node moving in Stay-At, Oval and Scan is high compared to eight and way-point.

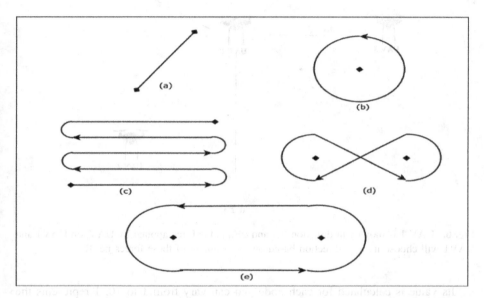

**Fig. 5.** PPRZM path schemes: (a) Way Point, (b) Stay-At, (c) Scan (d) Eight, and (e) Oval

### 3.4    Alpha Based Mobility Model

Apart from considering connectivity and coverage, energy is also accounted to be an important factor while making the decision of next move in alpha based model [8] (ABM). After every t update second, each node update its neighbor table based on the information gathered. In the next step, value of followship ($\alpha$) weight is calculated for each UAV based on (i) hop count (ii) number of neighbors (iii) energy. Followship weight ($\alpha$) represents node's willingness to be followed by adjacent nodes.

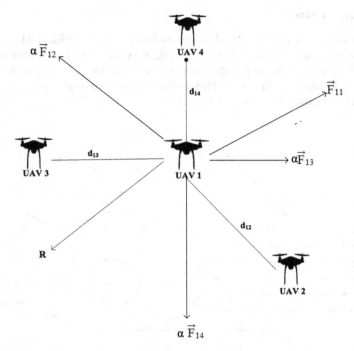

**Fig. 6.** UAV 1 is moving in direction $F_{11}$ and $\alpha F_{1i}$ is the force applied by $UAV_i$ on UAV1 and UAV1 will choose its next direction based on the resultant of these forces i.e. R

Its value is calculated for each node and can vary from 1 to 10. 1 represents the least recommended node to be followed while 10 represents most recommended node. Three evaluation levels, big, small and medium are proposed based on the three parameters mentioned above. Big and small represents the neighbors high and low level of followship weights and medium corresponds to undecided or neutral level of followship weights. The value of $\alpha$ is computed locally based on information of neighbor nodes and a force of magnitude equivalent to $\alpha$ is applied on UAV by each of the neighboring nodes. Node with high $\alpha$ value has greater impact than node with smaller value and each UAV also applies a force vector in the direction of current movement. Resultant of all the forces will determine the next direction and movement of UAV. Figure 6 depicts the forces applied on UAV1. Length of force vector indicates the magnitude of force. UAV 1 will choose its next direction based on resultant of forces i.e. R.

### 3.5  Connected Coverage Model

Connected Coverage model (CCM) [2] is based on three steps (i) neighborhood selection (ii) computing viable alternative (iii) selecting best alternative. In the first step appropriate neighborhood is selected based on the hop count. In step (ii) the future position of the node is considered and if the node is going out of transmission range in any of the scenarios that alternative is discarded. Out of available situations best alternative is selected in (iii) step based on pheromone based approach i.e. by moving in a direction where number of node is less to maintain appropriate density of nodes at a particular region. As depicted in Fig. 7. Circles represent the radio range of nodes and their next direction of movement is shown with arrows. Node in consideration can go up, down and right. The alternative which becomes out of range is discarded and out of available alternatives one which will not go out of range in near future is chosen as best alternative and is followed by node in consideration.

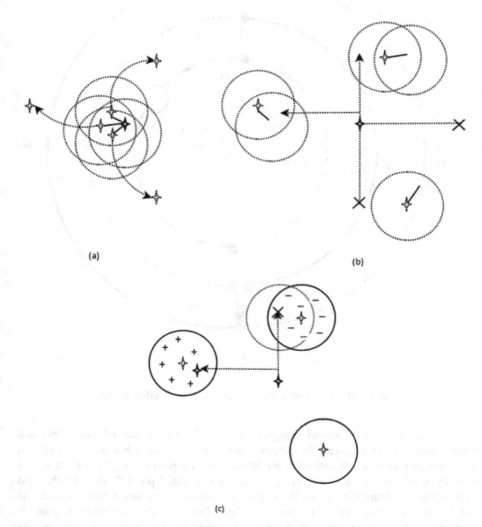

**Fig. 7.** Computation of viable alternative and Selection of best alternative in CCM.

### 3.6   Semi Random Circular Movement

Semi Random Circular Movement (SRCM) [18] model is designed especially for FANET network and is an enhancement to its predecessor models Random Direction Model and Random Way Point Model.

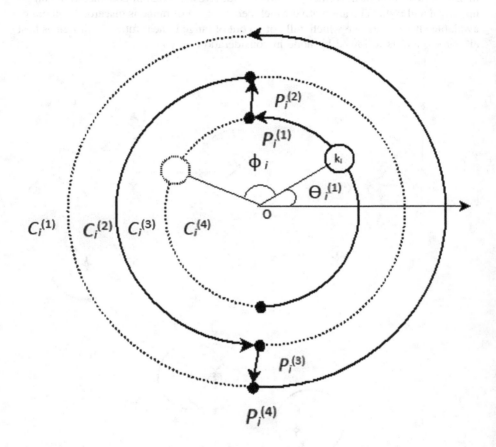

**Fig. 8.** Movement of a node $k_i$ using SRCM mobility model.

It is an entity based model designed specifically for the curved movement and sudden turns of flying nodes. As the name suggests, this model has both random and non-random properties. It follows a predefined circular path say $C_i$ for a node $k_i$ at a velocity $v_i$ which can vary from $[v_{min}, v_{max}]$ from an initial point $P_i$ and take the next step on the same circle based on the step point, step time and step length calculations but it can choose to move to different radius after reaching a destination point $P_i$ randomly. The starting point of the current period is the destination point of the previous period. Figure 8 represents a node $k_i$ travelling in circles centered O with step angle $\phi_i$.

### 3.7 Comparative Study and Discussion

FANET has different constraints compared to other Adhoc network. Table 2 shows analysis of new FANET mobility models based on connectivity, coverage, and energy and collision avoidance and neighbor awareness.

1. Connectivity Awareness: Because of its high mobility and unstable nature connection between the nodes tends to break. Pheromone, Particle Swarm, ABM, CCM and SRCM models are designed to avoid sudden connection breakage.
2. Energy Awareness: Energy is an important resource for flying network because of its application in hostile environments where battery cannot be charged. ABM keeps constraint of energy in consideration while deciding next move [8].
3. Collision Avoidance: Nodes should avoid collision with each other while moving. PSMM model avoids collision in collision adjustment phase [16] and due to circular movement SRCM model also partially avoids it [3].

**Table 2.** Comparison of new FANET specific Mobility Models.

| Mobility model | Connectivity awareness | Energy awareness | Collision avoidance | Global coverage | Neighbor awareness |
|---|---|---|---|---|---|
| Pheromone [9, 15] | No | No | Yes | Yes | Partially |
| Particle Swarm [16] | Yes | No | Yes | Yes | Yes |
| Paparazzi [3, 11] | No | No | No | No | No |
| Alpha Based [8] | Yes | Yes | No | Yes | Yes |
| Connected Coverage [2] | Yes | No | No | Yes | Yes |
| Semi Random Circular Movement [18] | No | No | Partially | Yes | Partially |

4. Global Coverage: Main aim of UAV network is to cover whole area of interest. Except PPRZM [11] all other models described above try to cover the maximum area.
5. Neighbor Awareness: Most of the application scenarios for FANET require group movement which in turn demands knowledge about neighboring nodes. MPU, DPR and SRCM models have partial neighbor awareness. ABM, PSMM and CCM are having full neighbor awareness.

## 4 Conclusion and Future Research

This paper presented a useful insight towards study of mobility models for flying adhoc networks. A mobility model helps in accurately handling the movement of nodes in real time scenario. An inappropriate model can give inaccurate results and can be misleading. In this paper, we have discussed about various mobility models for FANET, some of them are adapted from traditional mobile ad hoc networks and some are new models designed for flying network. Mobility models are classified based on different characteristics such as a model's ability to cover or connect, movement in group or as single entity, motion in fixed trajectory or randomly etc. The mobility models are compared with their contribution and limitations in tabular form. New FANET models are examined in length and a comparative study of models is discussed based on its ability to cover an area, maintenance of connectivity, energy efficiency etc.

As per our study, very few/limited work is found which considers energy as a decision parameter. Hence, in the future, more mobility models can be designed keeping energy consumption in consideration. Also, some clustering based approaches can be applied for movement of UAV network. Furthermore, applying machine learning approaches a good mobility models can be designed or existing models can be improved while considering sudden change of direction, acceleration and deceleration of moving nodes. Finally, work can be carried out using concept of virtual forces and controlled mobility. An attempt towards designing a mobility model which will consider coverage, connectivity and energy constraint while moving can be appreciated.

# References

1. Yanmaz, E., Yahyanejad, S., Rinner, B., Hellwagner, H., Bettstetter, C.: Drone networks: communications, coordination, and sensing. Ad Hoc Netw. **68**, 1–15 (2018)
2. Schleich, J., Panchapakesan, A., Danoy, G., Bouvry, P.: UAV fleet area coverage with network connectivity constraint. In: Proceedings of the 11th ACM International Symposium on Mobility Management and Wireless Access, pp. 131–138. ACM (2013)
3. Bujari, A., Calafate, C.T., Cano, J.C., Manzoni, P., Palazzi, C.E., Ronzani, D.: Flying ad-hoc network application scenarios and mobility models. Int. J. Distrib. Sens. Netw. **13**(10), 1–17 (2017). https://doi.org/10.1177/1550147717738192
4. Xie, J., Wan, Y., Kim, J.H., Fu, S., Namuduri, K.: A survey and analysis of mobility models for airborne networks. IEEE Commun. Surv. Tutor. **16**(3), 1221–1238 (2014)
5. Manimegalai, T., Jayakumar, C.: A conceptual study on mobility models in MANET. Int. J. Eng. Res. Technol. (IJERT) **2**(11), 3593–3598 (2013)
6. Bettstetter, C., Hartenstein, H., Pérez-Costa, X.: Stochastic properties of the random waypoint mobility model. Wirel. Netw. **10**(5), 555–567 (2004)
7. Liang, B., Haas, Z.J.: Predictive distance-based mobility management for PCS networks. In: Proceedings of the Eighteenth Annual Joint Conference of the IEEE Computer and Communications Societies, INFOCOM 1999, vol. 3, pp. 1377–1384. IEEE (1999)
8. Messous, M.A., Sedjelmaci, H., Senouci, S.M.: Implementing an emerging mobility model for a fleet of UAVs based on a fuzzy logic inference system. Pervasive Mob. Comput. **42**, 393–410 (2017)
9. Atten, C., Channouf, L., Danoy, G., Bouvry, P.: UAV fleet mobility model with multiple pheromones for tracking moving observation targets. In: Squillero, G., Burelli, P. (eds.) EvoApplications 2016, Part I. LNCS, vol. 9597, pp. 332–347. Springer, Cham (2016). https://doi.org/10.1007/978-3-319-31204-0_22
10. Yanmaz, E.: Connectivity versus area coverage in unmanned aerial vehicle networks. In: IEEE International Conference on Communications (ICC 2012), pp. 719–723. IEEE (2012)

11. Bouachir, O., Abrassart, A., Garcia, F., Larrieu, N.: A mobility model for UAV ad hoc network. In: 2014 International Conference on Unmanned Aircraft Systems (ICUAS), pp. 383–388. IEEE (2014)
12. Guillen-Perez, A., Cano, M.D.: Flying ad hoc networks: a new domain for network communications. Sensors 18(10), 3571 (2018)
13. Zhao, H., Wang, H., Wu, W., Wei, J.: Deployment algorithms for UAV airborne networks toward on-demand coverage. IEEE J. Sel. Areas Commun. 36(9), 2015–2031 (2018)
14. Jawhar, I., Mohamed, N., Al-Jaroodi, J., Agrawal, D.P., Zhang, S.: Communication and networking of UAV-based systems: Classification and associated architectures. J. Netw. Comput. Appl. 84, 93–108 (2017)
15. Kuiper, E., Nadjm-Tehrani, S.: Mobility models for UAV group reconnaissance applications. In: International Conference on Wireless and Mobile Communications, ICWMC 2006, p. 33. IEEE (2006)
16. Li, X., Zhang, T., Li, J.: A particle swarm mobility model for flying ad hoc networks. In: 2017 IEEE Global Communications Conference on GLOBECOM 2017, pp. 1–6. IEEE (2017)
17. Cheng, X., Dong, C., Dai, H., Chen, G.: MOOC: a mobility control based clustering scheme for area coverage in FANETs. In: 2018 IEEE 19th International Symposium on "A World of Wireless, Mobile and Multimedia Networks" (WoWMoM), pp. 14–22. IEEE (2018)
18. Wang, W., Guan, X., Wang, B., Wang, Y.: A novel mobility model based on semi-random circular movement in mobile ad hoc networks. Inf. Sci. 180(3), 399–413 (2010)
19. Kumari, K., Sah, B., Maakar, S.: A survey: different mobility model for FANET. Int. J. Adv. Res. Comput. Sci. Softw. Eng. 5(6), 1170–1173 (2015)
20. Wan, Y., Namuduri, K., Zhou, Y., Fu, S.: A smooth-turn mobility model for airborne networks. IEEE Trans. Veh. Technol. 62(7), 3359–3370 (2013)
21. Kumari, K., Sah, B., Maakar, S.: A brief survey of mobility model for FANET. In: Proceedings of National Conference on Innovative Trends in Computer Science Engineering (ITCSE) (2015)
22. Bai, F., Sadagopan, N., Helmy, A.: The IMPORTANT framework for analyzing the impact of mobility on performance of routing protocols for Adhoc networks. Ad Hoc Netw. 1(4), 383–403 (2003)
23. Williams, S.A., Huang, D.: Group force mobility model and its obstacle avoidance capability. Acta Astronaut. 65(7–8), 949–957 (2009)
24. Sanchez-Garcia, J., Garcia-Campos, J.M., Toral, S.L., Reina, D.G., Barrero, F.: A self organising aerial ad hoc network mobility model for disaster scenarios. In: 2015 International Conference on Developments of E-Systems Engineering (DeSE), pp. 35–40. IEEE (2015)
25. Biomo, J.D.M.M., Kunz, T., St-Hilaire, M.: An enhanced Gauss-Markov mobility model for simulations of unmanned aerial ad hoc networks. In: 2014 7th IFIP Wireless and Mobile Networking Conference (WMNC), pp. 1–8. IEEE (2014)
26. Kennedy, J.: Particle swarm optimization. In: Encyclopedia of Machine Learning, pp. 760–766. Springer, Boston (2011)

# Evolutionary Computing Through Machine Learning

# Role of Artificial Intelligence and Machine Learning in Resolving the Issues and Challenges with Prosthetic Knees

Deepali Salwan[1](✉) ⓘ, Shri Kant[2], and G. Pandian[1]

[1] Pt. Deen Dayal Upadhyaya National Institute for Persons
with Physical Disabilities, New Delhi, India
deepalisalwan@gmail.com, gpnbpoiph@yahoo.co.in
[2] Research and Technology Development Center,
Department of Computer Science, School of Engineering and Technology,
Sharda University, Greater Noida, India
shrikant.ojha@gmail.com

**Abstract.** From the ancient times until today, in the age of Artificial Intelligence, field of prosthetics is where human is continuously striving to do better. Upper limb amputation deprives the person from routine activities, which the amputee never thought was important, on the other hand lower limb amputation restricts the person's physical movements to a great extent. In this article we will highlight the issues and challenges with the active lower limb prosthetic from the socket to the knee, the material, the sensors and the algorithms used for controlling the movement. How each of them plays a pivotal role in providing a comfortable gait which resembles the natural human gait, this article throws light upon where are we in terms of advancement in lower limb prosthetic and the issues and challenges which are still there even with the finest active artificial knee available.

**Keywords:** Prosthetic-knee · Artificial-knee · Lower limb amputation · Microprocessor prosthetic knee · Sensors · Classifier algorithm · Sockets and liners

## 1 Introduction

Prosthetic knees or artificial knees are the replacement of lower limb which may have been removed. The removal of limb through surgical procedure is called amputation. This may happen due to several factors such as accidents, diseases like peripheral vascular disease, diabetes in old age. Injured soldiers from war front have been a constant to amputation since ages. History of amputation goes back to Egyptian era [1] where a sign of first prosthetic toe has been found attached to a mummy. From then until now there have been numerous inventions ranging from a prosthetic leg with a locking knee joint in sixteenth century to the most advanced Microprocessor Prosthetic Knee (MPK). Figure 1 below shows the types of lower limb amputation (LLA) level like: (1) Hip Disarticulation (2) Above Knee (AK) or as Transfemoral (3) Through knee (4) Below Knee (BK) or Transtibial. In this article we will focus extensively on

A. B. Gani et al. (Eds.): ICICCT 2019, CCIS 1025, pp. 147–155, 2019.
https://doi.org/10.1007/978-981-15-1384-8_12

Transfemoral Prosthetic Device since the simulating knee mechanism is an important part of prosthetic device.

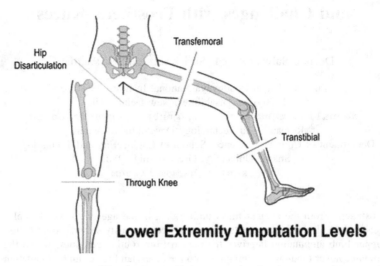

**Fig. 1.** Lower limb amputation level

## 2   Prosthetic Knee

An ideal prosthetic knee is the one which provides stability, comfort, natural gait, cosmoses and durability [2]. Figure 2 below shows part of a basic prosthetic knee which is (1) Socket (2) Knee Joint (3) Pylon (4) Foot. Stability is necessary for weight

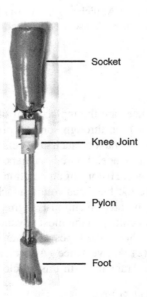

**Fig. 2.** Parts of prosthetic knee

bearing which can be obtained by one or combination of these three: (1) The stump (residual part of human limb) may hold the knee stable at the heel strike and while standing (2) By antero-posterior alignment (3) By manual or automatic lock.

# 3 Classification of Prosthetic Knee Joints

Today prosthetic knees are available in wide variety of designs and with broad spectrum of capabilities that it is difficult to determine universally acceptable classification system. However, we have tried to present a basic level of classification issues and challenges with each one of them. There are two functional groups i.e. Mechanical Controlled [3] and Microprocessor Controlled Knees. Further subdivision may be specified with regard to the complexity of stance phase stability and swing phase control.

## 3.1 Mono-centric Articulated Knees

This type of knee has a single pivot point or single axis. This type of knee joint is recommended for the users who have very limited movements or cannot afford a sophisticated and expensive knee joint.

The advantages of mechanical knees are the following:

(1) They are very simple mechanically
(2) Requires servicing rarely
(3) Very inexpensive
(4) Preferable where follow-up care is difficult or impossible especially when the amputee is residing in remote rural area.

## 3.2 Polycentric Articulated Knee

This design is also known as polycentric knee. It provides increased stance stability and ease of flexion (bending of a knee) during pre-swing. This also reduces the risk of stumble. These types of knees are suitable for the individuals who walk at a constant speed like elderly amputees.

## 3.3 Fluid Controlled or Variable Cadence Knees

These are Hydraulic (compressed oil) or Pneumatic (compressed gas) Controlled Knee. Pneumatic Dampers are the ones which are not affected by the temperature changes, so it is most preferred by the amputees who reside in the variable temperatures, on the contrary hydraulic dampers have the advantages over pneumatic in shock absorption during variable speed from slow walk to running, here the knee resistance compensate automatically. Therefore, Pneumatic are known for their controlled force while hydraulic are used for reduced energy cost and more resistance.

### 3.4   Microprocessor Controlled Knees (MPKs)

These are the most sophisticated prosthetic knees. These knees use sensors to detect the movement of the limb and control the swing accordingly with the help of micropro-cessor. This is majorly achieved by controlling the resistance while walking on dif-ferent terrain. Other advantages to MPK are it requires minimal efforts from the amputee and provide suitable gait symmetry with the help of hydraulic or pneumatic damper.

The figure below shows (a) Pneumatic Knee (b) Hydraulic Knee (c) MPK as stated above has the microprocessor system which works with the help of input of different sensors and apply sophisticated AI algorithms to give a better gait pattern to an amputee (Fig. 3).

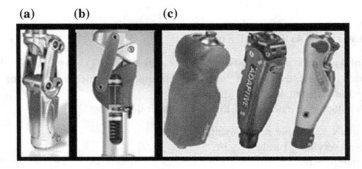

**Fig. 3.**   (a) Pneumatic knee (b) Hydraulic knee (c) Microprocessor controlled knee

## 4   Detail Study of Microprocessor Controlled Knee

### 4.1   Material Used in Making the Joints

Materials like Aluminum Alloy 6061, Aluminum Alloy 7075 and AISI4130 Steel [4] were compared to develop a knee joint to check the flexion. Flexion was as much as 90° considering properties of the materials (bears heavy weight) and cost effectiveness. It was found out that Aluminum Alloy 6061 was best of the all for developing a prosthetic joint as it gave more stability and comfort to the amputee and is safe and affordable.

### 4.2   Materials Used in Making the Socket

Socket is that part which integrates the prosthesis with the amputee. It has been found out that socket's fit is utmost important factor else even the most sophisticated knee is not able to provide promised comfort. As a result, the material should be lightweight yet rugged sockets for the fittings. Sockets are either fabricated manually or CAD/CAM [5] systems are used. Polypropylene, epoxy and acrylic as used for fab-rication; thermostat polymer laminate is used because of its properties. The socket must be designed perfectly and should not restrict blood circulation. Which socket will be applied to which amputee is by and large depend on the prosthetist (one who fits the prosthetic limb) after preparation of the fitting. The prosthetist must be skilled enough to suggest a suitable socket for the amputees.

## 4.3 Sensors

MPKs use various sensors to provide a better gait pattern compared to other mechanical knees. Studies shows that various sensors [5] such as Force Sensing Resistor (FSR), knee angle sensors, potentiometer (moment sensors), torque sensors [6], strain gauge, hall position sensors and piezoelectric sensors could identify the movement like flexion and extension, the issue with these sensors are that wearing of these sensors leads to disruption of signals malfunctioning of the device. Various articles show that [7, 8] 3 axis accelerometer and 3 axis gyroscopes embedded together provides all the necessary and accurate data for calculating accelerations and angular rate for mimicking natural human gait. Also, it shall be noted that magnetometers should not be use as it affects the other readings.

## 4.4 Algorithms

Algorithms are everywhere, in this world of Machine Learning and Artificial Intelligence. Prosthetic Limb driven by sensors and electronics backed up by a perfect control algorithm is the need of the hour. We are thriving to develop not an instrument which is a replacement of the lost limb but an extension of the body which feels real. As mentioned above MPKs use various algorithms to bring gait symmetry, motion analysis, stumble control and comfort with ease of walking.

(a) *Control Logic:* Control Logic/Control Loop is everywhere. Even while someone is driving a vehicle, the feedback given through the eyes to the brain and from the brain to the hands and legs, and these in turn responding as to move the vehicle in the right direction with the correct velocity is an example of Control Logic. Control algorithms work on PID [9], the principle of Proportional, Integral and Derivative and hence the name. There are 3 different type of PID Controls Algorithms like: Interactive, Noninteractive and Parallel Algorithms.

A PID controller is used in multiple applications where continuous control is required. It calculates an error in the process called an error value e(t) as the difference between Set Point (SP) and the Process Variable (PV). Thereafter, Proportional, Integral and Derivative (P, I and D respectively) based correction is applied. The oldest controllers were based on Interactive Algorithms (IA). These are derived from Ziegler-Nichols PID Control Algorithm.

The second algorithm is called as Non Interactive Algorithm. This algorithm is used when the intent is to use robust and stable control systems which is capable of absorbing disturbances as well. If the time difference is Td = 0, interactive and non-interactive algorithms works in the similar manner. Parallel Algorithm is generally detailed in textbooks but this algorithm is not practically used to tune. Interactive and Non Interactive algorithms are recommended to be used in Controllers. As stated in the

**(a)**

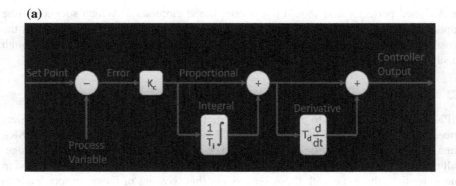

**(b)**

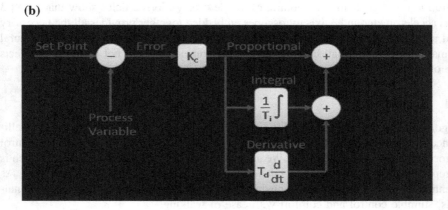

**(c)**

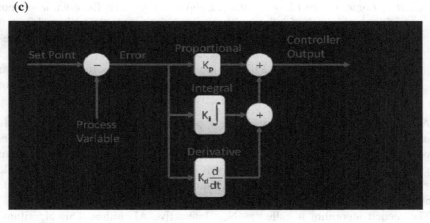

**Fig. 4.** (a) Interactive algorithm (Proportional mode), (b) Non interactive algorithm (Integral mode), (c) Parallel algorithm (Derivative mode)

Fig. 4, the use of Proportional and Integral and Derivative Mode is explained. These are also different in the way output is calculated and how these calculations act on Set Point Changes. Control Algorithm [6] are used wherein the controller supervises the trajectory of the joint position, however the output of these algorithms are not always guaranteed.

(b) *Intent Detection Algorithm:* This type of algorithm comes under Machine Learning for utterance/text analysis and is used in Chat Bots as in the case of Amazon's Alexa that interprets a text command which the computer understands. For instance, "Call Dad" would be interpreted as extract number fed for "Dad" and then dial the number to make a call. Intent Detection Algorithm can be done in two ways: unsupervised and supervised machine learning. Supervised Machine Learning includes entering millions of data records and the machine analyses it under two further subcategories i.e. Classification and Regression. While unsupervised Machine Learning is not supervised, there are no correct answers and machines are left on their own to derive meaning out of the data records. Intent Detection Algorithm in prosthetic knee is unsupervised Machine Learning and is based on the reference pattern. Dependent prosthetic knee also has a bypass system in case the algorithm [5] does not provide the required functionality. This prosthetic knee joint is specifically programmed for elderly amputees which needs a strong locking mechanism.

(c) *Genetic Algorithm:* Physical parameters in a prosthetic knee joint were also optimized through Genetic Algorithm [7], which decreases the amount of damping coefficient and dissipative energy. Genetic Algorithm as the name suggest is based on Charles Darwin theory of "Survival of the fittest". As per Darwin "It is not the strongest of the species that survives, nor the most intelligent, but the one most responsive to change." Therefore, the Genetic Algorithm can be explained as (1) We take initial population (Initialization) (2) We define function to classify whether the data from the population is Fit or Unfit by assigning a fitness score to them (Fitness Function) (3) The Fittest Individuals are selected so that they pass their properties(genes) to the next generation (Selection). (4) Crossover happens randomly resulting in creation of new offspring (Crossover) (5) Mutation happens in the offspring when the properties are exposed to certain kind of changes. Therefore, given certain situations as to how amputee wants the movement of the prosthetic part the solutions are derived from the large datasets which the sensors have gathered over a period of time (Fig. 5).

(d) *Fuzzy logic-based classifier:* Fuzzy logic-based classifier [8, 9] work on the logic which classifies the amputees into groups like young: tall: lean, old: short: fat, old: tall: slim etc. Amputated population falls into groups like ones mentioned above and hence the input data is classified as per the logic created in the algorithm. The data such as lean, old, short are not significant but precise. The term Fuzzy Logic was coined by Lotfi Zadeh in 1965 and revolves around the significant and non-significant values being assigned to real numbers ranging from 0 to 1 capturing "degrees of truth" ranging between completely true to completely false.

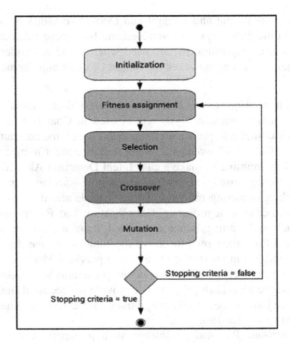

**Fig. 5.** Genetic algorithm

(e) ***Expectation Maximization Algorithm:*** Expectation Maximization Algorithm [10] is used in an active knee where the dataset is roughly clustered, this is done using k-means algorithm which helps in stability while standing and sitting, further work is needed on different postures. The EM algorithm [11] maximize the likelihood functions that arises in statistical estimation problems. It maximizes the conditional likelihood of a complete data space which is not measured rather than maximizing the function of the incomplete or measured data. This type algorithm is especially useful even if only a few data points are present and all do not cover all the categories of classification, it shows that this approach provides such data mapping which in turn improves the efficiency and accuracy.

(f) ***Impedance Control Algorithm:*** Impedance Control Algorithm [12–14] is used in MPKs but have inconsistencies between the framework and the real-world model. This algorithm is used in majorly controlling the movement of lower limbs in the amputee and is a part of special branch called Robotic Prosthetic Control. In this algorithm the variables like force or velocity or position are not regulated separately but their relationship is modified in such a way that a particular calculated torque required at knee angle is applied [15].

$$\tau = k(\theta - \theta_0) + b\theta$$

Here, k (spring stiffness), $\theta_0$ (equilibrium angle), and b (dampening coefficient) which provides a stable gait to the amputee, therefore providing a stumble free walk.

# 5   Conclusion and Future Work

The discussion above shows that there have been significant developments in the field of Lower Limb Amputation. Microprocessor Controlled Knees in combination with hydraulic dampers are best suited for the amputees and sensors like gyrometer and accelerometer provides the required input to the microprocessor in order to attain a natural gait for the prosthetic limb. These artificial limbs are comfortable enough owing to its socket material, artificial limb material and the lining. There are multiple algorithms which have been used in various MPKs and most of them are working fine for a specific category of amputees. There is still a need of writing exhaustive algorithm on the top of these sensors which best suits the need of amputees of all the ages and still provides the stability, comfort and varying need. Future work includes developing a Machine Learning based algorithm with a holistic approach to target and do away with all the challenges of current MPKs.

# References

1. Finch, J.L.: Assessment of two artificial big toe restorations from ancient Egypt and their significance to the history of prosthetics. J. Prosthet. Orthot. **24**(4), 181–191 (2012)
2. Hafner, B.J.: Evaluation of function, performance, and preference as transfemoral amputees transition from mechanical to microprocessor control of the prosthetic knee. Arch. Phys. Med. Rehabil. **88**(2), 207–217 (2007)
3. Khadi, F.M.: Design and manufacturing knee joint for smart transfemoral prosthetic. IOP Conf. Ser. Mater. Sci. Eng. **454**, 012078 (2018)
4. Kishore Kumar, P., Subramani, K.: Trends and challenges in lower limb prosthesis. IEEE Potentials **36**(1), 19–23 (2017)
5. Krut, S.: Secure microprocessor-controlled prosthetic leg for elderly amputees: preliminary results. Appl. Bionics Biomech. **8**(3–4), 385–398 (2011)
6. Awad, M., Abouhussein, A.: Towards a smart semi-active prosthetic leg: preliminary assessment and testing. IFAC-Pap. Online **49**(21), 170–176 (2016)
7. Karimi, G., Jahanian, O.: Genetic algorithm application in swing phase optimization of AK prosthesis with passive dynamics and biomechanics considerations. IntechOpen (2012). https://doi.org/10.5772/38211
8. Alzaydi, A.A., Cheung, A.: Active prosthetic knee fuzzy logic - PID motion control, sensors and test platform design. Int. J. Sci. Eng. Res. **2**(5), 1–17 (2011)
9. Arzen, K.-E.: A simple event-based PID controller. In: IFAC World Congress (1999)
10. Fessler, J.: Space-alternating generalized expectation-maximization algorithm. IEEE Trans. Signal Process. **42**(10), 2664–2666 (1994)
11. Nandi, G.: Biologically inspired CPG based above knee active prosthesis. In: International Conference on Intelligent Robots and Systems, pp. 1–6. IEEE (2008)
12. Frank, C.S.: Design and control of an active electrical knee and ankle prosthesis (2008). https://doi.org/10.1109/BIOROB.2008.4762811
13. Frank, S., Bohara, A.: Design and control of a powered transfemoral prosthesis. In: IEEE International Conference on Robotics and Automation, pp. 263–273 (2007)
14. Herr, H., Wilkenfeld, A.: User-adaptive control of a magnetorheological prosthetic knee. Ind. Robot. **30**(1), 42–55 (2003)
15. Hogan, N.: Impedance control: an approach to manipulation. In: American Control Conference. IEEE (1984)

# A Noise Tolerant Auto Resonance Network for Image Recognition

Shilpa Mayannavar[1]([✉]) and Uday Wali[2]

[1] C-Quad Research, Belagavi 590008, India
mayannavar.shilpa@gmail.com
[2] KLE DR MSS CET, Belagavi 590008, India
udaywali@gmail.com

**Abstract.** Auto Resonance Network (ARN) is a biologically inspired generic non-linear data classifier with controllable noise tolerance useful in control applications like robotic path planning. It appears that multi-layer ARN can be useful in image recognition but such capabilities have not been studied. This paper presents a multi-layer ARN implementation for an image classification system for handwritten characters from MNIST data set. ARN nodes learn by tuning their coverage in response to statistical properties of the training data. This paper describes the effect of adjusting the resonance parameters like coverage and threshold on accuracy of recognition. ARN systems can be trained with a small training set. It has been possible to attain accuracy up to 93% with as few as 50 training samples. ARN nodes can spawn secondary nodes by perturbing the input and output similar to a mutated cell in a biological system. Use of input transformations to achieve required perturbation of the network is discussed in the paper. Results show that the network is able to learn quickly without any stability issues. The present work shows that reasonable level of image classification can be obtained using small sample training sets using resonance networks.

**Keywords:** Artificial Intelligence · Artificial Neural Networks · Auto Resonance Network · Character recognition · Deep learning · Handwritten digit recognition · Image classification · Neural network hardware

## 1 Introduction

Artificial Neural Networks (ANNs) have been successful in image recognition and classification in recent years. Multi-layered Convolution Neural Network (CNN) has been very successful in image classification. Most of the progress we have seen in Artificial Intelligence (AI) is due to success of CNN and other deep learning network architectures. Success of such systems is partly due to invention of these new algorithms and partly due to the availability of high performance Graphic Processing Units (GPUs) with large number of compute cores on single chip. These two developments go hand-in-hand: ANNs require massively parallel processing capability and GPUs offer such capability at low cost. However, GPUs were primarily designed to handle display tasks which are different than ANN tasks. Therefore, many organizations are developing special purpose hardware to address the massively parallel computer architectures to meet the computational demand posed by deep learning ANNs. Ultimate aim of deep learning technology is to achieve a generalized intelligence that is comparable to biological systems, especially that of humans. Biological systems use thousands of types of neurons to achieve highly efficient sensory and motor control. Therefore, it is imperative that we need to explore many more ANN architectures on our quest to reach anything close to General Artificial Intelligence (GAI). This will require both evolutionary and exploratory approaches to find new architectures.

As any other technology, deep learning systems have scope for improvements. Some of the leading researchers including Geoffrey Hinton et al. [1] point out some such issues. For example, pooling layers that perform some approximation and image scaling incur loss of features. CNNs serialize the image and therefore loose spatial relations among the image features. Guanpeng et al. [2] have noted that errors and their propagation in deep neural networks like CNN depend on data, learning process and number of layers. Capsule networks were introduced by Hinton et al. [1] to overcome some of these issues and improve the performance of deep neural networks.

Auto Resonance Network (ARN) proposed by Aparanji et al. [3] take a different approach to the AI problem. An ARN node behaves very similar to a biological neuron: When the input matches a specific combination of values to which it is tuned, the neuron fires. Such a node is said to be in resonance. Output of an ARN node varies as a continuous value between well defined limits: generally, 0 and 1. Low precision arithmetic is often enough to implement ARN [4–6]. Every node in ARN represents a feature or a sequence of features, similar to capsules described by Hinton et al. [1]. Range of values to which the node resonates can be controlled using node parameters to match statistical properties of incoming data. Multiple layers of ARN nodes can be used in a sequence or hierarchy, each higher layer representing an improvement in recognition of input. Each node in a layer responds to a specific combination of inputs, independent of the other nodes in the same layer. A group of such nodes can cover any convex or concave set of patterns. Perturbation techniques are used by ARN to reinforce the training and reduce the number of actual inputs. Perturbations are similar to biological mutations. Perturbation of input, output and the system parameters are all supported by ARN. Mechanisms to distinguish between real and perturbed nodes are described in [3]. As the perturbed nodes can respond to similar input, we can say that they can perform an affine transform on the input. These properties of ARN are useful in building image recognition and classification systems. The recognition performance of such a network can be improved by increasing the number of training samples.

It is possible to implement various network structures based on ARN. Some of these possibilities include the choice of resonator, spatial spread of neural input, neuronal density, selection of threshold and coverage parameter $\rho$ (rho), perturbation methods, number of layers, etc. It is also possible to inter-spread resonators and other nodes like constrained Hebbian nodes [7] and logical decision making layers to improve overall quality of recognition.

In this paper, we have presented methods to use ARN for image recognition and classification task. This work was conducted with an idea to test generalization capabilities of ARN. Section 2 presents brief literature review on recent history of image recognition techniques, shifting from an algorithmic approach to a deep learning approach. Section 3 gives an overview of ARN structure, methods and its application to image processing. Section 4 gives details of how ARN was used to implement a hand written character recognition system for a specific public dataset. Results and discussions on the implementation are in Sect. 5. Conclusion and future work are in the Sect. 6. An adequate list of references is provided at the end of the paper.

## 2  Literature Survey

Till recently, image recognition and classification was performed by specialized signal processing algorithms. For example, Giuliodori et al. [8] performed experiments on MNIST and USPS datasets to recognize hand written characters. Algorithmic and statistical methods have been studied and compared. Thresholding (called Binarization method in the paper) and other methods including Hough transform for circle detection have been used to extract the features using shape and structural characteristics of digits.

Many algorithms based on geometric features have been reported to be useful for specific classes of image recognition problems. For example, second derivative of image intensity expressed by the Hessian matrix has been used to identify blood vessels by Frangi et al. [9] and to identify embryo cells by Patil et al. [10].

Everything in image processing changed in the last few years with the availability of CNN and web services that support remote implementation of such networks. Convolution networks have been used extensively for image classification in recent years. Low cost commercial applications using image recognition have been possible due to availability of these technologies. Alex Krizhevsky et al. [11] reported an error rate less than 0.3% with deep convolutional neural networks using ImageNet database with more than a million images. Traditional image recognition methods have been replaced by machine learning and CNN to improve the accuracy of results. Babu et al. [12] reported an accuracy of 96.94% (k = 1) using K-nearest neighbor method. Patil et al. [13] report an accuracy of 96 to 99.9% for various grades of embryo images using CNN as a web service.

Further improvements to CNN are being suggested. Juefei-Xu et al. [14] suggest replacing a convolutional layer with perturbation layer to reduce the size of the learning sample set. Many CNN based image classification systems have been built using millions of images, which is an achievement in itself. However, the authors show that new input images can be perturbed with a set of pre-defined random additive noise masks, achieving accuracy comparable to larger input image. In this way, computational complexity of convolution operation is replaced by noise addition. It is reported that the method is more efficient than the convolution layer. Simonyan and Zisserman [15] report how the accuracy of a model can be improved by increasing the depth of a neural network. In general, it is now recognized that the depth (number of layers) of neural network increases the quality of recognition.

There are many applications that can benefit from the current state-of-art technologies for image recognition. However, there is ample scope for further development in terms of neuronal structure, perturbation methods, hardware acceleration, etc.

## 3  Brief Introduction to ARN

An ARN *node* is similar to a neuron. ARN is an aggregation of neurons. A node has several input ports and one output port. Every input port implements a resonator, generating a near maximal output when the input is in the neighborhood of a tuned input value. The node will generate maximal output when all the input ports are resonating. Generally, each node has a single point in the input space at which it resonates. If some of the inputs match but not all, the output is proportional to the number of matched inputs. Each input port produces a value-limited output and hence there is a well known upper bound on the value attained at the output port. This computation is akin to neuronal firing. The accuracy of an ARN based classifier depends on the number of inputs $N$, value of *resonance control parameter* ($\rho$) and *threshold* ($T$). These parameters control the *coverage* of an ARN node (see Fig. 1). The effect of $\rho$ and $T$ on resonance depends on the statistical properties of input and size of sample set. For simplicity, these values are held constant at each layer in the work reported here. The network works as follows: When a new input is applied one of the three possibilities occurs, viz.,

(a)  one or some of the nodes are at resonance or
(b)  one or more of the nodes are near resonance or
(c)  none of the nodes are in resonance.

If there is a single resonating node and its output is above threshold value, it is the winner. Two nodes cannot be at peak resonance simultaneously by design. If one or more are near resonance, one with the highest output and above threshold is selected as winner. If two nodes are near resonance and have same output, one of them is selected and its tuning is shifted towards the input. If none of the nodes are resonating or not near resonance, a new node is appended to the network such that it will resonate at the newly applied input. Alternately, we can say that a new node is pre-tuned to resonate at current input.

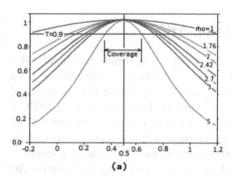
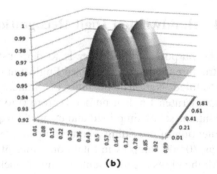

(a)                                            (b)

**Fig. 1.**  (a)Tuning curve of ARN node with single input port, resonating at input value of 0.5., (b) 3-D view of coverage of two-port 3-node ARN

Several ARN nodes may recognize same class of data. Therefore, labeling a node with a class indicates a learning process. Coverage of nodes belonging to a class may form continuous set of values or isolated set(s) of values in input/output space. Therefore the overall coverage of a data class can be convex, concave, contiguous or isolated. Coverage of a class is described as union of coverage of all nodes belonging to the same class. ARN nodes can spawn new nodes by perturbation of existing nodes and may suffer attrition if not used. Therefore, the number of output nodes in ARN varies in time. Repeated application of data closer to the resonance value of the node results in reduced coverage of the node, effectively reducing the range of input values that bring the node to a resonance. This implies an improved tuning, indicating a highly specialized node. On the other hand, node coverage may increase in response to data with high variance, representing a de-tuned node.

Therefore, learning can be described as the ability to generate, spawn and merge nodes with similarly labeled nodes into a single class. When none of the nodes produces an output above their acceptance threshold, a new node is added to the ARN, similar to adding a new case to an existing knowledge base.

ARN resonators (synapses) can be implemented with any of the several symmetric resonator functions. However, resonator function described by following Eq. (1) is preferred because of its controllability, symmetry and range.

$$y = 4 * \left( \frac{1}{1 + e^{-\rho(x - x_m)}} \right) * \left( \frac{1}{1 + e^{\rho(x - x_m)}} \right) \tag{1}$$

Figure 1(a) shows the output of a single input ARN node tuned for an input value of 0.5 for different values of resonance control parameter $\rho$. As the value of $\rho$ increases, the output becomes more focused. Therefore, for a given threshold value of say 0.9 and $\rho$ of 5, the coverage extends from 0.37 to 0.63. Input values outside this range will have output that will be less than the threshold. It can be noticed that, as the value of $\rho$ increases, resonance becomes sharper, indicating the reduced coverage or more focused learning. Figure 1(b) shows 3-D view of two port 3-node ARN.

# 4  Handwritten Digit Recognition Using ARN

The purpose of this paper is to demonstrate the ability of a simple two-level ARN structure to identify rasterized English numeric patterns. We have used the dataset from Modified National Institute of Standards and Technology (MNIST) database of handwritten English numerals from 0 to 9, established by LeCun et al. [16]. MNIST contains set of simplified and cleaned images of hand written characters created with an intention of providing a common platform to experiment with deep neural networks. It has 70,000 samples in total consisting of 60,000 training and 10,000 testing samples. All the digits are represented using a black background and white/grey foreground. The pixel intensities vary between 0 to 255, with 0 representing black and 255 as white. The size of each image in MNIST data set is 28 × 28 pixels. That means, each image has 784 pixel values and a label corresponding to each image.

## 4.1    Feature Extraction

The original data set is in 16-bit Unicode. We first need to convert this into 8 bit values and normalize the pixel values to be between 0 and 1 as (pixel value/255). These normalized pixel values are given as inputs to ARN. We have used image tiling to divide the image into smaller segments. The original image of $28 \times 28$ pixels is divided into 16 tiles of $7 \times 7$ pixels. Image tiling is shown in Fig. 2.

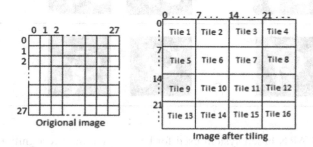

**Fig. 2.**  Image tiling for feature extraction

## 4.2    Input Perturbation

A neural network is said to be robust if it is able to classify the data even in presence of noise. Such a network should be able to distinguish between noise and novel input [17]. It is necessary that the system should recognize slight changes to orientation and scaling of input. Input perturbation refers to performing an affine/projective transformation on the input image, e.g., rotation by a small angle, translation, skew, etc. ARN supports perturbation of input as well as output for continuous function. For discrete output values, as in our case, output perturbation does not hold good. The neural networks are supposed to be noise tolerant; therefore if the input is perturbed by a small factor and applied as repeated input, some improvement in the performance is expected. However, if the input is perturbed by a large factor then it may mislead the network to completely misclassify the data. We are performing the rotation of image towards left and right by an angle $\theta + \Phi$ and $\theta - \Phi$. The sample images after applying perturbation are shown in Table 1.

## 4.3    Architecture

To keep the network simple, we have assumed only two ARN layers, first layer performing a labeling task and the second layer performing aggregation and decision making. We have used only one type of perturbation to keep the number of nodes low. Input image is symmetrically segmented in to $4 \times 4$ array, each element of the array containing $7 \times 7$ pixel part of the image. These were found to be adequate for the data set used to test the learning ability of the ARN structure. We have been able to achieve about 93% accuracy with as little as 50 samples per set.

**Table 1.** Perturbed image samples

| Original Image | Rotate left by 5 degree | Rotate right by 5 degree |
|---|---|---|
| 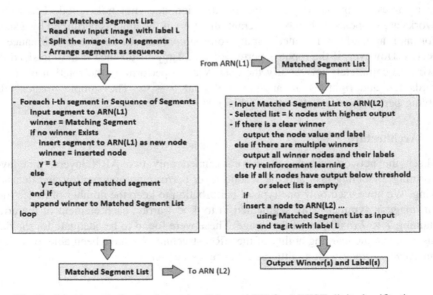 | | |

Working of ARN based system used for handwritten digit recognition is shown in Fig. 3. Each image is divided into 16 tiles of $7 \times 7 = 49$ pixels. First ARN layer has nodes with 49 input ports, corresponding to each tile (or segment) of image. When a tile is presented at the input port of the ARN, it will try to approximately match it to an existing node in the ARN. Matching node with highest output will be selected. If there is no matching node, a new node is created. One node will be created for a set of tiles that are similar to each other. Therefore, the number of nodes will equal the number of distinct tile sets, i.e. 16 in this case. As it is mentioned in earlier sections, the number of output nodes in ARN is not known apriori and it varies dynamically depending on the values of $\rho$, T and the training samples.

```
- Clear Matched Segment List
- Read new Input Image with label L
- Split the image into N segments
- Arrange segments as sequence
```

From ARN(L1) ⇒ Matched Segment List

```
- Foreach i-th segment in Sequence of Segments
    Input segment to ARN(L1)
    winner = Matching Segment
    if no winner Exists
        insert segment to ARN(L1) as new node
        winner = inserted node
        y = 1
    else
        y = output of matched segment
    end if
    append winner to Matched Segment List
loop
```

```
- Input Matched Segment List to ARN(L2)
- Selected list = k nodes with highest output
- If there is a clear winner
    output the node value and label
else if there are multiple winners
    output all winner nodes and their labels
    try reinforcement learning
else if all k nodes have output below threshold
    or select list is empty
    if
    insert a node to ARN(L2) ...
        using Matched Segment List as input
        and tag it with label L
```

Matched Segment List ⇒ To ARN (L2)

Output Winner(s) and Label(s)

**Fig. 3.** Pseudo-code for implementing 2-layer ARN for MNIST digit classification

The 16 nodes recognized by the first layer forms input layer of second layer of ARN, which has nodes with 16 input ports. Working of second layer is also similar to that of first ARN layer, adding to the node list when a new pattern is input. Label of the winner node indicates the recognized digit. Therefore, nodes in first layer are labeled with an index while the nodes in the second layer are labeled with the digit being trained. Winning digit is selected at the output of second layer of ARN. We have taken equal number of training samples for each digit.

### 4.4 Effect of Tuning on Learning

We have trained the ARN using MNIST data set for different values of $\rho$ and T. We have selected 500 samples per digit from the available set of 60,000 samples (6000 per digit, approximately). The network is trained for different sample sizes viz., 50, 100, 200, 300 and 500. Around 30% of samples from test data are used for testing. We have studied the effect of tuning for all these sample sizes. It is noticed that recognition accuracy was better with values of around 2 (>94%). For $\rho = 1$ and $\rho = 5$, the recognition rate was not high ($\sim 85\%$). For $\rho = 1$, the curve is almost flat and therefore, the coverage of each node is very large. It results into a very few number of nodes but there is a lot of overlap among nodes. On the other hand, with $\rho = 5$, the curve is very narrow and the coverage of each node is very small. Therefore, the significant values of $\rho$ can be chosen to be between 2 to 3. Similar to $\rho$, the value of T also affects the coverage and recognition rate of the network. Learning curves for $\rho = 2.42$ at different values of T are shown in Fig. 4. Learning curves for $T = 0.9$ at different values of $\rho$ are shown in Fig. 5. From these diagrams, we can observe that $T = 0.9$ and above yields slow and stable learning. Similarly $\rho = 2.42$, representing a half power point gives good learning.

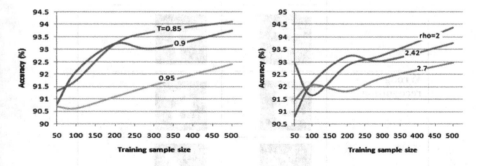

**Fig. 4.**  Effect of T on accuracy, for $\rho = 2.42$    **Fig. 5.**  Effect of $\rho$ on accuracy, for $T = 0.9$

## 5   Results and Discussions

A Python program was developed to implement the models and structures discussed in the above sections. The program was run on a typical laptop with no specific hardware accelerators like graphics processor. We have trained and tested the network for different sample sizes, different values of $\rho$ and T. Samples in MNIST test set are different than the training set. The network is able to recognize the digits even when the test image is somewhat different than the actual trained image. As any other neural network, there are times when it does wrong recognition or does not recognize or does multiple recognition. In-case of multiple recognition, the tested image is assigned equal probability for all the digits recognized. Higher level layers can resolve such cases. Some examples of each case are shown in Table 2. Notice that wrong recognitions are due to strong similarities with the recognized digit. Often, (3, 5), (3, 8), (4, 9) etc. (3 5  8 8  4 9) are wrongly recognized because of the way they are written (apriori probability). The accuracy of ARN varies with the value of $\rho$, $T$ and sample size. The confusion matrix for sample size = 200 and $\rho$ = 2.42, $T$ = 0.9 is given in Table 3 as an illustration. Notice that digit 1 has highest true positive and 3 has lowest true positive, indicating that a higher layer is required to resolve issues with recognition of digit 3.

**Table 2.** Results for $\rho$ = 2.42, $T$ = 0.9 and TrainingSize = 200 × 10

| | Test image | Matched image from training set |
|---|---|---|
| Correct recognition | | |
| | | |
| | | |
| Wrong recognition | | |
| | | |
| Multiple recognition | | |

**Table 3.** Confusion matrix for $\rho = 2.42$, $T = 0.9$ and TrainingSize $= 200 \times 10$, TestSize $= 60 \times 10$

|   | 0 | 1 | 2 | 3 | 4 | 5 | 6 | 7 | 8 | 9 | Σ |
|---|---|---|---|---|---|---|---|---|---|---|---|
| 0 | **41.75** | 1.00 | 2.50 | 2.75 | 0.00 | 3.75 | 2.00 | 0.00 | 6.25 | 0.00 | 60 |
| 1 | 0.00 | **59.00** | 0.00 | 0.00 | 0.00 | 0.00 | 0.00 | 1.00 | 0.00 | 0.00 | 60 |
| 2 | 1.31 | 4.14 | **35.97** | 8.03 | 1.61 | 2.11 | 1.11 | 2.61 | 1.98 | 1.11 | 60 |
| 3 | 0.00 | 6.17 | 3.33 | **25.83** | 0.333 | 13.33 | 1.17 | 5.00 | 4.50 | 0.33 | 60 |
| 4 | 2.00 | 5.92 | 1.42 | 0.00 | **37.58** | 0.17 | 0.16 | 2.17 | 0.48 | 10.16 | 60 |
| 5 | 3.25 | 1.70 | 0.70 | 3.25 | 1.67 | **37.95** | 3.50 | 0.20 | 4.78 | 3.00 | 60 |
| 6 | 3.19 | 1.36 | 2.20 | 0.00 | 4.69 | 2.36 | **44.30** | 0.11 | 1.61 | 0.11 | 60 |
| 7 | 0.00 | 4.03 | 0.00 | 1.33 | 3.20 | 0.00 | 0.00 | **40.03** | 0.70 | 10.70 | 60 |
| 8 | 5.37 | 2.50 | 5.20 | 2.70 | 1.58 | 4.37 | 1.20 | 1.58 | **32.41** | 3.08 | 60 |
| 9 | 0.25 | 1.00 | 0.00 | 1.25 | 8.67 | 0.25 | 0.00 | 2.83 | 2.58 | **43.16** | 60 |

# 6  Conclusions and Future Scope

A simple 2-level ARN network has been demonstrated to be able to efficiently classify MNIST data set with a small number of training samples. The network is able to learn fast without loss of stability. The ARN nodes have good noise tolerance necessary for implementing artificial intelligence related experiments. We have implemented the network on general purpose hardware. It will be very interesting to see how this system will behave with high performance MIMD processors like NVidia CUDA graphics accelerators. Processing time would reduce to reasonable limits, which in turn can allow larger training sets to be used. Increasing the number of layers would also yield much better accuracy. It will be interesting to see how this system will scale up when real life input is applied. ARN structures can add several features to further improve the accuracy of recognition with low demand on computational hardware. Attempts are also being made to implement low precision hardware accelerators specifically designed for implementing ARN structures [4–6].

# References

1. Sabour, S., Frosst, N., Hinton, G.E.: Dynamic routing between capsules. In: 31st Conference on Neural Information Processing Systems (NIPS 2017), Long Beach, CA, USA (2017)
2. Li, G., Hari, S.K.S., Sullivan, M., et al.: Understanding error propagation in deep learning neural network (DNN) accelerators and applications. In: ACM/IEEE Supercomputing Conference SC 2017, Denver, Co, USA, 17 November 2017
3. Aparanji, V.M., Wali, U., Aparna, R.: A novel neural network structure for motion control in joints. In: ICEECCOT-2016, IEEE Xplore Digital Library, pp. 227–232 (2016). Document no. 7955220
4. Mayannavar, S., Wali, U.: Hardware implementation of activation function for neural network processor. In: IEEE EECCMC, Vaniyambadi, India, 28–29, January 2018

5. Mayannavar, S., Wali, U.: Fast implementation of tunable ARN nodes. In: Abraham, A., Cherukuri, A.K., Melin, P., Gandhi, N. (eds.) ISDA 2018 2018. AISC, vol. 941, pp. 467–479. Springer, Cham (2020). https://doi.org/10.1007/978-3-030-16660-1_46
6. Mayannavar, S., Wali, U.: Performance comparison of serial and parallel multipliers in massively parallel environment. In: 3rd ICEECCOT, Mysore, 14–15 December 2018. IEEE Xplore Digital Library, December 2018, in press
7. Aparanji, V.M., Wali, U.V., Aparna, R.: Pathnet: a neuronal model for robotic motion planning. In: Nagabhushan, T.N., Aradhya, V.N.M., Jagadeesh, P., Shukla, S., M. L., C. (eds.) CCIP 2017. CCIS, vol. 801, pp. 386–394. Springer, Singapore (2018). https://doi.org/10.1007/978-981-10-9059-2_34
8. Giuliodori, A., Lillo, R., Pena, D.: Handwritten digit classification. Working paper 11–17, Statistics and Economics Series 012, pp. 11–17, June 2011
9. Frangi, A.F., Niessen, W.J., Vincken, K.L., Viergever, M.A.: Multi-scale vessel enhancement filtering. In: International Conference on Medical Image Computing and Computer-Assisted Interventions – MICCAI 1998, pp. 130–137 (1998)
10. Patil, S., Wali, U., Swamy, M.K.: Application of vessel enhancement filtering for automated classification of human In-Vitro Fertilized images. In: 2016 International Conference on Electrical, Electronics, Communication, Computer and Optimization Techniques (ICEEC-COT), Mysuru, India, 9–10 December 2016. IEEE Xplore (2017). https://doi.org/10.1109/iceeccot.2016.7955180
11. Krizhevsky, A., Sutskever, I., Hinton, G.E.: ImageNet classification with deep convolutional neural networks. In: Advances in Neural Information Processing Systems 25, pp. 1097–1105 (2012)
12. Babu, U.R., Venkateswarlu, Y., Chintha, A.K.: Handwritten digit recognition using K-nearest neighbour classifier. In: 2014 World Congress on Computing and Communication Technologies (WCCCT). IEEE (2014)
13. Patil, S., Wali, U., Swamy, M.K.: Deep learning techniques for automatic classification and analysis of human In-Vitro Fertilized (IVF) embryos. J. Emerg. Technol. Innov. Res. 5(2) (2018). ISSN 2349-5162
14. Juefei-Xu, F., Naresh Boddeti, V., Savvides, M.: Perturbative neural networks. arXiv digital library. Cornell University (2018)
15. Simonyan, K., Zisserman, A.: Very deep convolutional networks for large-scale image recognition. In: Conference Paper at ICLR (2015). https://arxiv.org/pdf/1409.1556.pdf
16. LeCun, Y., Cortes, C., Burges, C.J.C.: The MNIST database of hand written digits. http://yann.lecun.com/exdb/mnist
17. Stephen, G.: Competitive learning- from interactive to action to adaptive resonance. Cogn. Sci. 11, 23–63 (1987)

# Design of Hardware Accelerator for Artificial Neural Networks Using Multi-operand Adder

Shilpa Mayannavar[1(✉)] and Uday Wali[2]

[1] C-Quad Research, Belagavi 590008, India
mayannavar.shilpa@gmail.com
[2] KLE DR MSS CET, Belagavi 590008, India
udaywali@gmail.com

**Abstract.** Computational requirements of Artificial Neural Networks (ANNs) are so vastly different from the conventional architectures that exploring new computing paradigms, hardware architectures, and their optimization has gained momentum. ANNs use large number of parallel operations because of which their implementation on conventional computer hardware becomes inefficient. This paper presents a new design methodology for Multi-operand adders. These adders require multi-bit carries which makes their design unique. Theoretical upper bound on the size of sum and carry in a multi-operand addition for any base and any number of operands is presented in this paper. This result is used to design modular 4-operand, 4-bit adder. This module computes the partial sums using a look-up-table. These modules can be connected in a hierarchical structure to implement larger adders. Method to build a 16 bit 16 operand adder using this basic 4-bit 4-operand adder block is presented. Verilog simulation results are presented for both $4 \times 4$ and $16 \times 16$ adders. Design strategy used for the $16 \times 16$ adder may further be extended to more number of bits or operands with ease, using the guidelines discussed in the paper.

**Keywords:** Artificial Intelligence · Deep Learning · Hardware accelerators · Hardware optimization · Massive parallelism · Multi-operand addition · Neural computing · Neural network processor

## 1 Introduction

Accelerating developments in Artificial Intelligence (AI) present many new design challenges. Processor architectures for Deep Learning (DL) are characterized by large number of low precision parallel operations using distributed memory. Artificial Neural Networks (ANNs) are designed to behave like biological neuronal bundles. Each of these neurons in the bundle performs a non-linear transformation on input data and communicates the results to hundreds to thousands of other neurons. The number of synapses on a typical neuron in a human brain is about thirty five thousand while that of mice, another mammal, is said to be around fifteen thousand [1]. Each neuron performs scaling, summation and a non-linear transformation on the input data supplied by the synapses to the cell. Therefore, architectures supporting such operations are expected to be very different than Princeton or Harvard architectures generally used in

© Springer Nature Singapore Pte Ltd. 2019
A. B. Gani et al. (Eds.): ICICCT 2019, CCIS 1025, pp. 167–177, 2019.
https://doi.org/10.1007/978-981-15-1384-8_14

contemporary processors. The design goals for the neural processors will be influenced by the type and size of data being processed as well as the end application. Modern deep neural networks use several types of neural networks arranged in a hierarchy. For example, Convolutional Neural Network (CNN) has layers of convolution neurons, non-linear activation layers, pooling layers and fully connected layers in several iterations. Traditional processors are designed to consume more time in computation compared to the time taken for transfer of data. However, in modern deep learning networks, most of the time is spent in data communication between layers. As neural hardware becomes more capable, the number of synapses is bound to increase, directly increasing the time required to communicate among various neurons, over limited transfer lines. Accessing a central memory, to read and write from compute nodes (representing neurons) and distributing the output to nodes on the next layer, etc. can present several bottlenecks and hence design challenges. The problem becomes more severe as several layers of neural hardware compete to get data from a common pool of memory. Needless to say, design of communication channels to meet such demand poses several design issues. Neural processors tend to use segmented memory or compute-in-memory architectures to overcome these bottlenecks.

Neurons in a new type of neural networks called Auto Resonance Network (ARN), called *nodes*, register the input and respond to input in the neighborhood of the registered input. Hierarchical Auto Resonance Networks use resonators in multiple layers [2, 3]. Implementing ARN or other neural networks in hardware requires identification of basic sets of operations in popular neural computations. For example, Theano is a library in Python in which, mathematical computations can be carried out on multidimensional arrays. Some of the instructions available in Theano support machine learning operations such as activation functions (softmax, reLu, tanh and sigmoid), multi class hinge loss and tensor operations such as mean, sum, min, max, reshape, flatten and many more. It is reported that, Theano implementations of machine learning operations are 1.8 times faster compared to the implementations in NumPy [4], suggesting that efficiency concerns in machine learning are critical to success of a neural network processor. Auto Resonance Networks suggested by our working group require resonators, multi-operand adders and path search networks in several layers as required by the end application [2].

Most of the companies working on the design of custom neural processors have specific architectures in mind. For example Cambricon [5] supports simple vector and matrix operations at hardware level while the latest Google TPUs [6] are designed for the Tensorflow framework. Tensorflow is largely focused on Convolution Neural Network and Long Short Term Memory architectures. TrueNorth from IBM [7] is designed with a perceptron as a basic unit, suggesting its suitability for Multilayer Perceptron (MLP) implementations. There are many new types of neural structures being proposed, e.g., Generative Adversarial Networks [8], Capsule Networks [9] etc. and therefore, identification and design of modules required for neural hardware becomes necessary.

The number of parallel computations to be performed can be estimated by looking at how typical CNN systems are designed. For example, AlexNet [10] uses more than 650 thousand neurons. Each neurons in the first layer of convolutions use 363 input values ($11 \times 11 \times 3$) followed by a layer of filter neurons with 1200 inputs ($5 \times 5 \times 48$). Speed advantage using Graphic Processing Units (GPUs) has been reported to be 20–60 times the single CPU time. More modest MNIST digit classification using two layers of ARN use neurons with 49 inputs in first layer to achieve an accuracy of 93% with as little as $50 \times 10$ training samples [3]. Currently, most of AI cloud services are run on GPUs and TPUs configured as accelerators. On a rough estimate, a typical MLP network requires 10–20M multiplications. Processing demands of contemporary CNNs are at least an order higher [11, 12]. In general, the industry is moving towards compute intensive SIMD and MIMD streaming architectures that will reconfigure the hardware to suit the configuring instruction set or end application. This is clearly different than the issues addressed by RISC and CISC computers used today in everyday life. This implies that processor designs have to evolve beyond traditional processor design strategies and architectures like RISC/CISC.

This paper is divided into 5 sections. Discussion on the impact of massive parallelism of neural networks on hardware implementation and the need for multi-operand addition is presented in Sect. 2. Section 3 presents theoretic considerations for implementation of multi-operand adder. Implementation of a modular 4 bit 4 operand adder, its use in building a 16 bit 16 operand adder is discussed in Sect. 4. Results and conclusions are presented at the end in Sect. 5.

## 2  Impact of Massive Parallelism on Hardware Design

Each neuron in a neural network performs a weighted sum followed by a non-linear activation function. For a neuron with N-inputs, and one output, typically, N additions have to be performed to compute the output. In Deep Neural Networks (DNNs), the number of inputs is large. For example, in our implementation of ARN based MNIST image classifier, the input size N is 49 [3]. In Alexnet, the first convolution layer has 363 inputs [10]. Hence, computing the output involves multiple additions and multiplications. The total number of neurons in such networks is also very large. In Alexnet, the total number of neurons is in excess of 650,000. Implementing such networks on conventional hardware would yield a very slow performing network. Therefore, GPU implementations use sequential multiply-and-add operation to introduce pipelines that can speed such processing.

Linear memory organization can also severely impede the transfer of data between large number of neurons and their compute cores. Working with a single or a few busses to transfer large data sets would make it a bottleneck in the computation. On the other hand, it is possible to implement large number of high speed serial data transfer lines. This will improve the data transfer rate compared to conventional bus oriented architectures. In-core memory organization will allow local storage of values reducing overall need to move the data between memory and compute-cores. There could be several other memory and core organizations that become advantageous in massively parallel architectures. These need to be explored, studied and implemented.

Neural networks perform non-linear operations on the applied data. The case of vanishing or exploding numbers is common in such operations. For example, common activation functions like sigmoid ($1/(1 + exp(-x))$), hyperbolic tan ($tanh(x)$), softmax

$(x|\forall y : f(y) \leq f(x))$ have limited range of numbers where the output changes significantly. Therefore using limited accuracy is often sufficient in neural networks. Many of the neural network implementations therefore use low precision arithmetic over high precision. One such implementation is reported in [14].

Massive parallelism also has an impact on size of silicon required to implement necessary hardware while maintaining throughput. Earlier, we did mention that use of high speed serial transfer is a preferred solution instead of multiple parallel busses. In one of our earlier works, we have demonstrated that using large number of serial multipliers is more area-time efficient compared to parallel multipliers [13]. This is because the area required to implement a serial multiplier is much smaller than that of a parallel multiplier. So, while a parallel multiplier is performing a single computation, several serial multipliers are performing at a time. This ratio flips in favor of serial multipliers after a particular limit. That limit is fairly small when compared to the number of resources required for deep learning neural networks.

Further, in conventional processor architectures, math and logic operations are performed on two operands. As the number of operations increases, the operands have to be continuously moved between registers and cache, which reduces the overall speed. Therefore, it is necessary to explore multi-operand math operations. Some of the challenges in the design of multi-operand adder for Convolutional Neural Networks have been discussed in [15]. Multi-operand operations were simply an overhead in traditional computations, except in some scientific computing like multi-parameter simulations. Other requirement was in graphics, which was already addressed by the powerful MIMD GPUs. Therefore, most of the designers delegated such operations to software, following the RISC guidelines. However, the computational scenario in last couple years has changed so dramatically that immediate attention needs to be given to design of such operations.

Some of the issues related to design of multi-operand systems are (a) selection of number of operands, (b) modularity in design, (c) computation of carry, (d) area optimization, etc. Apart from these, complexity of computation in implementing multi-operand operation is also a factor. Systolic arrays for performing multiple operations over several steps have been described in literature. However, not enough attention has been given to multi-operand operations even in modern processor designs like that of Google TPU, possibly because of a technological bias in favor of two operand operations. In this paper, a modular multi-operand adder that can add 16 numbers as one integrated operation is discussed. The design is modular and uses 4-bit 4-operand addition as a basic operational unit. Hardware design and simulation results are also presented.

## 3   Multi-operand Addition

The basic design of a multi-operand adder follows a procedure similar to paper-pencil approach. Add the column and carry the overflow to left-side columns. In case of two operands, the overflow is limited to a single column (bit or digit). However, in case of multi-operand adder, carry can extend to more than one column on the left side. Designing a multi-operand adder therefore needs an estimate of how many columns will be affected by addition of numbers in a column. To illustrate the effect, consider 3-operand base-10 addition. The maximum number a column can generate is $3 \times 9 = 27$, with 2 as carry. However, if there are 15 operands, the maximum value is $15 \times 9 = 135$, with 13 as carry. Now, this means, the carry extends to two columns. The question now is how many operands will make the carry extend to 3 columns. This by induction can be continued to any number of columns and operands (digits and operands). In essence, it is necessary to know the upper bound on the size of sum and carry columns in a multi-operand addition. Such an upper bound on the size of carry is given by the following theorem.

**Theorem:** *An upper bound on value of the carry is numerically equal to the number of operands minus one, irrespective of the number of digits or the number system used. i.e., if there are N operands, the upper bound on the value of carry is $N - 1$.*

**Proof:** For $N = 2$, the result is obvious. We have been using a 1 bit carry in all two-operand adders. For any $N$, the maximum sum is sum of the maximum value of operands, i.e., for $N = 3$, maximum sum is $9 + 9 + 9 = 27$, with 2 as carry, and 7 as sum. Notice that the value of carry is independent of the number of digits. For 3 digits, the sum is 2997. This is because the LSB sum is less by the same amount as the value required to make higher digits equal to 9, i.e., $(N - 1)$. Therefore the sum of most significant digit will yield a sum equal to $N$ times 9 plus the $(N - 1)$ from carry.

The logic holds for any number of digits as well as number base used to calculate the sum. Figure 1 show the results for N = 3, 5, 8, 9 and 16, with upper bound on the carry given by N − 1, i.e., 2, 4, 7, 8 and 15. In Fig. 1, carry for N = 3 is $10_2, 2_{10}, 2_{16}$. Similarly, for N = 5 the bound is 4, i.e., $100_2, 4_{10}, 4_{16}$ and for N = 16, the upper bound is 15, i.e., $1111_2, 15_{10}$ and $15_{16}$. Notice that the carry for N = 16 in binary is $1110_2$, as this is a reduction in size of carry, the upper bound still holds. This occurs every time the sum columns add to full zero because the next addition will not produce a carry. This depends on the number of digits and N.

In general, for a base of k, and N rows, maximum value of column sum without carry is $N(k - 1) \bmod k$, or $(k - N) \bmod k$. Generated carry is

$$(N(k - 1) - (k - N) \bmod k)/k.$$

This value is always less than or equal to $(N - 1)$, depending on $N > k$ or $N \le k$. Therefore, the upper bound is $(N - 1)$.

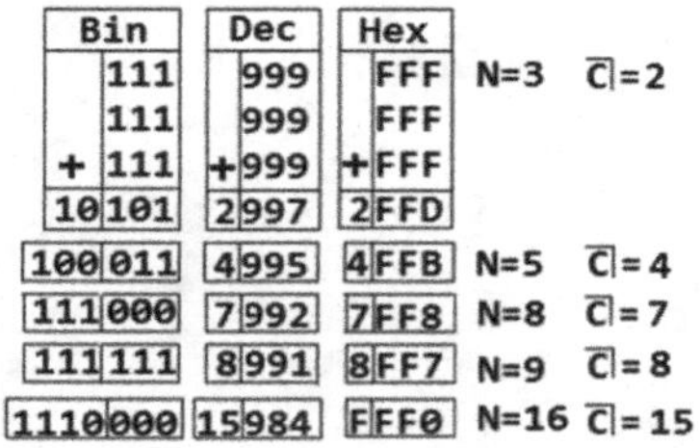

| Bin | Dec | Hex | | |
|---|---|---|---|---|
| 111 | 999 | FFF | N=3 | $\overline{C}$=2 |
| 111 | 999 | FFF | | |
| + 111 | +999 | +FFF | | |
| 10 101 | 2997 | 2FFD | | |
| 100 011 | 4995 | 4FFB | N=5 | $\overline{C}$=4 |
| 111 000 | 7992 | 7FF8 | N=8 | $\overline{C}$=7 |
| 111 111 | 8991 | 8FF7 | N=9 | $\overline{C}$=8 |
| 1110 000 | 15 984 | FFF0 | N=16 | $\overline{C}$=15 |

**Fig. 1.** Upper Bound on Carry using base 2, 10 and 16 number representation

In case we include the column sum with carry (full adder), maximum value in the column is $(k-1)$. Therefore, we can consider the column sum as corresponding to $N+1$ rows, giving an upper bound on the carry as less than $N$ or as $N-1$, when $N > k$. This proof is sufficient for our work now. A more rigorous proof using modular arithmetic will be given elsewhere. Knowing this upper bound on carry is useful in making sufficient resources available for implementation of multi-operand adder.

As an example, consider a simple 4-bit 4-operand (4 × 4) adder. Corresponding to four operands, from the above theorem, maximum size of carry is 3 which can be represented by two bits in binary system, or two columns to the left. Therefore, for a 4 × 4 adder, the number of output bits 2 + 4 = 6. If we consider a single bit addition, there will be 3 output bits. The algorithm uses this observation to construct the 4 × 4 adder.

If we consider a single column of 4 operands, maximum number of 1's in the partial sum is 4 ($100_2$). From the above theorem, 2 bits carry is sufficient. So, the carry will be of the form $xx_2$ which we can write as $0xx_2$. Actual values of the carry bits will depend on the data bits. There are two observations to be made here. First observation is that three bits are sufficient to represent a 4-operand column sum. Second observation comes from the fact that an upper bound on the sum of these two numbers is $100_2 + 011_2 = 111_2$. This is an important result, which shows that 3 bits are always sufficient to perform any 4 operand single column addition.

As there are only 4 operands or 4 bits per column, a 4-input 3-output lookup table (LUT) can be constructed to implement column addition quickly. Table 1 gives the details of this LUT. Essentially, the algorithm performs addition of bits in one column of multiple operands at a time and repeats the process for all the columns. The algorithm is simple, modular and straight forward. A 4 × 3 look-up-table is used as the basic block of multi-operand adder. This LUT is used to count the number of one's in the input. A 3 bit 2-operand adder is used to compute the carry to next column.

**Table 1.**  Input and output combination of a LUT

| Input | Output | Input | Output | Input | Output | Input | Output | Input | Output |
|-------|--------|-------|--------|-------|--------|-------|--------|-------|--------|
| 0 0 0 0 | 0 0 0 | 0 0 0 1 | 0 0 1 | 0 0 1 1 | 0 1 0 | 0 1 1 1 | 0 1 1 | 1 1 1 1 | 1 0 0 |
|  |  | 0 0 1 0 |  | 0 1 1 0 |  | 1 1 1 0 |  |  |  |
|  |  | 0 1 0 0 |  | 1 1 0 0 |  | 1 1 0 1 |  |  |  |
|  |  | 1 0 0 0 |  | 1 0 0 1 |  | 1 0 1 1 |  |  |  |
|  |  |  |  | 0 1 0 1 |  |  |  |  |  |
|  |  |  |  | 1 0 1 0 |  |  |  |  |  |

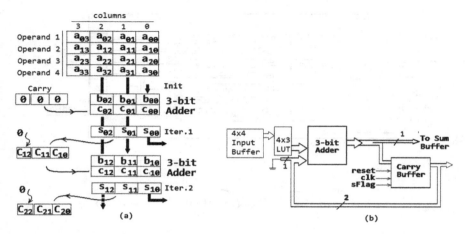

**Fig. 2.** (a) Illustration of the adder algorithm (b) Block diagram of single column adder

# 4 Implementation of Multi-operand Adder

## 4.1 4 × 4 Adder Operation

The implamentation details of 4 × 4 adder is shown in Fig. 2. The algorithm of the 4 × 4 adder is illustrated in the Fig. 2(a). From the earlier discussion it is clear that the sum of column data is computed separately and added to the pending carry. This will generate 1 bit column sum and 2 bit carry. The carry is used to calculate next column sum. The process is repeated for four columns.

Figure 2(b) shows the structure of one-column adder for 4 operands. The two carry bits and the sum bit are shown as output. On initialization, the carry buffer is cleared and data from first column is applied to the LUT. The three bit output as given in Table 1 is available at one of the inputs to the 3-bit adder. The other input comes from the carry buffer. The MSB of this operand to 3-bit adder is always zero. Two MSB bits of the 3-bit adder are copied to carry buffer and LSB is shifted to sum register. The iterations can be implemented in software or in hardware. A hardware implementation of a 4 × 4 adder is shown in Fig. 3. Organization of the column adder and percolation of carry bits is illustrated in Fig. 3(a). The control of data flow in the 4 × 4 adder is shown in Fig. 3(b).

A verilog simulation model was developed to test the correctness of concept. The simulation results are shown in Fig. 4. The timing diagram shows loading of four numbers and the output. The numbers are 4hA + 4hF + 4h3 + 4 × 9 = 6h25 corresponding to $10 + 15 + 3 + 9 = 37$ which is seen in the simulation output. The computation takes 5 clocks. The critical delay is due to propagation of carry between columns.

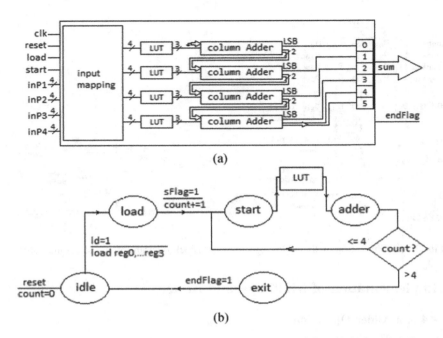

(a)

(b)

**Fig. 3.** (a) Hardware implementation of 4 × 4 adder module and (b) its operation

## 4.2  Optimized Logic Implementation of LUT

Generally, LUT is implemented as a 4 input 3 output memory. In such a design, the address decoder and 16 × 3 bit static RAM occupies lot of space. Hence a direct optimized logic implementation can easily replace such LUT. This implementation is much faster and occupies less space. One such circuit is given in Fig. 5

## 4.3  Implementation of 16 Bit 16 Operand Adder

Jumping back to the neural network implementation, the number of operands for addition depends on the type of the network. However, implementing 16 operand adder may be considered to speed up the addition process at the output of the neuronal cell. In this section, we will discuss the implementation of a 16 bit 16 operand adder, suitable for the number format described in [14].

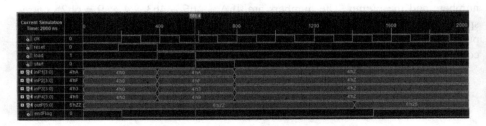

**Fig. 4.** Simulation results of 4 × 4 adder unit in verilog

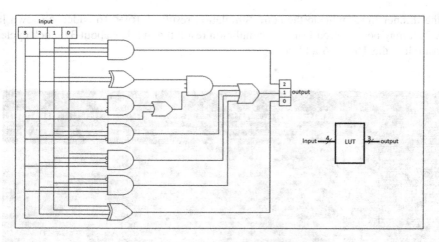

**Fig. 5.** Hardware implementation of LUT

## 4.4   16-Bit 16-Operand Adder

The idea of 4-operand, 4-bit adder explained in earlier section can be easily extended to any number of bits and to any number of operands. We will explain the implementation of $16 \times 16$ adder. The $4 \times 4$ module is extended to $4 \times 16$(4-operand, 16-bit) by increasing the number of iterations to 16 (see Fig. 2(a)). The $16 \times 16$ input is divided into four groups of $4 \times 16$ each. Each of $4 \times 16$ modules will produce a sum of 16-bits and a carry of 2 bits. These sum and carry of all the four modules are again grouped into two $4 \times 16$ modules as shown in Fig. 6. The thicker lines indicate the carry where

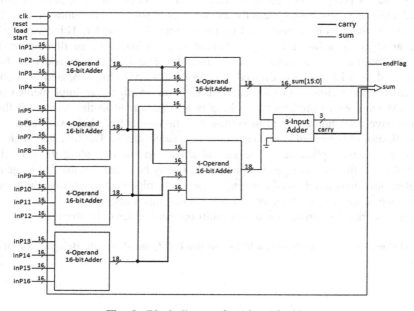

**Fig. 6.** Block diagram for $16 \times 16$ adder

as the thinner lines indicate the sum. Simulation result of $16 \times 16$ adder is shown in Fig. 7. It may be observed from the simulation result that it takes about 37 clock cycles to complete the $16 \times 16$ addition.

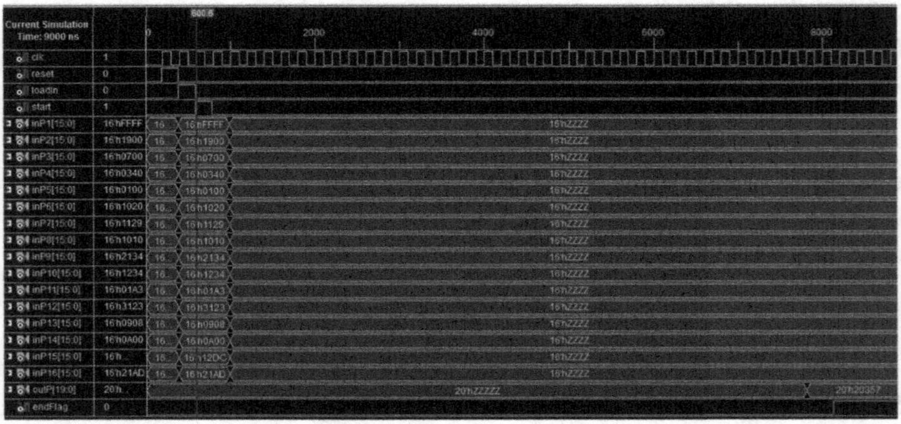

**Fig. 7.** Simulation result of $16 \times 16$ adder

## 5  Results and Conclusion

This paper presents design and implementation of multi-operand adders for use in neural processors and accelerators. A theorem to calculate an upper bound on the size of carry and its proof is presented in Sect. 2. The results can be applied to multi-operand addition in any base system for any number of rows and columns of data. The results of this theorem have been used to implement a $4 \times 4$ adder. This basic design can be arranged in a hierarchy to implement larger adders. As an illustration, the module has been used to implement a $16 \times 16$ adder. The basic column adder has been implemented as a LUT. An optimized version of this LUT, implemented as a combinational counting circuit is also presented. Column additions are implemented in a loop. FSM required to implement the loop is also presented in the paper. All these designed have been implemented in verilog and the results have been verified.

The algorithm presented here works in two separate stages: The first stage consists of column sum of applied data. The second stage consists of adding the carry bits generated in the first stage to get the final result. It may be noted that the first stage can be implemented in parallel and the second stage can be pipelined to improve the overall performance of the adder. Other optimization possibilities also exist.

Further work on optimization of the multi-operand adder is in progress.

**Acknowledgements.** The authors would like to thank C-Quad Research, Belagavi for all the support.

# References

1. Defelipe, J., Alonso-Nanclares, L., Arellano, J.: Microstructure of the neocortex: comparative aspects. J. Neurocytol. **31**, 299–316 (2002). https://doi.org/10.1023/A:1024130211265
2. Aparanji, V.M., Wali, U.V., Aparna, R.: Pathnet: a neuronal model for robotic motion planning. In: Nagabhushan, T.N., Aradhya, V.N.M., Jagadeesh, P., Shukla, S., Chayadevi, M.L. (eds.) CCIP 2017. CCIS, vol. 801, pp. 386–394. Springer, Singapore (2018). https://doi.org/10.1007/978-981-10-9059-2_34
3. Mayannavar, S., Wali, U.: A noise tolerant auto resonance network for image recognition. In: Gani, A.B., et al. (eds.) ICICCT 2019. CCIS, vol. 1025, pp. XX–YY. Springer, Cham (2019)
4. Bergstra, J., et al.: Theano: a CPU and GPU math compiler in Python. In: Proceedings of the 9th Python in Science Conference (SCIPY) (2010)
5. Liu, S., et al.: Cambricon: an instruction set architecture for neural networks. In: ACM/IEEE 43rd Annual International Symposium on Computer Architecture (2016)
6. Jouppi, N.P., et al.: In-datacenter performance analysis of a tensor processing unit. In: 44th International Symposium on Computer Architecture (ISCA), Toronto, Canada, June 2017
7. Akopyan, F., et al.: TrueNorth: design and tool flow of a 65mW 1 million neuron programmable neurosynaptic chip. IEEE Trans. Comput. Aided Des. Intergr. Circ. Syst. (2015). https://doi.org/10.1109/TCAD.2015.2474396
8. Goodfellow, I., Pouget-Abadie, J., Mirza, M., et al.: Generative Adversarial Nets, Achieves, Cornell University Library (2014). https://arxiv.org/pdf/1406.2661.pdf
9. Hinton, G.E., et al.: Dynamic routing between capsules. In: 31st Conference on Neural Information Processing Systems, NIPS 2017, Long Beach, CA, USA (2017)
10. Krizhevsky, A., Sutskever, I., Hinton, G.E.: ImageNet classification with deep convolutional neural networks. In: Pereira, F., et al. (eds.) Advances in Neural Information Processing Systems, NIPS 25, pp. 1097–1105. Curran Associates, Inc. (2012)
11. Hijazi, S., Kumar, R., Rowen, C.: Using Convolutional Neural Networks for Image Recognition. IP Group, Cadence, San Jose (2015)
12. Farabet, C., Poulet, C., Han, J.Y., LeCun, Y.: CNP: an FPGA based processor for convolutional networks. IEEE (2009)
13. Mayannavar, S., Wali, U.: Performance comparison of serial and parallel multipliers in massively parallel environment. IEEE Xplore Digital Library, December 2018, in press
14. Mayannavar, S., Wali, U.: Hardware implementation of activation function for neural network processor. IEEE Xplore Digital Library, January 2018, in press
15. Abdelouahab, K., Pelcat, M., Berry, F.: The challenge of multi-operand adders in CNNs on FPGAs; how not to solve it. In: SAMOS XVIII, Pythagorion, Samos Island, Greece, 15–19 July 2018. Association for Computing Machinery (2018)

# Performance Analysis of Collaborative Recommender System: A Heuristic Approach

Akanksha Bansal Chopra[1] (ID) and Veer Sain Dixit[2](✉) (ID)

[1] Department of Computer Science, SPM College,
University of Delhi, New Delhi, India
akankshabansal.asm@gmail.com
[2] Department of Computer Science, ARSD College,
University of Delhi, New Delhi, India
veersaindixit@rediffmail.com

**Abstract.** Security of Recommender systems from various type of attacks from non-genuine users are main point of research these days. The performance of Recommender systems get degraded due to these attacks from the non-genuine users. So exploration of the performance of Recommender Systems by considering possible attacks is the main issue of discussion in this paper. Here, a collaborative filtering based Recommender system is designed and implemented. Out of the generated recommendations, genuine and non-genuine recommendations are investigated. Push and Nuke attacks are considered, while testing matrix model. Various metrics are used to evaluate the performance of proposed Collaborative Recommender System.

**Keywords:** Recommender systems · Push attacks · Nuke attacks

## 1 Introduction

Nowadays, the amount of information is increasing tremendously. The surplus information available makes information retrieval a difficult task [1]. The Recommender systems help people to make decisions. Recommender systems predict user's interest on information, items or set of items and services from the pool of items available. These systems keep a track of users purchase, browse history and collect their recommendations in order to model user preferences for an item [9]. Collaborative Recommender systems and Content Based Recommender Systems are two main approaches of recommender systems that are used for information filtering [10]. In Content Based Recommender Systems (or Cognitive Recommender Systems), content of the item with a user profile is compared. An accurate and significant recommendation is then obtained by contextual information available in recommendation systems [11]. The recommendations are based on the information collected about content of an item and not on user's feedback. For example, if a Netflix user watches 'action' movies frequently, then it recommends movie from the movie dataset having genre 'action' [20]. Similarly, in mobile phones recommendations can be determined by comparing size, colour, brand, features, camera pixels etc. Another most widely used approach is Collaborative Filtering [10]. It has been the best and a successful approach [9]. In this

© Springer Nature Singapore Pte Ltd. 2019
A. B. Gani et al. (Eds.): ICICCT 2019, CCIS 1025, pp. 178–193, 2019.
https://doi.org/10.1007/978-981-15-1384-8_15

approach the items are rated by the users having similar tastes (or interest). This approach requires active participation of users. In this approach, recommendations are collected by many users. The items are then rated according to the given recommendations by these users for the items.

In our research, a Collaborative Recommender System is developed where user is allowed to rate the item even if he has not purchased or used it. The non-genuine user is also allowed to rate the item. This paper is structured in six sections. Sections 1 and 2 discusses in brief about Recommender systems and previous works done by various researchers. Section 3 explains the mathematical metrics used and algorithms that have been used to develop Recommender System. The dataset and result along with complete problem description and evaluations are discussed in Sect. 4. Section 5 gives analysis and interpretation. The conclusion and outlook for future work is discussed in Sect. 6.

## 2 Background and Related Work

Security is the utmost concern for any Recommender System. A Recommender System is secured only if its data cannot be altered, viewed, manipulated or used in any form by any non-genuine or any unauthorized person without the consent of its owner. Once the information gets online, its security becomes very sensitive and needs to be handled very carefully. Various types of attacks such as Shilling and Sybil attacks may lead to breach in the privacy of information. Sybil attack is a type of security threat in which multiple identities are detected by a node in a network. In this attack, an attacker creates fake identities to accomplish some purpose, like for example, increased recommendation of an item [5]. These attacks may be handled by developing attack-resistant algorithms [2, 3], or by identifying the increased identities [4]. These attacks may also be prevented by the use of CAPTCHAs [23]. This identifies the robotic interference, if any, and restricts the robotic user to make recommendation. Shilling attack may be executed by a group of users. In this type of attack, misleading recommendations are submitted to the system and system's recommendations are manipulated for an item [5]. Shilling attack may further be categorised in two attacks – RandomBot and AverageBot. These attacks may be executed with small datasets of users' recommendations. There are many ways to collect recommendations by users. K- means algorithm discussed in [16] and k-nearest-neighbour algorithm discussed in [5], are the two most popular and effective algorithms in such attacks to change the recommendations of the target item. The two more type of attacks are Push attack and Nuke attack. In Recommender systems, users give ratings for the items of interest. These ratings are used by the recommender systems to give recommendation for an item. It is always not necessary that the ratings are genuine and accurate. The ratings may be manipulated to change the existing recommendation of an item. A Push attack may be inserted to the recommender system to change the recommendations of items. In push attack number of higher and non-genuine ratings is inserted to the system. These ratings may be inserted by an individual or a group to increase the recommendation score of an item. Unlike push attack, in Nuke attack, lower and non-genuine ratings are inserted to the recommender system to change the recommendations. These ratings are inserted by an individual or a group to decrease the recommendation of any other item. It is contradictory to push attack in a way that, in push attack higher ratings are inserted to increase recommendation of own item or item of interest. But in nuke attack, lower ratings to other items are inserted to decrease the recommendations. In nuke attack, ratings are inserted for other items and not for self-items.

Shyong K. Tony Lam and others, in their paper have discussed three aspects of privacy and security: value and risks of personal information, shilling attacks and distributed recommenders [5]. Mobasher, B. and others, have shown that it is possible to mount successful attacks against collaborative recommender systems without substantial knowledge of the system or users [7]. Frank McSherry and Ilya Mironov have discussed that accuracy is inversely proportional to the available for a fixed value of privacy parameters [8]. Bo Zhang and others studied the psychological effects of user's privacy in recommender system [12]. They also determined that the perceived disclosure value of user can control the privacy of a recommender system. Shlomo Berkvosky and others investigated that when scattered user profiles are combined with data alteration methods, privacy issues are observed [13]. Authors of paper [9] have given empirical results by analysing various accuracy metrics. Recommender systems are categorised in many incomparable ways by the authors in paper [9]. The authors have shown that matrix factorization models yields better results than classic nearest-neighbour techniques [20].

As discussed, many researchers have focussed on security of recommender systems in their work. Majority has discussed the effect of attacks on individual ratings of an item. Whenever a user rates an item, a search query is processed by the algorithm to find other similar items, using similar keywords. In this approach the properties of the items are examined and not the users feedbacks. As an example [20], if a Netflix user watches action movies frequently, then it recommends movie from the movie dataset having genre 'action'. The performance of recommender system after being affected by inserting push attacks and nuke attacks is studied and analysed in the paper. The number of ratings for an item are considered and not the number of users giving these ratings. Push and Nuke attacks in different proportions are randomly inserted to the system and the pattern for performance of recommender system is analysed using various statistical methods. Validation of results is done by applying the developed algorithm on different sets of data.

## 3   Measure Metrics Used

A matrix is an array, arranged in rows and columns [21, 22]. The individual items in matrix A, with order m × n, denoted by $a_{ij}$, are called its elements [23]. To construct a matrix model for recommender system, let us assume that user 'u' gives rating 'r' for an item 'i'. Then,

$$r_{u_i} = x_u y_i$$

where,

$x_u = \{x_{u_1}, x_{u_2}, \ldots\ldots, x_{u_n}\}$ is a set of n number of user, and
$y_i = \{y_{i_1}, y_{i_2}, \ldots\ldots, y_{i_N}\}$ is a set of N number of items.

Let $\hat{R}$ be defined as an accumulation (total number of ratings) of user – item ratings, such that

$$\hat{R} = r_{u_i} = x_u y_i^T = XY^T \tag{A}$$

where,

$$X = [x_{u_1} x_{u_2} \ldots\ldots x_{u_n}];$$

$$Y^T = \begin{bmatrix} y_1 \\ y_2 \\ \cdot \\ \cdot \\ \cdot \\ y_N \end{bmatrix}$$

is possible only if, we have $n_{users} \times NmatrixX$ and $N_{items} \times NmatrixY$.

Number of items and users are getting increased on web day by day. Recommender systems are being used on millions of items. It is a challenge to ensure user satisfaction all time. These systems must cope up with increasing information with much accuracy, efficiency and scalability. Accuracy measures how closely recommendation matches a user's preferences. It is an important metric in improving quality of recommendation. It can be measured using **Mean Absolute Error** that compares numerical recommendation scores with actual user rating. Mean Absolute error (or MAE) is an average of the absolute errors between the expected and the actual value. Mathematically,

$$\text{MAE} = \frac{1}{n} \sum_{i=1}^{n} |e_i - a_i| \tag{1}$$

Where, $e_i$ is the expected rating and $a_i$ is the actual rating. Lesser the mean absolute error more is the performance.

A Recommender System needs to be efficient along with being accurate. The system is efficient if it processes in reasonable time, make good use of available resources and handle hundreds of requests per second. Memory and computation of time are used as measures of efficiency. In some situations, preferences of user changes, memory consumption reduces and computation time increases. This implies efficiency is dependent on the calculations required in an algorithm. These calculations can be speed up using data mining techniques such as hierarchical clustering.

Scalability is another important issue in recommender systems. The number of items and users are getting increased day-by-day. It is, therefore, necessary for a good recommender to manage this huge amount of information and predict relevant items of interest to the user. Measures such as **Euclidean Distance** and **Manhattan Distance** can be used as a solution to scalability. Euclidean Distance [24] and Manhattan

Distance [25] are the measures used to calculate the distance between two points. This calculated distance is the amount of disorder (or impurity or uncertainty) in the information. Mathematically, Euclidean Distance is calculated as

$$ED = \sqrt{\sum_{i=1}^{n} (e_i - a_i)^2} \qquad (2)$$

Where, $e_i$ is the expected rating of item and $a_i$ is the actual rating of item i. Another measure, Manhattan Distance is calculated as

$$MD = \sum_{i=1}^{n} |e_i - a_i| \qquad (3)$$

Recommender systems can use either implicit feedback or explicit feedback as input.

Implicit feedback is an indirect feedback of user. This includes purchase history, browsing history and search patterns of user. For instance, a user browsing for a mobile of particular brand, say X, on mobile sites may probably like to purchase mobile of brand X. This method of input is dependent on past history and is crucial when missing data is involved, which in turn may lead to negative feedback. User may be recommended with item of his interest using past history of preferences of similar users. Perhaps, the user does not like the item he purchased that was recommended. This is another drawback of implicit feedback because only guesses can be made using history of preferences of users but no opinion for positive feedback of the product can be made. Explicit feedback serves as better input feedback. Here, direct feedback of the users is gathered. Any missing information is omitted. For instance, in implicit feed back value is noted that how many times user A has purchased books of author A. Whereas, explicit feedback takes numerical values (or ratings), ranging between, say 1 (lowest) to 10(highest).

The book crossing dataset has 1.1 million ratings. It is quite tedious to work with individual values. To deal with such huge data, mean and standard deviation of accumulated ratings are calculated. Formula to calculate Mean of Accumulated Ratings is defined as

$$MAR = \frac{1}{n} \sum_{i=1}^{n} r_i \text{ and } SDAR = \frac{1}{n} \sum_{i=1}^{n} (r_i - \bar{r}_i)^2 \qquad (4)$$

where, $r_i$ is the total rating of a particular book ISBN.

Since users are not considered for inserting attacks, therefore, model matrix given by Eq. (A) is altered as

$$r_{u_i} = y_i,$$

implies

$$\hat{R} = r_{u_i} = y_i^T = Y^T$$

For every unique book ISBN total ratings are calculated. To optimize the algorithms [17], weights are assigned to total ratings of a particular ISBN. These total individual ratings given for an ISBN are multiplied with the probability of that ISBN book to give a weighted rating score (Si). Mathematically, rating score can be calculated using formula

$$Si = p_i \times \hat{R}' \tag{5}$$

where, $p_i$ is the probability of each book with unique ISBN and $\hat{R}'$ is the total rating for a particular book rating.

Mean of Rating Score and Standard Deviation Rating Score are defined as

$$\text{MSR} = \frac{1}{n}\sum_{i=1}^{n} S_i \text{ and } SDSR = \frac{1}{n}\sum_{i=1}^{n}(S_i - \bar{S}_i)^2 \tag{6}$$

where Si is the weighted rating score calculated using formula (5).

To understand the proposed work, the following algorithms are developed.

Algorithm 1: Genereate_recommendations
1. START
2. COLLECT $r_{u_i}$           \\ $1 \leq r_{u_i} \leq 10$
3. CLUSTER $r_{u_i}$           \\ using k-n-n algorithm
4. Evaluate $p_i$, $\hat{R}$
5. Evaluate $S_i$
6. Generate Recommendations

Algorithm 2: Push_attack
1. MERGE(EQUAL(ISBN)) && SUM($y_i$)
2. Evaluate (MAR , SDAR)
3. INSERT(PUSH_ATTACK)           \\ value = 10
   {
       for $\hat{R}$ = {0.1% , 0.2%,.......,0.9%,1.0%}of $\hat{R}$
                   Evaluate (S$_i$, MSR, SDSR, MAE, MD, ED)
   }
4. END

Algorithm 3: Nuke_attack
1. MERGE(EQUAL(ISBN)) && SUM( $y_i$ )
2. Evaluate (MAR , SDAR)
3. INSERT(NUKE_ATTACK)          \\value=1
  {

  for ( $\hat{R}$ = {0.1% ,0.2%,.......,0.9%,1.0%}of $\hat{R}$ )
                Evaluate (S$_i$, MSR, SDSR, MAE, MD, ED)

  }
4. END


## 4   Datasets and Results

In the experiment, a book ratings dataset namely Book-crossings is considered. It contains 1.1 million ratings of 270,000 books by 90,000 users. The ratings are on a scale from 1(lowest) to 10(highest). Implicit ratings are also included in the dataset. The dataset comprises of three tables. The table BX-Users, contains the users. Table BX-Books, contains the ISBN of books. The table BX-Book-Ratings contains the book rating information. Ratings are either explicit, expressed on a scale from 1 (lowest)–10 (highest) or implicit, expressed by 0.

The experiment is divided in four phases. In first phase, recommender system with actual book ratings is observed and analysed. In the second phase, push attacks at different proportions are randomly inserted. The effect of push attacks is studied at various proportions and the performance of recommender system with push attacks is compared with the recommender system with actual book ratings.

The aim is also to compare the algorithms at a common argument [15]. Performance is calculated in terms of mean, standard error, mean absolute error, Euclidean distance and Manhattan distance. In third phase, nuke attacks are randomly inserted at different proportions. The acceptance and rejection proportion is finally observed and stated. Lastly, in the fourth phase, repeatedly same algorithm (both for push and nuke attacks) is used to find the results for different number of observations in a dataset.

It is to note that, in the experiment, number of users is not of concern, and instead focus is only on the ratings given for the books. Rating '5' is considered as the average rating, whereas 10(highest) and rating 1(lowest) are considered suspicious and are used to insert push and nuke attacks, respectively. It is also to be noted that dataset contains implicit information expressed by 0.

- In the experiment, first user – item ratings ( $r_{u_i}$ ) from a Book – Crossing dataset are collected. The ratings for same ISBN are merged and summed up. Mean Accumulated Ratings (MAR) and Standard Deviation Accumulated Ratings (SDAR) using Eq. (4) are noted, for the recommender system with actual book ratings. In the next step, push attacks at different proportions (0.1% to 1.0% of total number of ratings, $\hat{R}$) are randomly inserted with value "10". The effect of push attacks is studied by evaluating Weighted Rating Score ($S_i$), MSR and SDSR, using Eqs. (5) and (6), at different proportions. Table 1 shows the evaluation results.

**Table 1.** Evaluation result table for actual book ratings and randomly pushed ratings at different Proportions

| Push attack proportions | MAR | MSR | SDAR | SDSR |
|---|---|---|---|---|
| Actual rating | 4.77088229 | 0.00010133 | 7.50356832 | 0.00016147 |
| 0.10% | 4.71788251 | 0.00010153 | 7.50534674 | 0.00016151 |
| 0.20% | 4.72692059 | 0.00010172 | 7.50484951 | 0.00016150 |
| 0.30% | 4.73623843 | 0.00010192 | 7.50905122 | 0.00016159 |
| 0.40% | 4.74553475 | 0.00010212 | 7.51277284 | 0.00016167 |
| 0.50% | 4.74564235 | 0.00010212 | 7.51305290 | 0.00016168 |
| 0.60% | 4.75496019 | 0.00010232 | 7.51832602 | 0.00016179 |
| 0.70% | 4.77854530 | 0.00010283 | 7.52793618 | 0.00016200 |
| 0.80% | 4.78875210 | 0.00010305 | 7.54119228 | 0.00016228 |
| 0.90% | 4.79855821 | 0.00010326 | 7.54645644 | 0.00016239 |
| 1% | 4.80867226 | 1.60293242 | 7.56483366 | 0.00016279 |

To compare the performance of recommender system with push attacks with a recommender system with actual book ratings- Mean Absolute Error, Euclidean Distance and Manhattan Distance are calculated. The three methods are used for same parameter so as to compare and find the best suitable method. Table 2 shows the evaluation results for Pushed Ratings to compare performance of recommender system at various levels of attacks.

**Table 2.** Evaluation result table for Mean Absolute Error, Manhattan Distance and Euclidean Distance for Randomly pushed ratings at different proportions

| Push attack proportions | MAE | MD | ED |
|---|---|---|---|
| 0.10% | 0.00000020 | 0.00000020 | 0.00000020 |
| 0.20% | 0.00000039 | 0.00000039 | 0.00000039 |
| 0.30% | 0.00000059 | 0.00000059 | 0.00000059 |
| 0.40% | 0.00000079 | 0.00000079 | 0.00000079 |
| 0.50% | 0.00000079 | 0.00000079 | 0.00000079 |
| 0.60% | 0.00000099 | 0.00000099 | 0.00000099 |
| 0.70% | 0.00000150 | 0.00000150 | 0.00000150 |
| 0.80% | 0.00000172 | 0.00000172 | 0.00000172 |
| 0.90% | 0.00000193 | 0.00000193 | 0.00000193 |
| 1% | 1.60283109 | 1.60283109 | 1.60283109 |

Next to analyse, nuke attacks are inserted at different proportions (0.1% to 1.0% of total number of ratings, $\hat{R}$) randomly with value "1". The effect of nuke attacks is studied by evaluating Weighted Rating Score, MSR and SDSR, by using Eqs. (5) and (6) at different proportions. Table 3 shows the evaluation results for nuke attacks.

**Table 3.** Evaluation result table for actual book ratings and randomly nuked ratings at different proportions

| Nuke attacks proportion | MAR | MSR | SDAR | SDSR |
|---|---|---|---|---|
| Actual rating | 4.77088229 | 0.00010133 | 7.50356832 | 0.00016147 |
| 0.10% | 4.70455620 | 0.00010111 | 7.44373110 | 0.00015998 |
| 0.20% | 4.70541586 | 0.00010113 | 7.47451146 | 0.00016064 |
| 0.30% | 4.70621105 | 0.00010114 | 7.47433717 | 0.00016064 |
| 0.40% | 4.70719966 | 0.00010116 | 7.47399640 | 0.00016063 |
| 0.50% | 4.70820976 | 0.00010119 | 7.47379579 | 0.00016062 |
| 0.60% | 4.70909091 | 0.00010121 | 7.47351238 | 0.00016062 |
| 0.70% | 4.71029443 | 0.00010123 | 7.47292644 | 0.00016060 |
| 0.80% | 4.71151945 | 0.00010126 | 7.47353318 | 0.00016062 |
| 0.90% | 4.71248657 | 0.00010128 | 7.47342235 | 0.00016062 |
| 1.00% | 4.71351816 | 0.00010130 | 7.47423264 | 0.00016063 |

Mean Absolute Error, Euclidean Distance and Manhattan Distance are calculated to study and analyse the effect of nuke attacks on the performance of recommender system. Table 4 shows the evaluation results for the same.

**Table 4.** Evaluation result table for Mean Absolute Error, Manhattan Distance and Euclidean Distance for randomly nuked ratings at different proportions

| Nuke attacks proportion | ABS(MAE) | MD | ED |
|---|---|---|---|
| 0.10% | 0.00000022 | 0.00000022 | 0.00000022 |
| 0.20% | 0.00000020 | 0.00000020 | 0.00000020 |
| 0.30% | 0.00000019 | 0.00000019 | 0.00000019 |
| 0.40% | 0.00000017 | 0.00000017 | 0.00000017 |
| 0.50% | 0.00000014 | 0.00000014 | 0.00000014 |
| 0.60% | 0.00000012 | 0.00000012 | 0.00000012 |
| 0.70% | 0.00000010 | 0.00000010 | 0.00000010 |
| 0.80% | 0.00000007 | 0.00000007 | 0.00000007 |
| 0.90% | 0.00000005 | 0.00000005 | 0.00000005 |
| 1.00% | 0.00000003 | 0.00000003 | 0.00000003 |

In evaluation of nuke attacks, absolute value for MAE is considered. It is to note that negative sign should be ignored. The parameter MAE finds the disagreement values between the parameters of interest. Hence negative sign is irrelevant.

# 5    Analysis and Interpretation

- In the previous section, MAR, MSR, SDAR and SDSR for actual ratings, push ratings and nuke ratings are evaluated. Corresponding to Tables 1 and 3, Figs. 1, 2, 3, 4, 5, 6, 7 and 8 are plotted. The Figures show the pattern for the values of MAR, MSR, SDAR and SDSR for push and nuke attacks, respectively. When compared, it is studied and analysed that randomly pushed attacks till 0.8% do not show any significant changes while from 0.9% onwards a significant drift in rating is noticed.

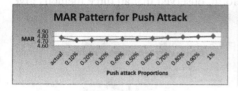

**Fig. 1.** MAR (Push attack)

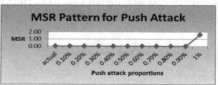

**Fig. 2.** MSR (Push attack)

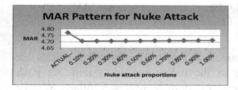

**Fig. 3.** MAR (Nuke attack)

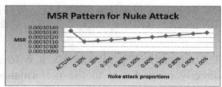

**Fig. 4.** MSR (Nuke attack)

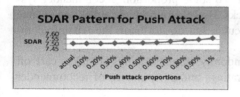

**Fig. 5.** SDAR (Push attack)

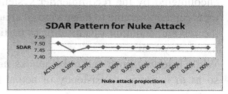

**Fig. 6.** SDAR (Nuke attack)

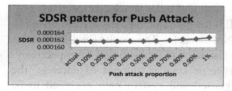

**Fig. 7.** SDSR (Push attack)

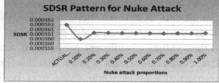

**Fig. 8.** SDSR (Nuke attack)

– The Tables 2 and 4, shows how performance of recommender system is being affected by inserting push attacks or nuke attacks at various levels of proportions. The three methods- Mean Absolute Error, Euclidean Distance and Manhattan Diustance are used to evaluate the measure of performance. The performance can be analysed by using these methods by measuring the difference between expected and actual values. More is the difference, shows more is the disagreement. It is observed that push attacks till 0.8% do not affect the performance of the system significantly but from 0.9% onwards, it is affected and needs action. Figures 9 and 10, shows the comparison between the three methods used, i.e. Mean Absolute Error, Euclidean Distance and Manhattan Distance. Henceforth, observed that all three methods give the same result values.

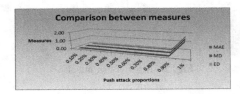

**Fig. 9.** Comparison between MAE, ED & MD measures for Push attack

**Fig. 10.** Comparison between MAE, MD & ED measures for Nuke attack

– To summarize Figs. 9 and 10, it is to conclude that push and nuke attacks are directly proportional to the performance of the recommender system. As push attacks increases, performance of recommender system decreases. Similarly as nuke attack increases, performance of recommender system also increases. But at the same time, it has been analysed that no action is required every time any attack is encountered by the recommender system. To some proportion, recommender systems can be tolerable to these kind of attacks. However any single attack can change the recommendation of an item, in particular, but do not harm the whole recommender system.
– While inserting push and nuke attacks randomly, implicit ratings are taken care off so as to maintain consistency of actual ratings and performance of recommender system
– As mentioned above, the same algorithm is run for different sets of data and MAR, MSR, SDAR, SDSR, MAE, ED and MD for one-half, one-fourth, one-fortieth and one-sixtieth values of original dataset are evaluated. Following are the evaluation results shown in Tables 5 and 6, for both push and nuke attacks, respectively.
– The amount of disagreement (MAE\ED\MD) for the same datasets are calculated to study and analyse the pattern of Push and Nuke attacks. The results for different sets of data values, at different proportions are also shown Tables 5 and 6, respectively.
– The evaluation result shows that for larger or smaller amount of datasets, the inserted attacks to some value may be ignored and system may not need to be modified or altered in terms of its performance. It is also observed that at least one-sixty fourth data values of original data values in a dataset to reach at a conclusion. Dataset smaller than one-sixty fourth data values in dataset, may not yield the

**Table 5.** Evaluation results for various measures for pushed ratings inserted at different datasets at different proportions

| Proportions of Push attacks | | Actual | 0.10% | 0.20% | 0.30% | 0.40% | 0.50% | 0.60% | 0.70% | 0.80% | 0.90% | 1% |
|---|---|---|---|---|---|---|---|---|---|---|---|---|
| 1/2th | MAR | 4.998428 | 5.004143 | 5.019858 | 5.035097 | 5.050812 | 5.066527 | 5.082242 | 5.097481 | 5.113196 | 5.128911 | 5.144626 |
| | SDAR | 8.619570 | 8.628026 | 8.629881 | 8.630605 | 8.636263 | 8.636873 | 8.646437 | 8.655968 | 8.662816 | 8.663586 | 8.667075 |
| | MSR | 0.000238 | 0.000238 | 0.000239 | 0.000240 | 0.000241 | 0.000241 | 0.000242 | 0.000243 | 0.000243 | 0.000244 | 0.000245 |
| | SDSR | 0.000410 | 0.000411 | 0.000411 | 0.000411 | 0.000411 | 0.000411 | 0.000412 | 0.000412 | 0.000413 | 0.000413 | 0.000413 |
| | MAE | 0.000000 | 0.000000 | 0.000001 | 0.000002 | 0.000003 | 0.000004 | 0.000004 | 0.000005 | 0.000006 | 0.000007 | 0.000007 |
| | ED | 0.000000 | 0.000000 | 0.000001 | 0.000002 | 0.000003 | 0.000004 | 0.000004 | 0.000005 | 0.000006 | 0.000007 | 0.000007 |
| | MD | 0.000000 | 0.000000 | 0.000001 | 0.000002 | 0.000003 | 0.000004 | 0.000004 | 0.000005 | 0.000006 | 0.000007 | 0.000007 |
| 1/4th | MAR | 4.844586 | 4.859145 | 4.873703 | 4.889172 | 4.903731 | 4.918289 | 4.933121 | 4.947680 | 4.963148 | 4.977707 | 4.992266 |
| | SDAR | 8.508706 | 8.523065 | 8.525535 | 8.527671 | 8.529555 | 8.529707 | 8.531442 | 8.538474 | 8.542710 | 8.542760 | 8.544171 |
| | MSR | 0.000441 | 0.000442 | 0.000443 | 0.000445 | 0.000446 | 0.000448 | 0.000449 | 0.000450 | 0.000452 | 0.000453 | 0.000454 |
| | SDSR | 0.000774 | 0.000776 | 0.000776 | 0.000776 | 0.000776 | 0.000776 | 0.000776 | 0.000777 | 0.000777 | 0.000777 | 0.000777 |
| | MAE | 0.000000 | 0.000000 | 0.000003 | 0.000004 | 0.000005 | 0.000007 | 0.000008 | 0.000009 | 0.000011 | 0.000012 | 0.000013 |
| | ED | 0.000000 | 0.000001 | 0.000003 | 0.000004 | 0.000005 | 0.000007 | 0.000008 | 0.000009 | 0.000011 | 0.000012 | 0.000013 |
| | MD | 0.000000 | 0.000001 | 0.000003 | 0.000004 | 0.000005 | 0.000007 | 0.000008 | 0.000009 | 0.000011 | 0.000012 | 0.000013 |
| 1/40th | MAR | 4.919265 | 4.935252 | 4.943245 | 4.959233 | 4.975220 | 4.983213 | 4.999201 | 5.015188 | 5.031175 | 5.047162 | 5.079137 |
| | SDAR | 7.911958 | 7.990582 | 8.044525 | 8.044622 | 8.044687 | 8.044707 | 8.044725 | 8.044711 | 8.044665 | 8.044587 | 8.044335 |
| | MSR | 0.003932 | 0.003945 | 0.003951 | 0.003964 | 0.003977 | 0.003983 | 0.003996 | 0.004009 | 0.004022 | 0.004035 | 0.004060 |
| | SDSR | 0.006325 | 0.006387 | 0.006430 | 0.006431 | 0.006431 | 0.006431 | 0.006431 | 0.006431 | 0.006431 | 0.006431 | 0.006430 |
| | MAE | 0.000000 | 0.000013 | 0.000019 | 0.000032 | 0.000045 | 0.000051 | 0.000064 | 0.000077 | 0.000089 | 0.000102 | 0.000128 |
| | ED | 0.000000 | 0.000013 | 0.000019 | 0.000032 | 0.000045 | 0.000051 | 0.000064 | 0.000077 | 0.000089 | 0.000102 | 0.000128 |
| | MD | 0.000000 | 0.000013 | 0.000019 | 0.000032 | 0.000045 | 0.000051 | 0.000064 | 0.000077 | 0.000089 | 0.000102 | 0.000128 |
| 1/64th | MAR | 4.488010 | 4.500000 | 4.511990 | 4.523981 | 4.523981 | 4.535971 | 4.547962 | 4.559952 | 4.571942 | 4.583933 | 4.607914 |
| | SDAR | 5.642445 | 5.715389 | 5.808093 | 5.919632 | 5.919632 | 5.920585 | 5.921514 | 5.922418 | 5.923298 | 5.924153 | 5.925790 |
| | MSR | 0.005381 | 0.005396 | 0.005410 | 0.005424 | 0.005424 | 0.005439 | 0.005453 | 0.005468 | 0.005482 | 0.005496 | 0.005525 |
| | SDSR | 0.006766 | 0.006853 | 0.006964 | 0.007098 | 0.007098 | 0.007099 | 0.007100 | 0.007101 | 0.007102 | 0.007103 | 0.007105 |
| | MAE | 0.000000 | 0.000014 | 0.000029 | 0.000043 | 0.000043 | 0.000058 | 0.000072 | 0.000086 | 0.000101 | 0.000115 | 0.000144 |
| | ED | 0.000000 | 0.000014 | 0.000029 | 0.000043 | 0.000043 | 0.000058 | 0.000072 | 0.000086 | 0.000101 | 0.000115 | 0.000144 |
| | MD | 0.000000 | 0.000014 | 0.000029 | 0.000043 | 0.000043 | 0.000058 | 0.000072 | 0.000086 | 0.000101 | 0.000115 | 0.000144 |

**Table 6.** Evaluation results for various measures for NUKED ratings inserted at different datasets at different proportions

| Proportions of Nuke attacks | | Actual | 0.10% | 0.20% | 0.30% | 0.40% | 0.50% | 0.60% | 0.70% | 0.80% | 0.90% | 1% |
|---|---|---|---|---|---|---|---|---|---|---|---|---|
| 1/2th | MAR | 4.966890 | 4.968455 | 4.970020 | 4.971544 | 4.973109 | 4.974632 | 4.976197 | 4.977762 | 4.979327 | 4.980851 | 4.982416 |
| | SDAR | 8.570601 | 8.570055 | 8.569315 | 8.568900 | 8.568256 | 8.567941 | 8.567590 | 8.567128 | 8.566377 | 8.566100 | 8.566127 |
| | MSR | 0.000205 | 0.000205 | 0.000205 | 0.000205 | 0.000205 | 0.000205 | 0.000205 | 0.000205 | 0.000205 | 0.000205 | 0.000205 |
| | SDSR | 0.000353 | 0.000353 | 0.000353 | 0.000353 | 0.000353 | 0.000353 | 0.000353 | 0.000353 | 0.000353 | 0.000353 | 0.000353 |
| | MAE | 0.000000 | 0.000000 | 0.000000 | 0.000000 | 0.000000 | 0.000000 | 0.000000 | 0.000000 | 0.000000 | 0.000000 | 0.000000 |
| | ED | 0.000000 | 0.000000 | 0.000000 | 0.000000 | 0.000000 | 0.000000 | 0.000000 | 0.000000 | 0.000000 | 0.000000 | 0.000000 |
| | MD | 0.000000 | 0.000000 | 0.000000 | 0.000000 | 0.000000 | 0.000000 | 0.000000 | 0.000000 | 0.000000 | 0.000000 | 0.000000 |
| 1/4th | MAR | 4.844691 | 4.846147 | 4.847603 | 4.849149 | 4.850605 | 4.852061 | 4.853607 | 4.855063 | 4.856610 | 4.858066 | 4.859521 |
| | SDAR | 8.506102 | 8.505636 | 8.505117 | 8.504422 | 8.503849 | 8.503104 | 8.502547 | 8.502412 | 8.501940 | 8.501194 | 8.500587 |
| | MSR | 0.000441 | 0.000441 | 0.000441 | 0.000441 | 0.000441 | 0.000441 | 0.000442 | 0.000442 | 0.000442 | 0.000442 | 0.000442 |
| | SDSR | 0.000774 | 0.000774 | 0.000774 | 0.000774 | 0.000774 | 0.000774 | 0.000774 | 0.000774 | 0.000774 | 0.000773 | 0.000773 |
| | MAE | 0.000000 | 0.000000 | 0.000000 | 0.000000 | 0.000000 | 0.000000 | 0.000000 | 0.000000 | 0.000000 | 0.000000 | 0.000000 |
| | ED | 0.000000 | 0.000000 | 0.000000 | 0.000000 | 0.000000 | 0.000000 | 0.000000 | 0.000000 | 0.000000 | 0.000000 | 0.000000 |
| | MD | 0.000000 | 0.000000 | 0.000000 | 0.000000 | 0.000000 | 0.000000 | 0.000000 | 0.000000 | 0.000000 | 0.000000 | 0.000000 |
| 1/40th | MAR | 4.491039 | 4.492234 | 4.493429 | 4.494624 | 4.495818 | 4.497013 | 4.498208 | 4.499403 | 4.500597 | 4.501792 | 4.502987 |
| | SDAR | 5.592757 | 5.597462 | 5.596608 | 5.595755 | 5.594901 | 5.594046 | 5.593191 | 5.592336 | 5.591480 | 5.590624 | 5.589768 |
| | MSR | 0.005366 | 0.005367 | 0.005368 | 0.005370 | 0.005371 | 0.005373 | 0.005374 | 0.005376 | 0.005377 | 0.005378 | 0.005380 |
| | SDSR | 0.006682 | 0.006682 | 0.006687 | 0.006685 | 0.006684 | 0.006683 | 0.006682 | 0.006681 | 0.006680 | 0.006679 | 0.006678 |
| | MAE | 0.000000 | 0.000001 | 0.000003 | 0.000004 | 0.000006 | 0.000007 | 0.000009 | 0.000010 | 0.000011 | 0.000013 | 0.000014 |
| | ED | 0.000000 | 0.000001 | 0.000003 | 0.000004 | 0.000006 | 0.000007 | 0.000009 | 0.000010 | 0.000011 | 0.000013 | 0.000014 |
| | MD | 0.000000 | 0.000001 | 0.000003 | 0.000004 | 0.000006 | 0.000007 | 0.000009 | 0.000010 | 0.000011 | 0.000013 | 0.000014 |
| 1/64th | MAR | 4.422535 | 4.423816 | 4.414853 | 4.417414 | 4.411012 | 4.412292 | 4.413572 | 4.414853 | 4.416133 | 4.417414 | 4.418694 |
| | SDAR | 5.527387 | 5.532505 | 5.528665 | 5.526849 | 5.527909 | 5.527002 | 5.526094 | 5.525186 | 5.524277 | 5.523368 | 5.522458 |
| | MSR | 0.005663 | 0.005664 | 0.005653 | 0.005656 | 0.005648 | 0.005650 | 0.005651 | 0.005653 | 0.005654 | 0.005656 | 0.005658 |
| | SDSR | 0.007012 | 0.007084 | 0.007079 | 0.007077 | 0.007078 | 0.007077 | 0.007076 | 0.007075 | 0.007073 | 0.007072 | 0.007071 |
| | MAE | 0.000000 | 0.000002 | 0.000010 | 0.000007 | 0.000015 | 0.000013 | 0.000011 | 0.000010 | 0.000008 | 0.000007 | 0.000005 |
| | ED | 0.000000 | 0.000002 | 0.000010 | 0.000007 | 0.000015 | 0.000013 | 0.000011 | 0.000010 | 0.000008 | 0.000007 | 0.000005 |
| | MD | 0.000000 | 0.000002 | 0.000010 | 0.000007 | 0.000015 | 0.000013 | 0.000011 | 0.000010 | 0.000008 | 0.000007 | 0.000005 |

results. Even one attack can alter the recommendation value. Figures 11 and 12 shows the performance pattern of push and nuke attacks at different dataset proportions respectively. The figures show that Push and Nuke attacks equally contribute to the performance of recommender system.

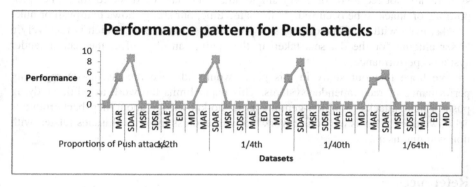

**Fig. 11.** Performance pattern for Push attacks inserted at different datasets

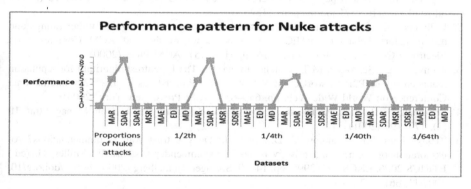

**Fig. 12.** Performance pattern for Nuke attacks inserted at different datasets

## 6 Conclusion and Future Scope

The issues of privacy and security are of rich concern in recommender systems. In this research paper, some aspects related to privacy and security are discussed and analysed. Some of the major factors such as leak of personal information and sale of information are discussed. Light on various attacks – Shilling attack, Sybil attack, Push attack and Nuke attack is also thrown. However focus is made only on Push attack and Nuke attack. To study Push and Nuke attacks, we constructed a matrix model and mathematical metrics. Various mathematical methods – Mean Absolute Error, Euclidean distance and Manhattan Distance are used to evaluate the effect of Push and Nuke attacks respectively, on the performance measure of recommender systems. It is to conclude that to some proportion, the attacks can be accepted and there is no need to make changes in the mechanism of performance measure of recommender system. We hope our study would contribute in high quality research in security of recommender system. Our study shows that alterations in performance measure of recommender systems are not required for every single attack and can be deferred till higher proportions of attack to be occurred. At the same time, our study shows that push or nuke attacks, even with small percentages of proportions (say, 0.9%), which is relatively a lesser amount for the datasets taken in the study, can also affect the recommender systems' performance.

We hope that our study in this paper would advance the discussion regarding performance of recommender systems. This paper limits the work as while studying push or nuke attacks from non-genuine users, no knowledge about their background is acknowledged. This is our future work that will address the critical issues related with non-genuine users.

# References

1. Haruna, K., Akmar Ismail, M., Damiasih, D., Sutopo, J., Herawan, T.: A collaborative approach for research paper recommender system. PLoS ONE **12**(10), e0184516 (2017). https://doi.org/10.1371/journal.pone.0184516
2. Dellarocas, C.: Immunizing online reputation reporting systems against unfair ratings and discriminatory behavior. In: EC 2000: Proceedings of the 2nd ACM Conference on Electronic Commerce, Minnesota, USA, pp. 150–157. ACM Press (2000)
3. Kamvar, S.D., Schlosser, M.T., Garcia-Molina, H.: The Eigentrust algorithm for reputation management in P2P networks. In: WWW 2003: Proceedings of the 12th International Conference on World Wide Web, pp. 640–651. ACM Press, New York (2003)
4. Friedman, E., Resnick, P.: The social cost of cheap pseudonyms. J. Econ. Manag. Strat. **10**(2), 173–199 (2001)
5. "Tony" Lam, S.K., Frankowski, D., Riedl, J.: Do you trust your recommendations? An exploration of security and privacy issues in recommender systems. In: Müller, G. (ed.) ETRICS 2006. LNCS, vol. 3995, pp. 14–29. Springer, Heidelberg (2006). https://doi.org/10.1007/11766155_2
6. Jeckmans, A., Beye, M., Erkin, Z., Hartel, P., Lagendijk, R., Tang, Q.: Privacy in recommender systems. In: Ramzan, N., van Zwol, R., Lee, J.S., Clüver, K., Hua, X.S. (eds.) Social Media Retrieval. Computer Communications and Networks. Springer, London (2013). https://doi.org/10.1007/978-1-4471-4555-4_12
7. Mobasher, B., Burke, R., Bhaumik, R., Williams, C.: Toward trustworthy recommender systems: An analysis of attack models and algorithm robustness. ACM Trans. Intern. Tech. **7**(4), Article 20, 38 pages (2007). http://doi.acm.org/10.1145/1278366.1278372
8. McSherry, F., Mironov, I.: Differentially private recommender systems: building privacy into the Netflix prize contenders. In: KDD 2009, Paris, France. ACM (2009). ISBN 978-1-60558-495-9/09/06
9. Herlocker, J.L., Konstan, J.A., Terveen, L.G., Riedl, J.T.: Evaluating collaborative filtering recommender systems. ACM Trans. Inf. Syst. (TOIS) **22**(1), 5–53 (2004)

10. Liu, F., Yuan, X., Yang, R., Liu, Y.: A recommender system based on artificial immunity. In: IEEE Ninth International Conference on Natural Computation (ICNC), pp. 639–643 (2013)
11. Champiri, Z.D., Shahamiri, S.R., Salim, S.S.B.: A systematic review of scholar context-aware recommender systems. Expert Syst. Appl. **42**(3), 1743–1758 (2015). https://doi.org/10.1016/j.eswa.2014.09.017
12. Zhang, B., Wang, N., Jin, H.: Privacy concerns in online recommender systems: influences of control and user data input. In: USENIX Association Tenth Symposium on Usable Privacy and Security, pp. 159–173 (2014)
13. Berkvosky, S., Eytani, Y., Kuflik, T., Ricci, F.: Enhancing privacy and preserving accuracy of a distributed collaborative filtering. In: RecSys 2007, Minneapolis, Minnesota, USA, 19–20 October 2007, Copyright 2007. ACM (2007). ISBN 978-1-59593-730-8/07/0010
14. O'Donovan, J., Smyth, B.: Is trust robust?: an analysis of trust-based recommendation. In: IUI 2006: Proceedings of the 11th International Conference on Intelligent User Interfaces, pp. 101–108. ACM Press, New York (2006)
15. Shindler, M., Wong, A., Meyerson, A.: Fast and accurate k-means for large datasets. In: 25th Annual Conference on Neural Information Processing Systems (NIPS 2011), pp. 2375–2383 (2011)
16. Castanedo, F.: A review of data fusion techniques. Sci. World J. **2013**, Article ID 704504, 19 pages (2013). https://doi.org/10.1155/2013/704504
17. Gipp, B., Beel, J., Hentschel, C.: Scienstein: a research paper recommender system. In: International Conference on Emerging Trends in Computing (ICETiC 2009), pp. 309–315 (2009)
18. Christensen, I., Schiaffino, S.: Matrix factorization in social group recommender systems. In: 12th Mexican International Conference on Artificial Intelligence. IEEE (2013). ISBN 978-1-4799-2604-6/13. https://doi.org/10.1109/micai.(2013)
19. Hu, Y., Koren, Y., Volinsky, C.: Collaborative filtering for implicit feedback datasets. In: 8th IEEE International Conference on Data Mining, ICDM 2008, pp. 263–272 (2008)
20. Association for Computing Machinery: Computer Figureics. Tata McGraw–Hill (1979). ISBN 978-0-07-059376-3
21. Anton, H.: Elementary Linear Algebra, 7th edn., pp. 170–171. Wiley, New York (1994). ISBN 978-0-471-58742-2
22. Krause, E.F.: Taxicab Geometry. Dover, New York (1987). ISBN 0-486-25202-7
23. Parra-Arnau, J., Rebollo-Monedero, D., Forné, J.: A privacy-protecting architecture for collaborative filtering via forgery and suppression of ratings. In: Garcia-Alfaro, J., Navarro-Arribas, G., Cuppens-Boulahia, N., de Capitani di Vimercati, S. (eds.) DPM/SETOP 2011. LNCS, vol. 7122, pp. 42–57. Springer, Heidelberg (2012). https://doi.org/10.1007/978-3-642-28879-1_4
24. Li, G., Cai, Z., Yin, G., He, Z., Siddula, M.: Differentially private recommendation system based on community detection in social network applications. Hindawi Secur. Commun. Netw. **2018**, Article ID 3530123, 18 pages (2018). https://doi.org/10.1155/2018/3530123
25. Mehta, H., Shveta, B.K., Punam, B., Dixit, V.S.: Collaborative personalized web recommender system using entropy based similarity measure. IJCSI **8**(6), 3 (2011). arXiv: 1201.4210 [cs.IR]

# Plant Disease Detection by Leaf Image Classification Using Convolutional Neural Network

Parismita Bharali[(✉)] [iD], Chandrika Bhuyan [iD], and Abhijit Boruah [iD]

Department of Computer Science and Engineering,
DUIET Dibrugarh University, Dibrugarh, India
parismitabharali44@gmail.com

**Abstract.** Plants are the source of food Plants are the source of food on the planet. Infections and diseases in plants are therefore a serious threat, while the most common diagnosis is primarily performed by examining the plant body for the presence of visual symptoms. As an alternative to the traditionally time-consuming process, different research works attempt to find feasible approaches towards protecting plants. In recent years, growth in technology has engendered several alternatives to traditional arduous methods. Deep learning techniques have been very successful in image classification problems. This work uses Deep Convolutional Neural Network (CNN) to detect plant diseases from images of plant leaves and accurately classify them into 2 classes based on the presence and absence of disease. A small neural network is trained using a small dataset of 1400 images, which achieves an accuracy of 96.6%. The network is built using Keras to run on top of the deep learning framework TensorFlow.

**Keywords:** Convolutional Neural Network · Deep learning · Plant disease detection · Image classification

## 1 Introduction

Majority of the leading economies in the world depend primarily on the agricultural sector. With modern methods and technology, the growth of crop production and quality has increased to a significant number. As per recent studies, agricultural production, however, needs to increase in higher percentage to meet the requirements of the growing population. Despite the latest advances, agriculture is not alien to challenges. One of the major challenges that the agricultural factor has to face almost inevitably is the different types of diseases that affect crops from time to time. A lot of these diseases are caused by pests like fungi, bacteria, and others. Figure 1 [12] shows some symptoms on a leaf of a plant affected by fungus. It has become a major cause of concern globally, with a contribution of around 16% of loss in global crop harvesting every year [23]. Several efforts to prevent plant diseases have been in practice over time. One of the earliest and rustic methods is the use of pesticides in the agricultural fields. However, excessive use of these chemicals has started to show its adverse

© Springer Nature Singapore Pte Ltd. 2019
A. B. Gani et al. (Eds.): ICICCT 2019, CCIS 1025, pp. 194–205, 2019.
https://doi.org/10.1007/978-981-15-1384-8_16

effects, proving fatal for both plants and animals including humans. In recent years, search for alternatives to use of excess and unnecessary chemicals in food has been gaining attention worldwide.

In contrary to the former, accurate and timely detection of plant diseases is recognized as crucial for precision agriculture [13]. Early detection and elimination of affected samples limit the unnecessary usage of chemicals thus promoting healthier production [22]. A majority of the symptoms of deadly diseases are prominent on the visible parts of the affected plants such as the leaves. Expert observation and analysis of the leaves alone, therefore, can simplify the whole scenario. A proper and timely examination of the leaves is sufficient for an early detection of disease in the plant. Naked-eye observation to detect the presence of any disease is a prominent practice among farmers. However, it requires a person to have all the necessary knowledge about every possible plant disease, for him to perform the diagnosis himself. Also, this manual process has been proved to be highly time consuming and prone to error. With the constant increase in the amount of crop variety and production, a human-eyed diagnosis of the plant diseases is definitely not appropriate.

**Fig. 1.** Fungus affecting plants

Instead, automatic techniques to detect diseases in the leaves are accurate and efficient for all types of small and large farms and a wide variety of crops altogether. Advances in technology have paved the way for the development of such techniques. With the growing efficiency of artificial intelligence, it is possible to get smart assistance in protecting the food source of the planet. Machine learning and its applications allow manipulating digital images so as to exploit information hidden in them. Deep Neural Networks have established success in the field of image classification beyond its peers. Modeled around the human brain, neural networks process information like the biological nervous system.

This paper proposes a method of classification of healthy plants from those affected by disease using Deep Convolutional Neural Network (CNN). A model is built to fit images of plant leaves, both healthy as well as disease affected, as inputs. The model is trained in a supervised learning method to produce an output indicating the presence of disease in a plant or not.

The paper further unfolds as Sect. 2 observes some of the works relatable in this field, Sect. 3 describes the methodology proposed for this study, Sect. 4 looks at the results obtained and Sect. 5 draws the final conclusion.

## 2    Related Work

Many researchers across the globe have proposed a number of approaches to check the threat to plants. Some early works in machine learning technologies have been successful in their attempt.

A method to identify and grade diseases in plants with the help of image processing and computer vision technology is described [18]. Leaves of Maple and Hydrangea were chosen for the study. These leaves suffered from diseases like Leaf Spot and Leaf Scorch. They first recognized the plant by pre-processing of the leaf images. The paper presented classification of diseases in the leaf, using segmentation based on K-Means, and classification of disease based on ANN. Finally, segmentation of separate areas of leaf, background and affected parts was done by implementing euclidean distance and K-Means clustering to classify the leaf infection into respective classes.

Ravindra Naik et al. [16] presented a classification model which involved Support Vector Machine, Genetic Algorithm and Neural Networks. The work focused on increasing the speed of classification by involving Genetic Cellular Neural Networks for image segmentation. Several preprocessing and feature extraction processes were employed on the input images captured with devices such as digital camera. Separated into training and testing sets, 5% of the images comprised of the training set, leaving the rest to constitute the testing set. The classification performed using SVM gained an accuracy of 86.77% which then was improved to 95.74% by a Neural Network Classifier.

Muthukannan et al. [15] tried to classify images of disease affected leaves only. Extraction of the relevant features like shape and texture of the affected leaf images was important for the study. The classification of the images was done by three different neural network algorithms - Feed Forward Neural Network, Learning Vector Quantization and Radial Basis Function Networks. The performance of each algorithm is measured in terms of four different classification parameters, and FFNN yielded better results based on them.

The deep neural networks have been able to be successfully growing in the areas of image identification and classification problems. CNNs are reputed to have given the best performance in pattern or image recognition problems, even outperforming humans [7]. Extending the implementation of deep learning in the field of agriculture has been successful so far.

Cortes [3] attempted to use a generative adversial network as a classifier. He designed a semi-supervised learning approach by implementing a loss function in the adversial network which acted as a classifier. The generalized discriminator classified the input images into two categories. A dataset of 86,147 images was segmented to remove the background from each image. The segmented images were resized along with the original images before feeding them as input to the network. The multi-label categorical generative network achieved its highest training accuracy of 80% and made it to 78% on the testing phase.

Detection of the presence of leaves along with distinguishing them between healthy and 13 different types of diseases was suggested by [22]. The dataset comprised of almost 3000 leaf images. For better extraction of features, preprocessing was done by manually cropping and making squares around them to highlight the images. Augmentation involved affine transformation, perspective transformation, and simple rotations which were done by particular application developed in C++. Models built in reference to CaffeNet were used for training followed by fine tuning to get better results. CaffeNet is reportedly a deeper CNN with a number of layers that extract features from input images progressively [19]. In particular, it comprises of a total of 8 learning layers, out of which, 5 constitute of convolutional layers and the rest 3 are fully connected layers [10]. The training performed on Caffe achieved an average precision of 96.3%.

Training CNN models to identify a large number of plant variants and to detect the presence of diseases in their dataset is suggested by Mohanty et al. [14]. They specifically considered the two predefined architectures of CNN, AlexNet and GoogleNet for their study. Training these architectures from scratch as well as using transfer learning, this paper compares the performance of both the architectures. Though this paper shows promising results in the use of CNN, yet it involves several pre-processing of the input images which requires time.

Another paper [26] describes the scope of CNN in plant disease classification by training a model to detect soybean plant diseases. Wallelign et al. in their work, designed a CNN on the basis of the architecture of LeNet, which had 3 conv layers, 1 max pooling layer following each of them. A ReLu activation function for each layer. A fully connected multilayer perceptron (MLP) comprised of the last layer. This model was trained with 12,673 images of soybean leaves that were collected in a natural environment and segmented to remove the background. The model achieved a classification accuracy of 99.32% over the dataset.

An experiment [1] was carried out to test and analyze as many as 6 conv net architectures for classification of diseases in plants. Brahimi et al. trained AlexNet, DenseNet-169, Inception v3, ResNet-34, SqueezeNet-1.1 and VGG13 using three learning strategies. Python deep learning framework called pyTorch was used for the training. Using pre-training and fine-tuning, the Inception v3 network gave the best results of 99.76%. This experiment also observed that the most successful learning strategy for all the CNNarchitectures was the deep transfer learning, which fine-tunes all layers in a network and starts backpropagation optimization from the pre-trained network.

The available literature provides important insights into the possible approaches. Convolutional Neural Networks, so far, has given promising results, but not without prior processing of the input data. Preprocessing of image data like segmentation requires time and can end up losing important data in the process. Involvement of huge datasets increases the level of complexity.

In this paper, Convolutional Neural Network is used for a binary classification of images of plant leaves according to the presence and absence of disease. The work considers a small dataset of healthy and affected plant leaves, without preprocessing, to analyze the performance of CNN on relatively fewer data.

## 3   Methods and Materials

Deep Convolutional Neural Network is employed in this work for the diagnosis of plant disease in a dataset of affected and healthy leaf images and accurately classifying them into two classes (affected and not affected) using supervised learning. This section elaborates the entire procedure from scratch. Figure 2 illustrates the entire method employed in this work.

**Fig. 2.** Methodology

### 3.1   Dataset Collection

All the machine learning systems start with data preparation. Both the quantity and quality of the data are highly stressed upon. The dataset for this paper is collected entirely from Google images, searched by the name of plants, healthy and diseased leaves downloaded separately Fig. 3.

The method for making deep learning dataset out of Google images described in [20] is used to create the data. Insignificant and blurry images are removed from the set. The whole dataset contains a total of over 1400 images, split into sets of 80:20 for training and validation.

Since this work focuses on the supervised method of learning, the data has been labelled accordingly. First, the whole dataset is separated into two directories -Train and Validate, inside a common Data directory. Second, both the Train and Validate directories consisted of two directories each, Disease affected and Not Disease affected, with all the images stored accordingly. This model trains on the images in the Train directory and is tested on the images in the Validate directory.

**Fig. 3.** Sample of the dataset

## 3.2 Data Preprocessing and Augmentation

Unlike most research works, the data collection is not followed by pre-processing in the study. No pre-processing is performed prior to training the model. The raw downloaded images are fed as input to the built CNN model. The data collected, however, are rotated into different angles between 0–360° to enlarge the dataset and to reduce over fitting during training.

## 3.3 CNN Structure

A number of standard pre-trained CNN architectures are available like AlexNet, GoogLeNet to name a few. The design of all these CNNs is motivated by the mechanism of visual cortex in the biological brain [7]. A CNN consists of a stack of convolution layers, each followed by a pooling layer, further followed by a number of fully connected or dense layers. Every layer in the network has a non-linear activation function associated [14].

A simple CNN model is built; a network comprising 3 convolutional layers Fig. 4. The top 2 layers have 32 neurons each of size 3 × 3 and the third layer has 64 neurons of size 5 × 5. Each layer has a ReLU activation function. Maxpooling layers of kernel size 2 × 2 follow each of the three layers Table 1. The network is flattened, after the third convolutional layer, to one dimension. This small network is then followed by two fully-connected layers activated with ReLU activation. One final dense classification layer is then added to the network which is a single unit with a sigmoid activation function generally used for binary classification.

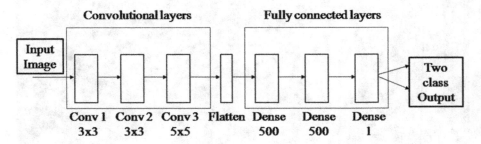

**Fig. 4.** CNN model

**Table 1.** Model architecture

| Layers | Activation | Units |
|---|---|---|
| Convolution 1 | ReLU | 32 filters, $3 \times 3$ |
| Maxpooling | – | $2 \times 2$ |
| Convolution 2 | ReLU | 32 filters, $3 \times 3$ |
| Maxpooling | – | $2 \times 2$ |
| Convolution 3 | ReLU | 64 filters, $5 \times 5$ |
| Maxpooling | – | $2 \times 2$ |
| Flatten | – | – |
| Dense | ReLU | 500, Dropout 0.2 |
| Dense | ReLU | 500, Dropout 0.2 |
| Dense | Sigmoid | 1 |

### 3.4 Training

The built CNN model is trained to classify the dataset into disease affected leaves and healthy leaves. The training is performed on an NVIDIA 940mx equipped with NVIDIA cuDNN, the deep neural network library, containing as many as 384 CUDA cores Table 2. Ubuntu 16.04.2 LTS (64-bit) served as the Operating System. The CNN model is built using Keras, a deep learning package written in python, on top of TensorFlow GPU backend. There are a number of frameworks like TensorFlow, Caffe, CNTK, PyTorch to name a few, each one built to support different deep learning purposes. Created by the Google Brain team, TensorFlow supports languages to build deep learning models [25]. Written in Python, Keras is a high-level neural networks API [2]. Here, Keras is used on top of TensorFlow so as to enable fast experimentation.

For this experiment, the whole dataset of images is split into the ratio of 80:20, where 80% were training images and 20% served for evaluating the trained model. The model, trained using Root Mean Square Propagation (RMSProp) with a batch size of 32 for 100 epochs, ran for over 10 h in the 940mx GPU. The parameters such as the kernel size, filter size, learning rate, batch size, no of epochs were all determined by trial and error method. The performance of the CNN is analyzed by training the model on the dataset using supervised learning. In supervised learning, annotations of data are used as references to perform the training.

**Table 2.** Basic Hardware & Software

| Hardware/Software | Characteristic |
| --- | --- |
| Processor (CPU) | Intel core-i5 (7th Gen) |
| Graphics (GPU) | NVIDIA GeForce 940mx |
| Operating System | Ubuntu 16.04.2 LTS (64 bits) |
| Environment | Keras, Tensorflow |

## 4 Result and Discussion

With the help of deep CNN, this paper tried to detect plant disease by classifying the leaf images based on the presence and absence of symptoms on them.

The acquired result of the CNN training on the dataset to classify leaf images into two categories (disease affected and healthy) is graphically represented in terms of accuracy and loss count for both training and validation processes Figs. 5 and 6. Binary cross entropy is used to return the model loss. This function measures the performance of a classification model when the output is a probability of 0 or 1 i.e., for binary output classes. The CNN model achieved an overall accuracy of about 96.6% when evaluated on a set of validation data. The result showed a minimized model loss after 100 epochs, which signifies the excellence of the model.

Figures 5 and 6. Binary cross entropy is used to return the model loss. This function measures the performance of a classification model when the output is a probability of 0 or 1 i.e., for binary output classes. The CNN model achieved an overall accuracy of about 96.6% when evaluated on a set of validation data. The result showed a minimized model loss after 100 epochs, which signifies the excellence of the model.

Using deep CNN, this paper successfully achieved the goal of detecting disease in plants and classifying leaf images on the basis of visible symptoms of diseases. The proposed CNN promises to overcome the shortcomings of traditional methods used in detecting disease in plants. It demonstrates its ability to take leaf images as input and this proves the scope and excellence of Convolutional Neural Networks in the field of agriculture and its growth.

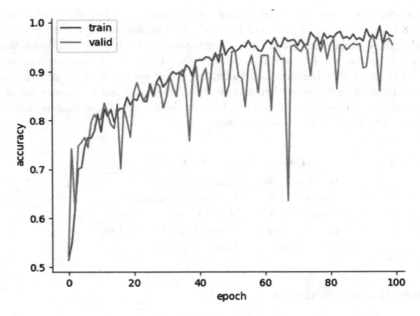

**Fig. 5.** Model accuracy

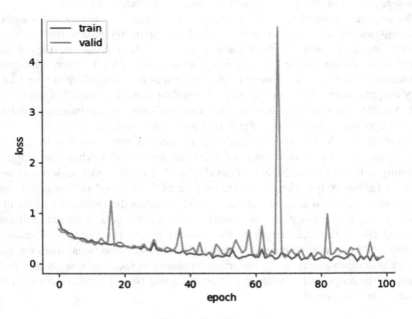

**Fig. 6.** Model loss

# 5  Conclusion

There are several methods in computer vision for the object detection as well as image classification process, although they are not yet efficiently utilized in solving the problems of the agricultural sector. This paper, basically, is an attempt to extend an opportunity to use neural networks in the field of agriculture. It tried exploring deep learning techniques of image classification with an aim to automatically detect disease in plants. This approach is concerned with classifying leaves of different plants into two categories of healthy leaves or disease-affected leaves, based on the visible symptoms in the leaves.

The success of CNN in terms of object recognition and image classification undoubtedly multiplied the chances of efficient solutions to plant health crisis. Therefore, a convolutional neural network model is developed for this study and it is successful in clearing the path for developing practical solutions to the early detection of plant disease. This work also created a new dataset of original plant leaf images, collected from the internet and augmented to fit into our developed CNN model. The model is trained on a set of train data and it finally achieves an accuracy of 96.6% over a set of validation data.

There is no compared analysis of this work with other related works because no other experiment has used the exact technique. However, in comparison with the related works presented in Sect. 2, the results are promising, especially when the data and the network architectures are taken into account. This experiment approaches with a smaller dataset, in contrary to the huge datasets used in image classification tasks by convolutional neural networks. In addition, it is an effort to build a simple CNN model with comparatively less number of layers, to simplify the task. The performance of the CNN shows the possibility of simpler ways of dealing with image classification tasks.

To extend this work further, larger datasets and more complicated networks can be developed to evaluate the performance and improve the accuracy to a whole. Also, lately image classifiers can be built in android platforms using TensorFlow. Development of effective practical applications focused to detect plant disease in real time can be a very important step in the field of agriculture. With smartphones taking over the world, the possibility of its involvement to detect plant diseases can only be beneficial.

# References

1. Brahimi, M., Arsenovic, M., Laraba, S., Sladojevic, S., Boukhalfa, K., Moussaoui, A.: Deep learning for plant diseases: detection and saliency map visualisation. In: Zhou, J., Chen, F. (eds.) Human and Machine Learning. HIS, pp. 93–117. Springer, Cham (2018). https://doi.org/10.1007/978-3-319-90403-0_6
2. Chollet, F., et al.: Keras: The python deep learning library. Astrophysics Source Code Library (2018)
3. Cortes, E.: Plant disease classification using convolutional networks and generative adverserial networks (2017)

4. Ertam, F., Aydin, G.: Data classification with deep learning using tensorflow. In: 2017 International Conference on Computer Science and Engineering (UBMK), pp. 755–758. IEEE (2017)
5. for General Microbiology, S.: Combating plant diseases is key for sustainable crops (2011). https://www.sciencedaily.com/releases/2011/04/110411194819.htm
6. He, K., Zhang, X., Ren, S., Sun, J.: Deep residual learning for image recognition. In: Proceedings of the IEEE Conference on Computer Vision and Pattern Recognition, pp. 770–778 (2016)
7. Hijazi, S., Kumar, R., Rowen, C.: Using convolutional neural networks for image recognition. Cadence Design Systems Inc., San Jose (2015)
8. Huang, G., Liu, Z., Van Der Maaten, L., Weinberger, K.Q.: Densely connected convolutional networks. In: Proceedings of the IEEE conference on computer vision and pattern recognition. pp. 4700–4708 (2017)
9. Iandola, F.N., Han, S., Moskewicz, M.W., Ashraf, K., Dally, W.J., Keutzer, K.: Squeezenet: Alexnet-level accuracy with 50x fewer parameters and¡ 0.5 mb model size (2016). arXiv preprint arXiv:1602.07360
10. Krizhevsky, A., Sutskever, I., Hinton, G.E.: Imagenet classification with deep convolutional neural networks. In: Advances in neural information processing systems. pp. 1097–1105 (2012)
11. Krizhevsky, A., Sutskever, I., Hinton, G.E.: Imagenet classification with deep convolutional neural networks. Communications of the ACM 60 (2017)
12. McKenney, M.A.: Apple scab: what causes it and what can be done about it? (2018). https://hubpages.com/living/How-to-Recognize-Manage-and-Prevent-Apple-Scab-on-Your-Apple-Trees
13. Miller, S.A., Beed, F.D., Harmon, C.L.: Plant disease diagnostic capabilities and networks. Ann. Rev. Phytopathol. **47**, 15–38 (2009)
14. Mohanty, S.P., Hughes, D.P., Salathé, M.: Using deep learning for image-based plant disease detection. Front. Plant Sci. **7**, 1419 (2016)
15. Muthukannan, K., Latha, P., Selvi, R.P., Nisha, P.: Classification of diseased plant leaves using neural network algorithms. ARPN J. Eng. Appl. Sci. **10**(4), 1913–1919 (2015)
16. Naik, M.R., Sivappagari, C.M.R.: Plant leaf and disease detection by using hsv features and SVM classifier. Int. J. Eng. Sci. **6**, 3794 (2016)
17. Pratt, H., Coenen, F., Broadbent, D.M., Harding, S.P., Zheng, Y.: Convolutional neural networks for diabetic retinopathy. Procedia Comput. Sci. **90**, 200–205 (2016)
18. Rastogi, A., Arora, R., Sharma, S.: Leaf disease detection and grading using computer vision technology & fuzzy logic. In: 2015 2nd International Conference on Signal Processing and Integrated Networks (SPIN), pp. 500–505. IEEE (2015)
19. Reyes, A.K., Caicedo, J.C., Camargo, J.E.: Fine-tuning deep convolutional networks for plant recognition. CLEF (Working Notes) **1391**, 467–475 (2015)
20. Rosebrock, A.: How to create a deep learning dataset using google images. https://www.pyimagesearch.com/2017/12/04/how-to-create-a-deep-learning-dataset-using-google-images
21. Simonyan, K., Zisserman, A.: Very deep convolutional networks for large-scale image recognition (2014). arXiv preprint arXiv:1409.1556
22. Sladojevic, S., Arsenovic, M., Anderla, A., Culibrk, D., Stefanovic, D.: Deep neural networks based recognition of plant diseases by leaf image classification. Comput. Intell. Neurosci. **2016**, 11 (2016)
23. Strange, R.N., Scott, P.R.: Plant disease: a threat to global food security. Annu. Rev. Phytopathol. **43**, 83–116 (2005)

24. Szegedy, C., Vanhoucke, V., Ioffe, S., Shlens, J., Wojna, Z.: Rethinking the inception architecture for computer vision. In: Proceedings of the IEEE Conference on Computer Vision and Pattern Recognition, pp. 2818–2826 (2016)
25. Techlabs, M.: Top 8 deep learning frameworks. https://www.marutitech.com/top-8-deep-learning-frameworks/
26. Wallelign, S., Polceanu, M., Buche, C.: Soybean plant disease identification using convolutional neural network. In: The Thirty-First International Flairs Conference (2018)

# Soft Modeling Approach in Predicting Surface Roughness, Temperature, Cutting Forces in Hard Turning Process Using Artificial Neural Network: An Empirical Study

Navriti Gupta[1(✉)] 🆔, A. K. Agrawal[1], and R. S. Walia[2]

[1] DTU, Delhi 110042, India
navritiguptadtu@gmail.com
[2] On Lien, PECTU, Chandigarh, India

**Abstract.** Hard Turning means turning of steel having hardness more than 50 HRC (Rockwell Hardness C). It is used for metal removal from hardened steels or difficult to machine steels. In this research hard turning of EN-31 steel (tool steel) 48 HRC was done with Carbon Nano Tubes based coated insert and two tools. Taguchi L27 orthogonal array was used for design of experiments. The input parameters taken in this research were cutting speed, feed, depth of cut, type of coating and cutting conditions. The output responses were surface roughness, temperature and cutting forces. 5-5-1 Feed forward artificial neural network was used in simulation of actual cutting conditions and prediction of responses before actual machining was done. The simulation results of ANN were in unison to those predicted by actual experimental procedure. It was concluded that relative error in values of surface finish, temperature and cutting forces as predicted with ANN versus those achieved during experimental procedure were 1.04%, 2.889% and 1.802% respectively, which were very close to actual values.

**Keywords:** Hard turning · Soft modeling · Carbon Nano tubes · ANN · Feed forward · Surface roughness · Temperature · Cutting force · Relative % error

## 1 Introduction and Literature Review

Hard turning means turning of steels having hardness more than 50 HRC [1]. Hardened steels find their usage in several industries as tool and die, heavy machine parts and automobile components. Hard turning is rapidly emerging as promising field of machining of hardened steel. It is rapidly replacing grinding as the material removal method from hard steel, due to added advantages as higher material removal, closer tolerances, higher doc, complex shape and sizes of the component [2].

Surface roughness (SR) is one of the important outcomes of the process [3]. It greatly depends on the temperature at tool tip while machining. The adverse conditions in hard turning can be addressed through temperature predictive model. Many researchers have applied different models of outcome response prediction using different control parameters. In earlier research [4] SR of AISI 304 stainless steel

© Springer Nature Singapore Pte Ltd. 2019
A. B. Gani et al. (Eds.): ICICCT 2019, CCIS 1025, pp. 206–215, 2019.
https://doi.org/10.1007/978-981-15-1384-8_17

specimen using multi-layer coated cutting tool was studied using integrated adaptive neuro-fuzzy particle swarm optimization (PSO). In another research [5] minimum surface roughness was predicted using Artificial Neural Network (ANN) and Genetic Algorithm (GA) integrated methodology. In another research [6] tool wear estimation was done using ANN approach. In a separate research [7] turning of D2 steel was done to optimize Temperature in cutting using Taguchi Technique. An important finding was effectiveness of Carbon Nano Tubes (CNTs) based cutting fluid helped in high rate of heat transfer.

ANN model for prediction of SR was presented while CNC turning AISI1030 steel [8]. Another research work elaborated on ANN model in simulation of surface milling of hardened AISI4340 steel with minimal fluid application [9].

In another research machine learning models were applied in determination of SR [10]. The advantage of applying ANN and other machine learning methods in experimental approach is that an idea about yet to be started actual machining process can be achieved. Based on the inhand information, machinist can devise strategies to improve SR and reduce power consumption. The achieved advantages are superior surface finish, low tool temperatures, low wear rate, low cutting forces etc.

## 2 Materials and Methods

### 2.1 Experimental Set Up

HMT Centre versatile lathe type 22 is used to carry out the experiments. EN-31 steel (tool steel) is used as work piece material. Bars of diameter 50 mm, length 900 mm are used for turning. Hardness was 48HRC.

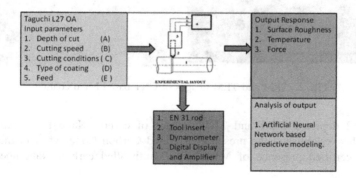

**Fig. 1.** Block diagram of experiment set up

Figure 1 depicted block diagram of the research conducted. The five input parameters taken in this research were depth of cut, cutting speed, cutting conditions, type of coating and feed. The output responses of experiments were SR, temperature and cutting forces.

In this research three tool bits viz. carbide, carbide + CNTs based Nano coating and Carbide + TiN.

**Table 1.** Cutting Parameters at different levels

| Input parameters | Level | | |
|---|---|---|---|
| | 1 | 2 | 3 |
| A Depth of cut (mm) | 0.53 | 0.81 | 1.06 |
| B Cutting speed (rpm) | 285 | 480 | 810 |
| C Cutting conditions | Dry | Wet | Cooled air |
| D Type of coating | WC | WC + CNT | WC + TiN |
| E Feed Rate (mm/rev) | 0.16 | 0.19 | 0.22 |

Table 1 was showing five different parameters used in the research with their levels.

Scanning Electron Microscopy and Tunnel Electron Microscopy tests were done for characterization of the tool coating.

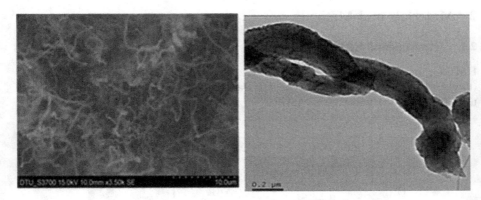

**Fig. 2.** SEM and TEM images of CNT based Nano coating

Figure 2 depicted SEM and TEM images of carbon Nano tubes based Nano coating. They established the presence of coiled Carbon Nano tubes in the coating. TEM test certified presence of MWCNTs (Multiwalled carbon Nanotubes) in the coating.

## 2.2 Method: Design of Experiments

Taguchi is well established and efficient approach in design of experiments. It reduces the total number of experiments without actually compromising on the quality of experimental output.

# 3 Artificial Neural Method

ANN can be used to model any non-linear process as it can approximate any functions more efficiently. Some of unique features are the learning ability, comprehensionability and generalizationability, which enable it to devise highly complex input–output [11, 12]. Accuracy of ANN model is determined by number of neurons, number of hidden layers, weight assigned to hidden layers, learning rate, transfer function and training function. Because of these calculation abilities, ANN is widely finding usage as predictive modeling technique of modeling in with very high accuracy. It expresses the predictive output as Relative % error. The relative % error is found out by calculating the percentage difference between the experimental values and ANN values.

The percentage relative error should be as low as possible. It depicts the deviation of predicted values from the experimental values. The lower % relative error symbolizes accuracy of prediction of responses. Accuracy is dependent on number of neurons and the epoch level selected. The ANN modeling is capable of modeling linear as well as non-linear processes and with higher accuracies than most of the other modeling techniques.

## 3.1 Artificial Neural Networks: Structure

The structure of artificial network principally consists of different layers of nodes or neurons. Each layer is different or same number of neurons. The input from previous layer node is going to each and every node/neuron in the next layer. This interaction basically is the artificial neural network calculation.

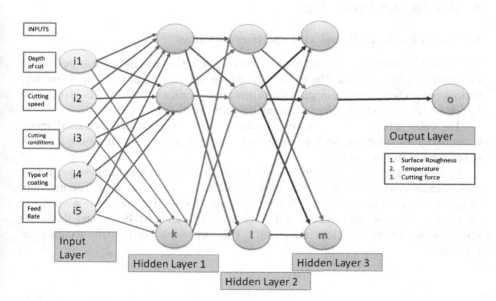

**Fig. 3.** 5-k-l-m-1 Artificial Neural Network used research

As shown in Fig. 3, a typical neural architecture of 5-k-l-m-1 was shown. It consists of 5 input neurons, k neurons in hidden layer 1, l neurons in hidden layer 2, and m neurons in hidden layer 3. And there was one output layer [13]. Five nodes/neurons in the input layer represents cutting speed v, feed rate f, depth of cut d, cutting condition ct, tool coating type t. Different researchers have used different ANN architectures in their research. In a research [14, 15] different structures of ANN were used and 4–1–1 network structure was labeled as accurate and reliable for the prediction of the SR. However, the good point in ANN is that researchers can choose any number of hidden layers with any number of nodes for each hidden layer [13]. But the general rule is more the number of hidden layers, more the number of node interactions. More the multiple interactions, more is the complexity of the mapping and computation time.

While no. of layers can be decided by taking a glimpse at the past work of researchers in the same field, no of nodes can be decided in accordance with earlier research [16], number of nodes in hidden layer can be approximated as "k/2", "1k", "2k", and "2k + 1" where k is the number of input nodes.

### 3.2    5-5-1 Feed Forward Network

The ANN calculation was done by nodes present in layer. It involves interaction among the neurons based on complex mathematical calculations. The feedforward type of network takes input and generates output after learning from the training data. The calculations of 5-5-1 feed forward learning type of Neural Network were done in Tiberius standard software of ANN soft prediction.

## 4    Results and Discussions

### 4.1    ANN Predicted Values vs. Experimental Values

80–20 training-testing rule is followed in this research. It means model was developed on training of 80% of data. The testing was conducted on 20% of rest data. However, various researchers have attempted using other configurations as 90–10, 70–30 etc. The Epoch of 100 is selected. Epochs are the iterations performed by ANN.

The Table 2 below was giving the absolute relative error between the experimental temperature response and ANN predicted SR, temperature and force response for each experimental run individually.

In Table 2 above the % relative errors between ANN predicted values and actual experimental values were shown for surface roughness and temperature. The numbers of epochs selected were 100. 80–20 training testing model was used. The normalized error values for SR, temperature and force were 0.07754, 0.049757 and 0.061359 respectively, which was less than 0.01. This represents very high accuracy in prediction.

**Table 2.** Absolute relative error in temperature by ANN Modeling

| S. No. | A | B | C | D | E | % Rel error SR | % Rel error Temp | % Rel error Force |
|--------|---|---|---|---|---|------|------|------|
| 1 | 1 | 1 | 1 | 1 | 1 | −0.01% | −1.06% | −0.53% |
| 2 | 1 | 1 | 2 | 2 | 2 | 6.41% | −0.81% | 0.25% |
| 3 | 1 | 1 | 3 | 3 | 3 | 2.95% | 0.98% | −3.04% |
| 4 | 1 | 2 | 1 | 2 | 2 | −1.93% | −1.31% | −0.67% |
| 5 | 1 | 2 | 2 | 3 | 3 | 0.56% | −0.61% | −1.81% |
| 6 | 1 | 2 | 3 | 1 | 1 | 14.22% | −0.98% | −0.88% |
| 7 | 1 | 3 | 1 | 3 | 3 | −1.17% | −1.26% | 3.16% |
| 8 | 1 | 3 | 2 | 1 | 1 | −3.28% | 1.21% | 1.39% |
| 9 | 1 | 3 | 3 | 2 | 2 | 1.13% | 0.39% | 3.77% |
| 10 | 2 | 1 | 1 | 2 | 3 | 1.90% | −2.29% | 1.80% |
| 11 | 2 | 1 | 2 | 3 | 1 | 0.96% | −0.29% | −3.35% |
| 12 | 2 | 1 | 3 | 1 | 2 | −10.31% | −0.12% | −1.63% |
| 13 | 2 | 2 | 1 | 3 | 1 | −0.28% | −1.46% | 0.92% |
| 14 | 2 | 2 | 2 | 1 | 2 | 0.93% | −4.29% | −0.28% |
| 15 | 2 | 2 | 3 | 2 | 3 | −0.83% | 0.86% | 2.98% |
| 16 | 2 | 3 | 1 | 1 | 2 | 0.92% | −0.59% | 0.62% |
| 17 | 2 | 3 | 2 | 2 | 3 | −1.58% | 0.48% | −5.57% |
| 18 | 2 | 3 | 3 | 3 | 1 | 7.47% | −0.51% | 2.73% |
| 19 | 3 | 1 | 1 | 3 | 2 | 0.28% | 0.53% | 0.42% |
| 20 | 3 | 1 | 2 | 1 | 3 | 0.88% | −0.52% | 1.39% |
| 21 | 3 | 1 | 3 | 2 | 1 | −2.83% | −0.89% | −5.82% |
| 22 | 3 | 2 | 1 | 1 | 3 | −3.31% | 0.50% | −1.04% |
| 23 | 3 | 2 | 2 | 2 | 1 | 4.84% | 0.21% | −0.42% |
| 24 | 3 | 2 | 3 | 3 | 2 | 4.16% | 0.71% | −0.40% |
| 25 | 3 | 3 | 1 | 2 | 1 | −0.91% | −3.63% | 0.19% |
| 26 | 3 | 3 | 2 | 3 | 2 | 3.18% | −0.46% | 2.57% |
| 27 | 3 | 3 | 3 | 1 | 3 | 0.79% | −1.23% | −1.07% |
| Average error | | | | | | 2.89% | 1.04% | 1.80% |
| ANN vs. Experimental | | | | | | | | |
| Epoch | | | 100 | | | | | |
| Hidden neurons | | | 5 | | | | | |
| Learning rate | | | 80% | | | | | |
| Normalised Error | | | | | | 0.07754 | 0.049757 | 0.061359 |

## 4.2    Graphical Interpretation of Results

Graphs were plotted for % ge relative error for all of 27 experimental runs for each response of SR, force and temperature. The graphs provide an overall picture of the process.

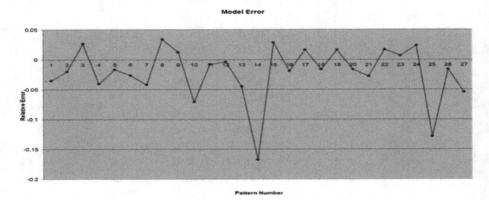

**Fig. 4.** Model error in Relative error in ANN Modeling value of SR

The Fig. 4 above is graphical representation of % ge relative error in ANN predicted values of SR.

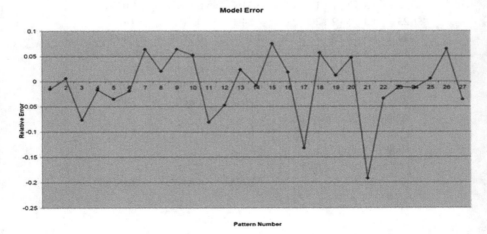

**Fig. 5.** Model error in Relative error in ANN Modeling value of force

The Fig. 5 above is graphical representation of % ge relative error in ANN predicted values of force.

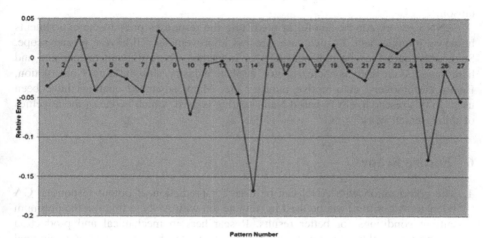

**Fig. 6.** Model error in Relative error in ANN Modeling value of force

The Fig. 6 above is graphical representation of % ge relative error in ANN predicted values of force.

## 5  Conclusions

Hard Turning is a versatile advanced metal removal especially in precision tooling machine industries as dies, molds, aerospace etc. The SR is an important indicator of machining performance. It itself is dependent upon various other factors as cutting force, tool tip temperature, wear rate, energy and power consumption, torque on tool etc.

So if advanced prediction of outcomes of machining can be done, it will help in improving the process capability. In this research work, Experiments were performed as per taguchi L27 model to predict SR, temperature and force. And advanced ANN prediction of surface roughness, temperature and force were done. The predicted values vs. the experimental results were in unison to each other.

**Table 3.** Average relative error in ANN based modeling and experimental investigation

|                         | Temperature | Surface roughness | Force  |
|-------------------------|-------------|-------------------|--------|
| Average relative error  | 1.04%       | 2.889%            | 1.802% |

Table 3 above was summarizing relative error between the predicted value of output response and actual response for Temperature, surface roughness and force.

The error value was in range 0–3%. This clearly shows that the two values were in unison.

ANN network can be helpful in predicting the responses prior to the experiments based on training data. In such case, process improvement will have a greater scope. Such techniques may be indispensable for industries where difficult to machine and hard materials are used as nuclear sector, airplane industry, ship building, Construction, Heavy earthmovers, auto-motive industry etc. Many advanced coatings have been used. In this research, CNTs based Nano coating is used, which is one of the novelties of this research work.

## 6  Future Scope

1. The optimization with ANN can be used for prediction of output responses. GA (Genetic Algorithm) can be used in conjunction with ANN to find out the optimum cutting conditions for better results. Researchers in mechanical and production engineering fields are not limiting themselves to the older and conventional methods of optimization techniques. Unconventional and modern optimization techniques are being used for testing older results and to arrive at accurate conclusions.
2. Predictive Modeling can be done for cost saving. As simulation results are closer to actual experimental results, process planning can be altered accordingly.
3. Reduction in cost of rejection, as simulation results will give us in sight of actual process.

## References

1. Tonshoff, H.K., Arendt, C., Ben Amor, R.: Cutting of hard steel. Ann CIRP **49**, 547–566 (2000)
2. Tonshoff, H.K., Wobker, H.G.: DBrandt: hard turning, Influence on the work piece prop. Trans. NAMRI/SME **23**, 215–220 (1995)
3. Kaladhar, M.: Evaluation of hard coating materials performance on machinability issues and MRR during turning operations. Measurement **135**, 493–502 (2019)
4. Aydın, M., Karakuzu, C., Uçar, M., Cengiz, A., Çavuşlu, M.A.: Prediction of SR and cutting zone temperature in dry turning process of AISI 304 stainless steel using ANFIS with Particle Swarm Optimization learning. Int. J. Adv. Manuf. Technol. **67**, 957–967 (2013)
5. Oktem, H., Erzurumlu, T., Erzincanli, F.: Prediction of minimum surface roughness in end milling mold parts using NN and GA. Mater. Des. **27**(9), 735–744 (2006)
6. Kaya, B., Oysu, C., Ertunc, H.M.: Force–torque based on-line tool wear estimation system for CNC milling of Inconel 718 using NN. Adv. Eng. Softw. **42**, 76–84 (2011)
7. Sharma, P., Sidhu, B.S., Sharma, J.: Investigation of effects of nanofluids on turning of AISI D2 steel using MQL. J. Clean. Prod. **108**, 72–79 (2015)
8. Nalbant, M., Gökkaya, H., Ihsan Toktaş, G.S.: The experimental investigation of the effects of uncoated, PVD- and CVD-coated cemented carbide inserts and cutting parameters on surface roughness in CNC turning and its prediction using artificial neural networks. Robot. Comput. Integr. Manuf. **225**, 211–223 (2009)

9. Leo, K., Wins, D., Varadarajan, A.S.: Prediction of surface roughness during surface milling of hardened AISI 4340 steel with Minimal cutting fluid application using Artificial Neural Network. Int. J. Adv. Prod. Eng. Manag. **7**(1), 51–609 (2012)
10. Çaydaş, U., Ekici, S.: Support vector machines models for surface roughness prediction in CNC turning of AISI 304 austenitic stainless steel. J. Intell. Manuf. **23**(3), 639–650 (2012)
11. Kosko, B.: Networks and Fuzzy Systems. Prentice-Hall of India, New Delhi (1994)
12. Schalkoff, R.B.: Artificial Neural Networks. McGraw-Hill International Edition. McGraw-Hill, New York (1997)
13. Kant, G.: Prediction and optimisation of machining parameters for minimizing surface roughness and power consumption during turning of AISI1045 Steel (2016)
14. Sanjay, C., Jyoti, C.: A study of surface roughness in drilling using mathematical analysis and neural networks. Int. J. Adv. Manuf. Technol. **29**, 846–852 (2006)
15. Sanjay, C., Jyoti, C., Chin, C.W.: A study of surface roughness in drilling using mathematical analysis and neural networks. Int. J. Adv. Manuf. Technol. **30**, 906 (2006)
16. Zhang, G., Patuwo, B.E., Hu, M.Y.: Forecasting with artificial neural networks: the state of the art. Int. J. Forecast. **14**(1), 35–62 (1998)

# Semi-automatic System for Title Construction

Swagata Duari$^{(\boxtimes)}$ and Vasudha Bhatnagar

Department of Computer Science, University of Delhi,
New Delhi 110007, India
sduari@cs.du.ac.in

**Abstract.** In this paper, we introduce a two-phase, semi-automatic system for title construction. Our work is based on the hypothesis that keywords are good candidates for title construction. The two phases of the system consist of - extracting keywords from the document and constructing a title using these keywords. The proposed system does not *generate* the title, instead it aids the author in creatively constructing the title by suggesting impactful words.

The system uses a pre-trained supervised keyword extraction model to extract important words from the text. Our KE approach gains from the advantages of graph-based keyword extraction techniques. We empirically establish the effectiveness of the proposed method, and show that it can be applied to any texts across domain and corpora. The keywords thus extracted are suggested to the author as potential candidates for inclusion in the title. The author can use creative transformations of the suggested words to construct an appropriate title for the manuscript. We evaluate the proposed system by computing the overlap between the list of title-words and the extracted keywords from the documents, and observe a macro-averaged precision of 82%.

**Keywords:** Title construction · Supervised keyword extraction · Graph-of-text

## 1 Introduction

The *title* of a scientific research article plays an important role in the process of literature review. It is the most important piece of information that assists the reader in sifting through vast amount of text in a repository. The title of a document conveys the central idea expressed in a document by establishing the premise of discussion contained in the text, and provides a clear yet simple one-line summary of the document content. Deciding on a title for scientific write-up or blog articles has always been a task of immense importance, because the reader often decides on the relevance of that document to his/her query just by observing the title. Writers often go through several iterations in order to decide upon the most satisfactory title for their article.

Studies have shown that the title of a scientific paper can influence the number of reads and citations for that article [10, 15, 21]. Paiva et al. reported that articles with short and result-oriented titles are more likely to be cited than those with long and method-describing titles [21]. Automatic generation of full-fledged title for scientific write-up is a complex process that requires natural language generation, which is still immature. We propose a semi-automatic method for *constructing* titles from scientific articles by identifying impactful words appearing in the abstracts. We hypothesize that

A. B. Gani et al. (Eds.): ICICCT 2019, CCIS 1025, pp. 216–227, 2019.
https://doi.org/10.1007/978-981-15-1384-8_18

keywords express the crux of the document and are therefore likely to be a part of the document title. Thus, we propose an application of automatic keyword extraction where extracted keywords are recommended to the author, which can be used for title construction after suitable transformation and by including other glue words. It is note worthy that our work is different from automatic title generation, where a full-fledged title is automatically generated for the document. Instead, the proposed systemaids in 'constructing' the title by automatically extracting the important words from the text and suggesting them to the author.

*Keyword Extraction* (henceforth, KE) is a classic data mining problem that addresses the task of automatically extracting special words from text documents to present a compact and precise representation of the document content. Typically embedded in document text, keywords for research articles not only convey the topics that the document covers, but are also used by search engines and document databases to efficiently locate information. We propose to identify keywords for title construction using an automatic keyword extraction method. The novelty of our approach lies in the fact that we design a generic supervised keyword extraction model that can be applied on any text without considering its domain or corpora. We aim to achieve the goal by exploiting the advantages of graph-based keyword extraction methods. Specifically, our contributions are as given below.

   i. We demonstrate that the properties of the words extracted from graph represen-
      tation of text are effective features to discriminate between keywords and non-
      keywords.
  ii. We note that simple classifiers perform well enough for the task of keyword
      extraction. Complex algorithms, such as deep learning, not necessarily yield better
      performance considering the small training sets and the training time.
 iii. We show that the extracted keywords appear $\approx 80\%$ times in the title of scientific
      articles in our dataset.

The paper is organized as follows. We discuss works related to our research in Sect. 2, followed by methodology of the proposed algorithm in Sect. 3. Section 4 covers experimental setup, dataset details, objectives of each experiment, and pre-liminary results and discussions. Finally, Sect. 5 concludes our paper.

## 2 Related Works

In this section, we discuss works related to automatic title generation and automatic keyword extraction, which are relevant to our study.

### 2.1 Automatic Title Generation

Various studies have been performed on automatic title generation from both spoken [5, 11, 12] and written text [13, 14, 22, 25, 27]. These works aim at converting the document into a 'title representation' by using either statistical, probabilistic, or machine learning methods. Kennedy et al. used an EM-based Bayesian statistical machine-translation method to identify title-word and document-word pairs that are most likely to constitute the document title [14]. Jin et al. proposed a probabilistic approach for automatic title generation that takes into account title word ordering [13].

Automatic title generation has also been viewed as an automatic summarization task by some researchers. Tseng et al. applied the task of automatic title generation to document clustering in order to identify generic labels for better cluster interpretation [27]. In a recent study, Shao et al. used a dependency tree based method to identify and filter candidate sentences that are likely to be titles [25]. In a similar fashion, Putra et al. used adaptive KNN to produce a single-line summary from article abstracts and argued that rhetorical categories (research goal, method, and not relevant) of sentences in the document have potential to boost the title generation process.

## 2.2  Automatic Keyword Extraction

Automatic KE methods fall under two categories - supervised and unsupervised [1, 3]. Supervised methods treat keyword extraction problem as a binary classification task ('keyword' and 'non-keyword' classes), whereas unsupervised methods use statistical or graph-theoretic properties to rank candidate words.

The primary task in any supervised KE methods is to construct the feature set. Identifying good quality features that effectively discriminate keywords from non-keywords is a challenging task. Some of the popular features that are used in literature are tf-idf, POS tags, n-gram features, etc. [4, 9, 20, 26]. Apart from these, topical information [30], linguistic knowledge [9], structural features of the document [17], knowledge about domain and collection [4, 20], expert knowledge [8], and external sources like Wikipedia links [18] are used to enrich the feature set. Moreover, the objective of supervised KE methods is to identify potential key-*phrases*, and not key-*words*. We, however, slightly differ from rest of the state-of-the-art supervised KE methods and focus on identifying keywords instead of phrases.

In general, supervised approaches for keyword extraction report better results compared to unsupervised counterparts. Unsupervised KE techniques largely comprise graph-based methods, which transform the text into a graph and use graph-theoretic properties to rank keywords. Local node properties like PageRank [19], PageRank along with position of the word in text [7], degree centrality [16], coreness [23], etc. have been studied extensively in the past. Unlike supervised methods, the primary advantage of unsupervised methods is that they are independent of the domain or corpus of the document.

## 3  Methodology

In this paper, we propose a semi-automatic system for title construction. The system works in two phases. In the first phase, the system automatically extracts stemmed keywords from the text document and presents them to the author. These words are the candidates for title construction. In the second phase, which requires manual intervention, the author can creatively weave the title by suitably transforming the stemmed candidates and using glue words.

We design a supervised keyword extractor to implement the first phase by exploiting graph-theoretic properties of candidate keywords. We hypothesize that certain node properties are capable of distinguishing keywords and non-keywords, and accordingly transform the document to a graph-of-text representation. The proposed algorithm comprises of the following steps.

i. Prepare the training set as follows.
   (a) Select candidate keywords from each document, and construct the corresponding graph-of-text (Sect. 3.1).
   (b) Extract select node properties from each graph-of-text and assign label to each candidate keyword based on the available gold-standard keywords list (Sect. 3.2).
   (c) Balance the training set, if required.
ii. Train a predictive model using the prepared training set and use it to predict keywords for target document (Sect. 3.3).
iii. Recommend top-$k$ extracted keywords as candidates for title construction.

The usage of the proposed system is summarized in Fig. 1. To construct the title, the text is converted to a graph. Node properties of each word are extracted and are supplied to a pre-trained model, which outputs the probability of each word being a 'keyword'. Top-$k$ keywords suggested to the user can be used for title construction ($k$ is user specified).

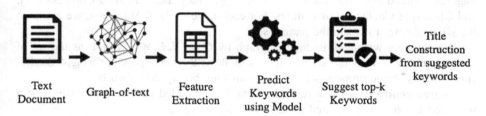

| Text Document | Graph-of-text | Feature Extraction | Predict Keywords using Model | Suggest top-k Keywords | Title Construction from suggested keywords |

**Fig. 1.** Diagrammatic representation of the semi-automatic system for title construction.

## 3.1 Candidate Selection and Modeling Text as Graphs

We follow the well-established convention to retain nouns and adjectives from the text as candidate keywords [7, 19, 23]. The text is then transformed to a graph representation, where the candidate keywords constitute the set of nodes and the set of links are defined based on a co-occurrence relation. Following Duari et al., we use a parameter-free approach for creating context aware graphs (CAG) [6], where links between nodes are forged if they co-occur within two consecutive sentences. The output graph is undirected, and is weighted by the number of times the adjacent nodes (words) co-occur in the original text. We exclude isolated nodes from computation. Please note that short texts (1–3 sentences) result into highly dense graphs, which are often complete graphs. Graph density decreases with increase in the number of sentences. Figure 2 shows CAG graph of a short example text. Edge width in the graph is proportional to corresponding edge weight.

As PDAs move beyond the personal
space and into the enterprise, you
need to get a firm grip on the
options available for your users.
What operating system do you
choose? What features do you and
your company need? How will these
devices fit into the existing
corporate infrastructure? What
about developer support?

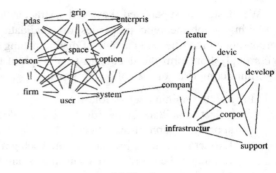

(a) Sample text.                                    (b) Graph.

**Fig. 2.** Graph created from sample text in (a) (document id 1450 from Hulth2003 dataset).

## 3.2  Training Set Construction

Node centrality measures quantify the importance of a node in the graph, and are well studied in graph theory. Centrality is a local node property that estimates a node's embeddedness in the network. Intuitively, nodes with high centrality value are more important in the network. We consider Degree Centrality, Eigenvector Centrality (Prestige), and PageRank [2] as features. Additionally, we consider an extension of PageRank called PositionRank [7], a graph degeneracy method called Coreness [24], and Clustering Coefficient as features. For each node in the CAG graph, we compute the six properties to create the feature set.

We denote a weighted, undirected graph by $G = (V, E, W)$, with $V$ as the set of vertices, $E \in V \times V$ as the set of edges, and $W$ as the corresponding weighted adjacency matrix. Node properties used as features are briefly described below.

**Degree centrality** of a vertex measures its embedded-ness at local level. For a weighted graph, it is computed as $deg(v_i) = \sum_j w_{ij}$ [28].

**Prestige** or *Eigenvector centrality* of vertex $v_i$ quantifies its embedded-ness in the graph while (recursively) taking into account the prestige of its neighbors. Starting with initial prestige vector $p_0$ where all nodes (words) are assigned equal *prestige, $p_k$* is computed recursively as follows till convergence is achieved [28].

$$p_k = W^T p_{k-1} = \left(W^k\right)^T p_0$$

According to this computation, a well-connected word attains more importance if it is connected to other well-connected words.

**PageRank** computes prestige in the context of Web search with an additional component called *random jump*. In case of text documents, this component relates to the concept of text cohesion [19]. We adopt the computation of word score (*WS*) from TextRank algorithm [19], with $d = 0.85$ as the probability of random jump.

$$WS(v_i) = (1 - d) + d * \sum_{v_j \in N_i} \left( \frac{w_{ji}}{\sum_{v_k \in N_j} w_{jk}} WS(v_j) \right)$$

**PositionRank** is an extension of PageRank that favors words occurring at the beginning of the document as keywords [7]. Node $v_i \in V$ is assigned a weight $p_i$ based on its positional information by taking the inverse of the sum of its positions of occurrences in the text. Subsequently, PageRank computation is performed on the weighted nodes of the graph to yield PositionRank scores for the candidate words. Mathematically, the PositionRank score of a node $v_i$ is computed as follows.

$$S(v_i) = (1 - \alpha).\tilde{p}_i + \alpha. \sum_{v_j \in N_i} \left( \frac{w_{ji}}{O(v_j)} S(v_j) \right)$$

Here, $\alpha$ is set to 0.85, $\tilde{p}_i = \frac{p_i}{\sum_{j=1}^{|V|} p_j}$ is the normalized positional weight of $v_i$, $N_i$ is the neighborhood of node $v_i$, $w_{ji}$ is the weight of edge $e_{ji}$, and $O(v_j) = \sum_{v_k \in N_j} w_{jk}$.

**Coreness** is a graph degeneracy property that decomposes graph $G$ into a set of maximal connected subgraphs $G_k$ ($k$ denotes the core), such that nodes in $G_k$ have degree at least $k$ within the subgraph and $G_k \subseteq G_{k-1}$ [24]. Coreness of a node is the highest core to which it belongs. Rousseau et al. [23] presume that words in the main (highest) core of the graph are keywords due to their dense connections. Though our findings differ, we are convinced that keywords tend to lie in higher cores. Hence, we choose to include coreness as a discriminating property.

**Clustering Coefficient** of a node indicates edge density in its neighborhood. Clustering coefficient for node $v_i$ is computed as the ratio of actual number of edges in the sub-graph induced by $v_i$ (excluding itself) to the total number of possible edges in that subgraph [28]. Mathematically, for an undirected graph $G$, clustering coefficient of node $v_i \in G$ is computed as below.

$$CC(v_i) = \frac{2|e_{jk} : v_j, v_k \in N_i, e_{jk} \in E|}{n_i(n_i - 1)}$$

Here, $n_i$ is the number of nodes in $N_i$, i.e., the subgraph induced by $v_i$. We speculate that nodes (words) with low clustering coefficient connect diverse contents together, and thus are likely to be important words.

All properties, except Clustering Coefficient, have been studied by state-of the-art unsupervised graph-based keyword extraction methods. To the best of authors' knowledge, complex interplay of these properties has not been explored for discriminating between keywords and non-keywords.

**Assigning Labels:** For each document, we consult the corresponding gold-standard keywords list and assign the label as 'positive' or 'negative' to the candidate words (nodes) depending on whether they are listed as a gold-standard keyword (as unigram) or not. The labels along with the feature set constitute the training set for our KE algorithm.

### 3.3    Model Training

We prepare three training sets using the steps described in Sect. 3.2 for the KDD, WWW, and Hulth2003 datasets (dataset details are given in Table 1). Training set for Hulth2003 dataset is relatively balanced. However, training sets for KDD and WWW datasets are imbalanced in nature. This is because on average, each document from KDD and WWW datasets is assigned $\approx 10$ gold-standard keywords (unigrams) out of $\approx 100$–$200$ words (columns $N_{avg}$ and $L_{avg}$ of Table 1, respectively). Since imbalanced dataset does not yield robust predictive model, we balance both these training sets by over-sampling the 'positive' class using Weka implementation of SMOTE filter[1]. Using the five training sets (two imbalanced and three balanced) as individual training sets, we train the predictive models.

Several classification algorithms have been explored in literature, including CRF and SVM [29], Bagged decision tree [18], Naïve Bayes [4, 26], gradient boosted decision trees [26], etc. However, we decided to use two classical algorithms - Naïve Bayes (NB) and Logistic Regression (LR) - because of their simplicity and fast execution time. Using these two algorithms and the five training sets, we induce ten (10) predictive models - five using NB and five using LR for each training set. Our empirical results validate that classical algorithms perform adequately for our experiments. We present the cross-validation and test results in Sect. 4.2.

## 4    Empirical Evaluation

In this section, we present our experimental setup and empirical results. We also discuss our findings, and empirically establish our claims.

### 4.1    Experimental Setup and Objectives

The proposed framework is implemented using R (version 3.4.0) and relevantpackages[2] (igraph, tm, openNLP, RWeka, caret and pROC). We use three publicly available datasets that have been used in similar studies. Hulth2003 dataset contains abstracts from medical domain, whereas KDD and WWW datasets contain abstracts from computer science domain published in these two well-known conferences. Each document in these datasets is accompanied by an associated gold-standard keywords list, which is used as ground truth for testing the classifier performance. Table 1 briefly describes the datasets along with relevant statistics. For KDD and WWW datasets, we consider only those documents which contain at least two sentences, and are accompanied by at least one gold-standard keyword. We create the individual training sets from Hulth2003, KDD, and WWW datasets using the methodology described in Sect. 3.

---

[1] We set 'percentage' parameter to 300%

[2] https://cran.r-project.org/web/packages.

**Table 1.** Overview of the experimental data collections. $|D|$: Number of docs, $L_{avg}$: average doc length, $N_{avg}$: average gold-standard keywords per doc, $K_{avg}$: average percentage of keywords present in the text.

| Collection | $|D|$ | $L_{avg}$ | $N_{avg}$ | $K_{avg}$ | Description |
|---|---|---|---|---|---|
| Hulth2003[a] [9] | 1500 | 129 | 23 | 90.07 | PubMed abstracts from *Inspec* |
| WWW [4] | 1248 | 174 | 9 | 64.97 | CS articles from WWW conference |
| KDD [4] | 704 | 204 | 8 | 68.12 | CS articles from KDD conference |

[a]We use Test and Training set, and uncontrolled gold standard list.

We designed experiments to:

i. evaluate the cross-validated performance of keyword classifiers trained on the individual training sets (Sect. 4.2).
ii. assess predictive power of the trained models over cross-collection and cross-domain datasets (Sect. 4.2).
iii. evaluate quality of extracted keywords for title construction of scientific papers (Sect. 4.2).

## 4.2 Results and Discussion

**Evaluating Cross-validated Performance of the Models:** We trained three models on the balanced training sets (Hulth, KDD-B, and WWW-B) and two models on the imbalanced training sets (KDD and WWW) using NaïveBayes (NB) and Logistic Regression (LR) algorithms (please see Sect. 3.3 for details). Since a well-written abstract contains the most important facts about scientific research and proxies well for the complete document, we empirically test the system on abstracts from scientific papers. Nevertheless, the system is extendable to full texts. We present 10-fold cross validation results in Table 2, showing precision, recall, and F1-score as performance evaluation metrics. Bold values represent best performance across all models in terms of the 'positive' class[3].

As expected, models trained on balanced training set yield better result as compared to the ones trained on imbalanced set. Thus, we discard the models trained on KDD and WWW training sets from further experiments. Although models trained on WWW-B training set turns out to be the best from cross-validation performance, we also retain models trained on Hulth and KDD-B for assessing the predictive power of the models over unseen documents. In subsequent experiments, we use a naming convention of M-X for all models, where M stands for the model, which is either NB or LR, and X stands for the training set. For example, NB-Hulth is the model trained on Hulth training set using Naïve Bayes classifier.

---

[3] Positive class is for keywords.

**Table 2.** Cross-validated classifier performance. NB: Naïve Bayes classifier results, LR: Logistic Regression classifier results, X-B: balanced training set for the corresponding dataset X.

| Training set | Naïve Bayes (NB) | | | Logistic Regression (LR) | | |
|---|---|---|---|---|---|---|
| | P | R | F1 | P | R | F1 |
| Hulth | 64.76 | 51.47 | 57.36 | 72.65 | 47.29 | 57.29 |
| KDD | 37.39 | 58.51 | 54.63 | 55.75 | 20.47 | 29.95 |
| WWW | 40.02 | 60.23 | 48.09 | 60.12 | 23.69 | 33.99 |
| KDD-B | 66.93 | 64.55 | 65.72 | 75.43 | 56.80 | 64.80 |
| WWW-B | 69.20 | **66.24** | 67.69 | **76.04** | 61.36 | **67.92** |

**Assessing Cross-collection Predictive Power of the Models:** We test the performance of the six models (three for each NB and LR) on cross-collection test sets from Hulth2003, KDD, and WWW collections. The test sets comprise the unbalanced training sets for Hulth, KDD, and WWW datasets as described in Sect. 3.2. We apply each model on all three test sets and report their performance in Table 3. For example, the model trained on Hulth training set is tested on all three training sets from Hulth, KDD, and WWW datasets. Table 3 shows results of the individual models on the corresponding test sets. Results are macro-averaged at the dataset level. Bold values indicate best performance in terms of precision, recall, and F1-score for the corresponding test sets.

**Table 3.** Performances of NB and LR models on test sets. P: Precision, R: Recall, and F1: F1-score.

| Models | Hulth2003 | | | KDD | | | WWW | | |
|---|---|---|---|---|---|---|---|---|---|
| | P | R | F1 | P | R | F1 | P | R | F1 |
| NB-Hulth | 66.7 | **58** | 57.5 | 36.6 | 66.4 | 44.7 | 38.5 | 70.4 | 47 |
| NB-KDD-B | 65.9 | 57.1 | 56.3 | 35.7 | 64.3 | 43.3 | 37.7 | 68.7 | 45.8 |
| NB-WWW-B | 66.5 | 55.4 | 55.8 | 36.8 | 63.5 | 43.9 | 38.4 | 67.5 | 46 |
| LR-Hulth | 74 | 52.6 | **57.7** | 41.5 | **66.9** | **48.7** | 43.7 | **70.4** | **51** |
| LR-KDD-B | 75.2 | 45.4 | 52.2 | **45.4** | 57.6 | 47.9 | **47** | 61.5 | 49.9 |
| LR-WWW-B | **75.3** | 47.8 | 54 | 44.9 | 59.1 | 48.1 | 46.6 | 62.9 | 50.3 |

Performance of LR models are relatively better than NB models in terms of F1-score, as reported in Table 3. We observe that models induced by Logistic Regression exhibit better results in terms of precision and models induced by Naïve Bayes exhibit better result in terms of recall. We also observe that the performance of the models are uniform across all datasets irrespective of the training set used. This indicates that the proposed method is independent of the domain or corpora of its training set, and is applicable to any text document. This experiment also establishes that node properties of graph-of-text are effective discriminators to distinguish keywords from non-keywords.

**Recommending Keywords for Title Construction:** We empirically validate our hypothesis that keywords are suitable candidates for generating titles for scientific documents. We experiment with Hulth2003 dataset that contains title and abstract for each document, where titles are clearly distinguishable from the rest of the text. Though KDD and WWW datasets contain title and abstract as well, titles are embedded in a manner that they are not clearly distinguishable. Thus, we present our results using only the Hulth2003 dataset. Moreover, to the best of the authors' knowledge, there is no work in literature that resembles our objective of semi-automatic title construction through automatic keywords extraction. Thus, we provide empirical validation for our experiment and present them below.

To compute the overlapping between keywords and words in title, we first tokenize the title text and remove stopwords from them. Since the proposed KE algorithm uses stemming, we stem the title-words for comparison. We compute macro-averaged precision and recall at a dataset level comparing both these lists of keywords and title-words. We rank the predicted keywords in decreasing order of their probability for the 'positive' class. We use the models trained using Logistic Regression, i.e., LR-Hulth, LR-KDD-B, and LR-WWW-B for our experiment as they clearly outperformed NB models in Table 3. We present our findings in Table 4, where bold values represent best result in terms of precision and recall.

**Table 4.** Overlapping of extracted keywords and title-words using precision and recall. @k: extracting top-k keywords, @lenW: extracting as many keywords as the number of corresponding title-words, P: Precision, R: Recall.

| Models | @5 | | @7 | | @10 | | @lenW | |
|---|---|---|---|---|---|---|---|---|
| | P | R | P | R | P | R | P | R |
| LR-Hulth | **82.12** | **63.37** | **74.41** | **69.51** | 70.61 | **72.07** | **79.24** | **69.22** |
| LR-KDD-B | 76.38 | 56.27 | 71.81 | 60.28 | 70.30 | 61.56 | 75.11 | 58.96 |
| LR-WWW-B | 77.72 | 58.56 | 72.64 | 63.08 | **70.79** | 64.53 | 76.67 | 62.01 |

We present four set of outcomes in Table 4. We extract top-k keywords (@k) with k being 5, 7, and 10 and we extract as many keywords as the number of title-words in the corresponding document (@lenW). We kept the number of extracted keywords to a low value, as effective titles tend to be shorter in length [15, 21]. Best precision is obtained when we extract top-5 keywords, and best recall is obtained when we extract top-10 keywords. LR-Hulth model outperforms other two models in all aspects. The results substantiate our claim that keywords are indeed good candidates for title construction.

## 5   Conclusion and Future Work

In this paper, we presented a semi-automatic system to suggest keywords for title generation. Our approach do not generate a title, instead it recommends impactful words for inclusion in the title. We design a supervised framework to automatically extract keywords from single documents. Our KE approach gains from advantages of graph-based keyword extraction techniques, which makes them applicable to texts from any domain or corpora.

The keywords extracted using predictions of the proposed model are then matched against the corresponding title-words from the document. Initial investigation shows a maximum macro-averaged precision of 82% for our dataset when we suggest top-5 extracted keywords, which supports our hypothesis that keywords are indeed good candidates for title construction. Please note that the extracted keywords constitute only nouns and adjectives (Sect. 3.1). Since title words are not restricted to only nouns and adjectives, our KE approach is expected to miss some words, which explains the 18% loss in precision. This can be improved by including more part-of-speech categories to the text graph after extensively studying the distribution of title-words. As we are reporting our initial investigation in this paper, this part is out of scope and can be considered as a future work.

Top-10 keywords (stemmed) extracted using LR-Hulth model for the abstract of this manuscript are – 'paper', 'system', 'extract', 'titl', 'construct', 'keyword', 'text', 'word', 'document', 'semiautomat'. We constructed the title for the manuscript using these suggested keywords.

# References

1. Boudin, F.: A comparison of centrality measures for graph-based keyphrase extraction. In: IJCNLP, pp. 834–838 (2013)
2. Brin, S., Page, L.: The anatomy of a large-scale hypertextual web search engine. Comput. Netw. ISDN Syst. **30**(1–7), 107–117 (1998)
3. Bulgarov, F., Caragea, C.: A comparison of supervised keyphrase extraction models. In: Proceedings of WWW, pp. 13–14. ACM (2015)
4. Caragea, C., et al.: Citation-enhanced keyphrase extraction from research papers: a supervised approach. In: Proceedings of EMNLP 2014, pp. 1435–1446 (2014)
5. Chen, S.C., Lee, L.S.: Automatic title generation for Chinese spoken documents using an adaptive k nearest-neighbor approach. In: Eighth European Conference on Speech Communication and Technology (2003)
6. Duari, S., Bhatnagar, V.: sCAKE: semantic connectivity aware keyword extraction. Inf. Sci. **477**, 100–117 (2019)
7. Florescu, C., Caragea, C.: A position-biased PageRank algorithm for keyphrase extraction. In: AAAI, pp. 4923–4924 (2017)
8. Gollapalli, S.D., Li, X.L., Yang, P.: Incorporating expert knowledge into keyphrase extraction. In: AAAI, pp. 3180–3187 (2017)
9. Hulth, A.: Improved automatic keyword extraction given more linguistic knowledge. In: Proceedings of the 2003 Conference on EMNLP, pp. 216–223. ACL (2003)
10. Jamali, H.R., Nikzad, M.: Article title type and its relation with the number of downloads and citations. Scientometrics **88**(2), 653–661 (2011)
11. Jin, R., Hauptmann, A.G.: Title generation for spoken broadcast news using a training corpus. In: Sixth International Conference on Spoken Language Processing (2000)
12. Jin, R., Hauptmann, A.G.: Automatic title generation for spoken broadcast news. In: Proceedings of the First International Conference on Human Language Technology Research, pp. 1–3. Association for Computational Linguistics (2001)

13. Jin, R., Hauptmann, A.G.: A new probabilistic model for title generation. In: Proceedings of the 19th International Conference on Computational Linguistics, vol. 1, pp. 1–7. Association for Computational Linguistics (2002)

14. Kennedy, P.E., Hauptmann, A.G.: Automatic title generation for EM. In: Proceedings of the Fifth ACM Conference on Digital Libraries, pp. 230–231. ACM (2000)

15. Letchford, A., Moat, H.S., Preis, T.: The advantage of short paper titles. R. Soc. Open Sci. **2** (8), 150266 (2015)

16. Litvak, M., et al.: DegExt-A language-independent graph-based keyphrase extractor. In: Mugellini, E., Szczepaniak, P.S., Pettenati, M.C., Sokhn, M. (eds.) Advances in Intelligent Web Mastering-3, vol. 86, pp. 121–130. Springer, Heidelberg (2011). https://doi.org/10. 1007/978-3-642-18029-3_13

17. Lopez, P., Romary, L.: GRISP: a massive multilingual terminological database for scientific and technical domains. In: LREC 2010 (2010)

18. Medelyan, O., et al.: Human-competitive tagging using automatic keyphrase extraction. In: Proceedings of EMNLP, vol. 3, pp. 1318–1327. ACL (2009)

19. Mihalcea, R., Tarau, P.: TextRank: bringing order into texts. In: Proceedings of the 2004 Conference on EMNLP, pp. 404–411. ACL (2004)

20. Nguyen, T.D., Kan, M.-Y.: Keyphrase extraction in scientific publications. In: Goh, D.H.-L., Cao, T.H., Sølvberg, I.T., Rasmussen, E. (eds.) ICADL 2007. LNCS, vol. 4822, pp. 317–326. Springer, Heidelberg (2007). https://doi.org/10.1007/978-3-540-77094-7_41

21. Paiva, C.E., Lima, J.P.d.S.N., Paiva, B.S.R.: Articles with short titles describing the results are cited more often. Clinics **67**(5), 509–513 (2012)

22. Putra, J.W.G., Khodra, M.L.: Automatic title generation in scientific articles for authorship assistance: a summarization approach. J. ICT Res. Appl. **11**(3), 253–267 (2017)

23. Rousseau, F., Vazirgiannis, M.: Main core retention on graph-of-words for single-document keyword extraction. In: Hanbury, A., Kazai, G., Rauber, A., Fuhr, N. (eds.) ECIR 2015. LNCS, vol. 9022, pp. 382–393. Springer, Cham (2015). https://doi.org/10.1007/978-3-319-16354-3_42

24. Seidman, S.B.: Network structure and minimum degree. Soc. Netw. **5**(3), 269–287 (1983)

25. Shao, L., Wang, J.: DTATG: an automatic title generator based on dependency trees. In: Proceedings of the International Joint Conference on Knowledge Discovery, Knowledge Engineering and Knowledge Management, pp. 166–173. SCITEPRESS-Science and Technology Publications, Lda (2016)

26. Sterckx, L., Demeester, T., Develder, C., Caragea, C.: Supervised keyphrase extraction as positive unlabeled learning. In: EMNLP 2016, pp. 1–6 (2016)

27. Tseng, Y.-H., Lin, C.-J., Chen, H.-H., Lin, Y.-I.: Toward generic title generation for clustered documents. In: Ng, H.T., Leong, M.-K., Kan, M.-Y., Ji, D. (eds.) AIRS 2006. LNCS, vol. 4182, pp. 145–157. Springer, Heidelberg (2006). https://doi.org/10.1007/11880592_12

28. Zaki, M.J., Meira Jr., W., Meira, W.: Data Mining and Analysis: Fundamental Concepts and Algorithms. Cambridge University Press, New York (2014)

29. Zhang, C.: Automatic keyword extraction from documents using conditional random fields. JCIS **4**(3), 1169–1180 (2008)

30. Zhang, Y., Chang, Y., Liu, X., Gollapalli, S.D., Li, X., Xiao, C.: MIKE: key phrase extraction by integrating multidimensional information. In: Proceedings of CIKM 2017, pp. 1349–1358. ACM (2017)

# Object Recognition in Hand Drawn Images Using Machine Ensembling Techniques and Smote Sampling

Mohit Gupta[(✉)] [ID], Pulkit Mehndiratta, and Akanksha Bhardwaj

Department of Computer Science, Jaypee Institute of Information Technology,
Noida, India
mohitatjammu@gmail.com,
{pulkit.mehndiratta, akanksha.bhardwaj}@jiit.ac.in

**Abstract.** With the increase in the advancement of technology, the size of multimedia content generated is increasing every day. Handling and management of this data to extract the patterns result in a more optimized and efficient way is the need of the hour. In this proposed work, we have presented techniques to classify the stroke-based hand-drawn object. We have used the Quick Draw dataset which is a repository of approximately 50 million hand-drawn drawings of 345 different objects. Our research presents an approach to the classification of these drawings created using hand strokes. We are converting the given raw image data, to much more simplified and concise data and then performed oversampling on data belonging to classes with fewer instances using Synthetic Minority Over-sampling Technique (SMOTE) method to balance the distribution of each class in the dataset. Finally, we are classifying the drawings using K-Nearest Neighbor (K-NN), Random Forest Classifier (RFC), Support Vector Classifier (SVC) and Multi-Layer Perceptron model (MLP) by working on their hyperparameters for the best-achieved classification result. The proposed solution attains the accuracy of 82.34% using best hyperparameter selection.

**Keywords:** SMOTE · Machine learning · Classification · Hyper-parameter selection · Ensembling

## 1 Introduction

In today's scenario, everyone is having easy access to unlimited internet connectivity in result producing large amounts of multimedia content every second. This data can contain text, video, speech, images, numerical values, and this is a never-ending list. As per Forbes, at the current pace of ours, about 2.5 quintillion data bytes are generating each day [25] and this rate is for sure accelerating as we speak of it, because of exposure to technology, knowledge, and growth of the Internet of Things (IoT). A massive portion of this user-generated data is image(s) data. Companies, governments, and researches are in continuous trials to extract useful information as much as possible thus, to make it useful for various applications.

© Springer Nature Singapore Pte Ltd. 2019
A. B. Gani et al. (Eds.): ICICCT 2019, CCIS 1025, pp. 228–239, 2019.
https://doi.org/10.1007/978-981-15-1384-8_19

Sketching or drawing is one of many ways of how human beings think, create or even communicate, and with the advancements, in the technology, these skills are just not pen-paper based anymore. Sketch recognition is the technique of automatic understanding of free-hand sketches on a digitizing screen with a stylus. This allows the user to draw or plot in an unrestrained manner without having any special skill set or knowledge with the help of training models to recognize the different style of drawing. The field of recognition of hand-drawn object has gained popularity and has drawn the attention of many types of research in the past few years. Effective recognition of hand drawing can perceive human thinking processes and making it useful in a number of applications like sketch-based search [1], hand-written bills reading by systems [2], game designing [3] and most importantly it can act as a digital medium for children education in various developing economies like India [3].

With all the knowledge base and literature available, we have come up with a unique approach to contribute to the on-going research in this field. We have proposed a solution to classify images of QuickDraw dataset provided by Google which contains the information about the strokes drawn by a user to draw any object which belongs to a particular class. This dataset, itself has posed major challenges to work upon because of its size and the distribution of data around 345 classes. Hence, in this paper, we have worked on simplifying the data to extract meaningful information in a more under-standable and workable manner. The balancing of the dataset class distribution has been done by using Synthetic Minority Over-Sampling Technique (SMOTE). After applying all these techniques, we have used machine learning models – K-Nearest Neighbor (K-NN), Random Forest Classifier (RFC), Support Vector Classifier (SVC) and Multi-Layer Perceptron model (MLP) along with their ensembling after comparing the hyperparameter results for the individual model.

## 2 Related Work

Researchers have been studying hand-drawn object recognition since the 1990s in the field of graphics and computer vision. With the increase in urge to extract useful information form the hand drawings for using in different applications, hand drawing recognition is one of the emerging topics for research purposes.

Many of the existing research work [4, 5] focuses on basic shape recognition in particular domain like UML (Unified Modeling Language) drawings. These approaches are limited to a particular domain knowledge only. In an article [6], researches have worked on a collection of one million clipart images extracted from the web. They have built up Sketch2Tag system that recognizes the object without any domain-specific restriction. Many researchers have tried to study free-hand objects recognition using graphical models and sketch features [7] or have used image mapping to the sketches [8]. People have tried to use deep neural networks like CNN [9–11] which work very well in the image or visual recognition task but to our knowledge, no significant work has been done regarding free-hand sketch recognition. Eitz et al. in their work [12] have tried to analyze the sketch drawings using a database containing around 250 classes of objects and achieved an accuracy of 54%. His work [12] was further improved by Rosalia et al. [13] using Fisher vector image recognition with an accuracy of 67%. Researchers in the paper [14], have used user's stroke and interpreted the pixels to produce geometrical descriptions such as ovals, lines, any arbitrary shape or point.

Working on our approach to recognize hand-drawn objects, this article is divided into various sections as follows - Sect. 3 discusses methodology which explains the dataset usage and techniques that have been implemented. Section 4 presents our idea of simplifying the data. SMOTE technique to balance the data is mentioned in Sect. 5. Section 6 describes the usage of machine learning models. The results of our implemented approach are mentioned in Sect. 7. Finally, Sects. 8 and 9 contains the conclusion and future scope of our experimental approach.

# 3    Methodology: Hand-Drawn Object Classification

## 3.1    Dataset

The dataset that has been used for this research is a Quick Draw dataset [26] provided by Google which is a repository of approximately 50 million hand-drawn drawings of 345 different objects.

We are using the dataset of 10 classes. Each class of object has different dataset repository, so we are randomly combining the dataset of these 10 classes (on which we are working). Finally, the combined data has about 13.5 lakh data instances belonging to the 10 chosen classes. The attributes of our data are key_id (uniquely identifies all the drawings), country code (a two-letter code of the country to which the player belongs), drawing (Co-ordinates of the strokes), recognized (whether the word was recognized by game, when the player made that drawing), timestamp (when the drawing was made), word (class to which drawing belongs) as shown in Fig. 1. For our classification task, we are only taking a word and drawing attributes into consideration.

| | countrycode | drawing | key_id | recognized | timestamp | word |
|---|---|---|---|---|---|---|
| 0 | BR | [[[0, 99, 118, 116, 59, 40, 25, 12, 12, 17, 31... | 5438082394357760 | True | 2017-03-17 16:37:44.609240 | hourglass |
| 1 | DE | [[[11, 38, 170, 154, 128, 95, 103, 132, 150, 1... | 4922317173948416 | True | 2017-01-30 14:51:30.634570 | hourglass |
| 2 | US | [[[53, 76, 144, 192], [3, 26, 25, 12]], [[45, ... | 6193290166665216 | True | 2017-03-04 12:12:23.979120 | hourglass |
| 3 | US | [[[12, 34, 63, 88, 102, 124, 137, 140, 140, 12... | 4776824523456512 | True | 2017-03-29 21:35:10.259580 | hourglass |
| 4 | US | [[[128, 0], [0, 2]], [[1, 57, 173], [247, 233,... | 4832689808998400 | True | 2017-03-24 15:29:52.056240 | hourglass |

**Fig. 1.** Depicts that dataset

## 3.2    Implementation

Initially, we have combined the Quick Draw dataset of different classes as each class has different data files and then performed the pre-processing on it. As the combined data is large, we try to extract the meaningful information from our data by converting the raw data to a much more informative simplified data and this can make our models efficient in terms of computations and predictions. Now such a large dataset along with a large number of classes brings the problem of imbalance distribution. This is highly undesirable for the training of any machine learning models as it will result in biased classification and misleading accuracy. So, we are applying Synthetic Minority Over-sampling Technique (SMOTE) [15, 16] to oversample the data instances of minority classes and thus, balancing the data distribution. After applying these techniques, our dataset is absolutely ready to be used for further classification. Figure 2 shows how the random hand-drawn sketch looks like in the dataset.

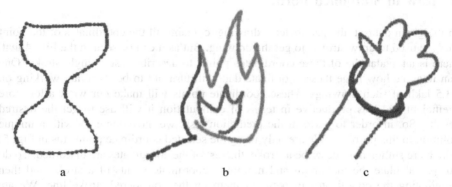

a         b         c

**Fig. 2.** a, b, c – Depicts the drawing of the classes - watch glass, campfire, hand respectively.

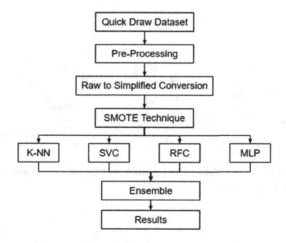

**Fig. 3.** Depicts the flow chart our work

We are aiming to classify hand-drawn object drawings to their corresponding classes based on the hand-drawn strokes (that have been used to draw that object). We are using machine learning models and training them using the coordinates of the point of joining the strokes. The co-ordinate dataset is the Quick Draw dataset that has been simplified and undergone SMOTE processing. We are using K-Nearest Neighbor

(K-NN), Random Forest Classifier (RFC), Support Vector Classifier (SVC) and Multi-Layer Perceptron model (MLP) as shown in (Fig. 3). Each of these ML models has been tested on different possible hyper-parameters. Finally, we have compared the performance of these models with their best-suited hyper-parameters and analyzed the performance that how well our trained models have classified for the hand-drawn object drawings.

## 4   Raw to Simplified Form

In the given dataset, the parameter – drawing, contains all the coordinates of the point that are used to draw strokes to get the drawing. But as it can be seen in the Fig. 4 that, there is an avalanche of these coordinates points to describe just a single stroke. One can imagine how huge these coordinate data will turn out to be if we are working on 13.5 lakh of such drawings. These coordinate points will make our work a lot more inefficient and less productive in terms of computation it will use to get the desired results. So, in order to perform the prediction task, we have come up with a unique solution to this problem. We are only using the selected coordinate points (as in Fig. 5) which are sufficient to describe a stroke, the rest of the coordinates are discarded. To do so we calculate the maximum and minimum coordinate points of a stroke and then eliminating the coordinates in between them on the considered stroke line. We are simplifying our data to make it much more informative and concise to perform our classification task efficiently because we are only concerned about the strokes of the drawings.

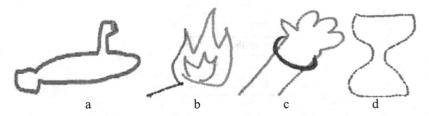

**Fig. 4.** 4, 5 – Depicts the raw data and simplified drawing of the classes – aeroplane, watch glass, campfire, hand respectively.

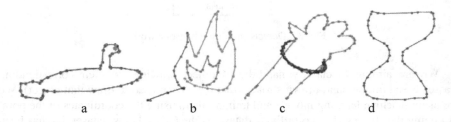

**Fig. 5.** Depicts the raw data and simplified drawing of the classes – aeroplane, watch glass, campfire, hand respectively.

# 5  Synthetic Minority Over-Sampling Technique (SMOTE)

We have a large dataset with many numbers of classes and hence, there is a high chance of data being an imbalance when we are dealing with classification. Imbalance in the dataset refers to the situation when we have a large number of instances belonging to a class i.e. a majority class compared to some other class with only a few instances i.e. a minority class. In such a case, the prediction results of the machine learning models being used can be biased and will result in misleading accuracies. So, we need to adjust the data distribution for ensuring unbiased prediction results by either increasing the minority class instances or decreasing the majority class instances. To resolve this issue, we are using Synthetic Minority Over-sampling Technique (SMOTE) [15, 16].

SMOTE randomly creates new minority class instances. It first identifies the feature vector in the dataset and its corresponding nearest neighbour data points. Once done it try to find out the differences between the two and multiply it with random numbers 0 and 1. Finally, it adds new data points by adding random numbers to the feature vector value on the line segment and the process repeats for other feature vectors. For example – Let (9, 6) be the feature vector value and (4, 2) be its nearest neighbour.

| Feature vector – | $f1\_1 = 9$ | $f2\_1 = 6$ | $f2\_1 - f1\_1 = -3$ |
| Nearest Neighbor – | $f1\_1 = 4$ | $f2\_1 = 2$ | $f2\_1 - f1\_1 = -2$ |

Then, new generated samples will be as $(f1', f2') = (9, 6) + rand(0,1) * (-3, -2)$.

# 6  Machine Learning Models

## 6.1  K-Nearest Neighbor (K-NN)

K- Nearest Neighbour (K-NN) [17–19] is an instance-based machine learning model. This model can be used for both classification and regression tasks. In both cases, the input is 'k' closest training instances in the feature vector space and output is different based on the task. Here, we have used it to perform the classification task where the output is the class membership voted out by the neighbour instances in feature space. The primal parameter of K-NN is the 'k' value i.e. the number of nearest neighbours to look for. While training our model, we set this value and our model searches for 'k' closest neighbour instances for the argument instance and thus assigns the class which is belonging to the majority of the neighbour instance, to the argument instance. By default, it is 5. So, in order to set the best parameter for K-NN model, we have tested it on different values of 'k' i.e. N neighbours. The result of these hyperparameters is mentioned in Table 1. As per the results, our K-NN is best performing at its default parameter for the given dataset and so, we are using K-NN with parameter – 'N_neighbor' value to be 5.

**Table 1.** Hyperparameter results of the K-NN model

| n_neighbors | Accuracy (%) | Precision (%) | Recall (%) | F-score (%) |
|---|---|---|---|---|
| 2 | 79.52 | 81.4 | 79.4 | 79.4 |
| 3 | 81.95 | 83.1 | 81.8 | 81.9 |
| 4 | 82.07 | 83.2 | 81.9 | 82.2 |
| 5* | **82.34** | **83.4** | **82.3** | **82.3** |
| 6 | 81.91 | 83 | 81.9 | 81.8 |
| 7 | 82.07 | 83.1 | 81.9 | 81.9 |

## 6.2    Support Vector Classifier (SVC)

Support Vector Machine (SVM) [20] is a supervised machine learning algorithm. We are using it to perform the classification task [21, 22] by finding the best parameters to find a hyperplane that differentiates the classes most optimally. Data points are marked in a feature vector space and SVC model works to fit the hyperplane in the best manner i.e. reducing the error. The parameter which can be altered is – 'kernel' i.e. the kind of hyperplane is required. The default kernel is 'RBF'. We have tried to analyze all the hyperparameters – linear, RBF and sigmoid kernels and the results of these hyperparameters are mentioned in Table 2. The choice of kernel depends entirely on the data because kernel adds non-linearity to the hyperparameters which are very essential when we are working on a large dataset like Quick Draw Dataset and which also has many classes. As per the results, we are setting our SVC model to 'RBF' kernel.

**Table 2.** Hyperparameter results of SVC model

| Kernel | Accuracy (%) | Precision (%) | Recall (%) | F-score (%) |
|---|---|---|---|---|
| Linear | 72.73 | 72.9 | 73.1 | 73.4 |
| RBF* | **78.15** | **78.3** | **77.9** | **78.1** |
| Sigmoid | 76.75 | 76.7 | 77 | 77.2 |

## 6.3    Random Forest Classifier (RFC)

Random forests [23] are an ensemble of decision tree algorithm. Random Forest uses multiple decision tree models known as forests to perform classification or regression task. In the case of random forest, the splitting is done randomly i.e. the features selected for decision making are chosen randomly as opposed to the concept of a decision tree. Random Forest model also overcomes the problem of overfitting by introducing randomness. It is a normal tendency that – the more the number of trees, the better the result or the better the accuracy (in case of classification) but it is not always true. By default, the number of trees RFC ensemble is 10. But we are testing on a different hyper-parameter value ranging from 10 to 2000 and getting the best fit value for a number of trees to take into consideration. So, we have set our RFC over 1200 number of trees to the ensemble, by proving the notion of increasing accuracy with more trees to be false. The result over different hyperparameters with RFC kernel are shown in Table 3.

## 6.4  Multi-layer Perceptron Model (MLP)

Multi-Layer Perceptron (MLP) Model [24] is a feed-forward artificial neural network. It mimics the brain in its working. The basic components of MLP are input layer, hidden layer and output layer, each of these layers consists of various nodes. These nodes in the layers are usually connected and the connections between these nodes have random weights (between 0 and 1) and bias values assigned to them. Also, the nodes of the hidden and output layer have activation functions assigned to them. These activation functions add non-linearity to the hypothesis. Information comes as input to a node, it gets multiplied with weight, added up with bias, then applied with activation function and further passed on to the next connected node in the layer. The most unique feature of MLP is the backpropagation. After reaching the output layer, MLP calculates the error and then backpropagate the information to update the earlier randomly assigned weights in order to reduce the error.

Our model performs these forward and backward passes for the set iterations and gets itself trained. We have worked on the combination of the hyperparameters – hidden layer, activation function and solver. The solver is a gradient descent algorithm (to update the weights) we can choose, that can be – 'sgd', 'Adam'. Activation functions under consideration are – 'logistic', 'tanh' and 'relu'. The results of all these hyperparameter combinations are mentioned in Table 4.

**Table 3.** Hyperparameter results of RFC model

| n_neighbors | Accuracy (%) | Precision (%) | Recall (%) | F-score (%) |
|---|---|---|---|---|
| 10* | 73.78 | 74.2 | 73.7 | 73.6 |
| 50 | 79.98 | 80.3 | 80 | 79.9 |
| 100 | 80.7 | 81.1 | 80.5 | 80.8 |
| 500 | 81.31 | 81.5 | 81.4 | 81.4 |
| 1000 | 81.53 | 81.8 | 81.5 | 81.6 |
| 1200 | 81.58 | 81.9 | 81.6 | 81.6 |
| 1300 | 81.52 | 81.7 | 81.5 | 81.4 |
| 1500 | 81.25 | 81.7 | 81.2 | 81.3 |
| 2000 | 81.42 | 81.8 | 81.5 | 81.4 |

**Table 4.** Hyperparameter results of the MLP model

| Hidden layer | Activation | Solver | Accuracy (%) | Precision (%) | Recall (%) | F-score (%) |
|---|---|---|---|---|---|---|
| 2 | logistic | sgd | 70.66 | 70.8 | 70.5 | 70.5 |
| 2 | logistic | adam | 80.17 | 80.2 | 79.9 | 80.3 |
| 2 | tanh | sgd | 81.8 | 81.9 | 81.8 | 81.8 |
| 2 | tanh | adam | 81.2 | 81.2 | 81.1 | 80.9 |
| 2 | relu | sgd | 81.77 | 81.9 | 81.7 | 82 |
| 2 | relu* | adam* | 80.53 | 80.5 | 80.6 | 80.5 |
| 3 | logistic | adam | 78.18 | 78.1 | 78.2 | 78 |
| 3 | tanh | sgd | 81.68 | 81.7 | 81.6 | 81.6 |
| 3 | tanh | adam | 81.44 | 81.6 | 81.5 | 81.3 |
| 3 | relu | sgd | 80.54 | 80.6 | 80.5 | 80.5 |
| 3 | relu | adam | 81.26 | 81.3 | 81.2 | 81.3 |

## 7   Results

In QuickDraw dataset, we have approximately 13.5 lakhs drawings. For every image, we are given with the coordinate point details of each stroke that is being used to draw each of these drawings. We have used 80–20% ratio of the dataset to train and test our prepared models. The choice of the models is made after going through the process of hyperparameter selection. We have compared the individual hyperparameters of each model based on the accuracy, prediction, recall and f-score as measures. Based on this selection results (as mentioned in Fig. 6), we have fixed our model with these best-suited parameters and this has helped us to achieve the state-of-the-art classification results on the given dataset.

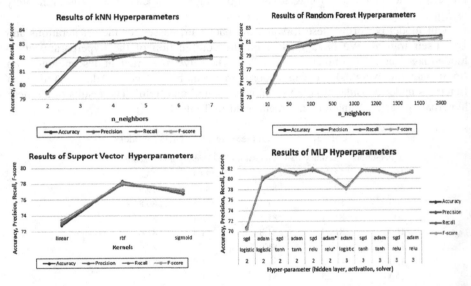

**Fig. 6.** Shows the graphical representation of hyper-parameter testing for models – K-NN, RFC, SVC and MLP respectively.

Each entry in the dataset has many numbers of coordinate points corresponding to the large number of strokes that are used to draw these drawing. So, we are refining the coordinate point information with SMOTE balancing using oversampling and this simplified version of the coordinate point information is given to our set models for training purposes. We have used K-NN with '5' n-neighbours, SVC with 'RBF' kernel, RFC with '1200' ensemble trees and MLP with 2 hidden layers (each consisting 100 nodes to have information) with *'tanh'* activation function and *'sgd'* as our gradient solver choice. We have ensembled the models in different combinations and thus, studied how the combination works on the coordinate information given to them based on the performance of these models. To our surprise, K-NN with '5' nearest neighbours has given the best results with 82.34% accuracy, 83.4% of precision, 82.3% of recall and 82.3% of f-score value, outperforming ensembling models (as mentioned in

Table 5). RFC is also performing very well and so the combination of both K-NN and RFC. K-NN being computationally very easy to perform and RFC being the ensemble, 1200 trees are generating the best results for our classification task.

**Table 5.** Classification Results of Machine Learning Models and their ensemble

| Models | Accuracy (%) |
|---|---|
| SVC | 78.15 |
| RFC | 81.58 |
| MLP | 81.8 |
| K-NN | 82.34 |
| RFC + SVC | 79.75 |
| K-NN + SVC | 80.8 |
| SVC + MLP | 81.2 |
| K-NN + MLP | 81.57 |
| RFC + MLP | 81.67 |
| K-NN + RFC | 81.75 |

# 8  Discussion

We have worked on the hand-drawn image classification task for the Quick Draw dataset provided by Google that has approximately 13.5 lakh drawings distributed over 10 different classes.

Even the single drawing instance consists of a large number of coordinate points depicting the strokes made to draw this drawing. So, the first challenge for us was to process a large amount of information. So, we came up with a very unique approach of simplifying our dataset to only have specific coordinate points which can give us all the information about the strokes as mentioned in Sect. 4. Then, we had to deal with the imbalance distribution of our data as it belongs to many classes. So, to resolve this issue, we used the SMOTE technique which oversampled the data for minority classes and balanced our data, this improved the training of our models used otherwise our machine learning models would have performed in a biased manner. We have used K-NN, SVC, RFC and MLP method for the classification task. Before applying these methods, we performed hyperparameter selection, where we tried different combinations and set our models to the best performing parameters. These models further used for the classification task and then ensembled to check for further improvements. Of all the models and ensembles, we used K-NN with the n_neighbor parameter as 5. This is because it performed well with 82.34% accuracy. The clear reason for K-NN to perform this well can be inferred as it is computationally less expensive approach and the data was large, the K-NN model approach got the advantage to predict the class for the drawings based on the stroke coordinate points.

Even the MLP model with 2 hidden layers (each consisting 100 nodes to have information) with 'tanh' activation function and 'sgd' gradient solver has given good results with an accuracy of 81.8%. The feed-forward network with backpropagation and the best parameter choice made it possible for MLP to work well as compared to the rest of the model choices we have made. Out of all the ensembles we used, KNN + RFC outperformed others with 81.75% accuracy because RFC itself is an ensemble approach and along with computationally less expensive K-NN.

## 9 Conclusion/Future Scope

We have successfully classified the hand-drawn object drawings of Quick Draw dataset by Google, using machine learning models like - K-NN, RFC, SVC and MLP with best selected hyperparameter and their ensembles. The given dataset is very large to handle. Thus, simplifying the coordinate points for the stroke information of the drawing and then balancing the simplified data with SMOTE technique worked very well in refining the data and retaining the meaningful information only. Training our models with this refined data gives us good results. All these techniques and hyper-parameter selection really played a significant role in deciding the performance of our models selected for our study. Our trained model can predict the strokes and guide the user as he/she starts drawing in real time. So, the user just needs to start and the model will tell which stroke to draw and the user whatever his age be can easily draw.

This work can be further extended and improved by using deep neural networks like Convolutional Neural Network (CNN). As CNN's [9–11] are known for their image recognition functioning. We can test and compare the working of stroke based and image-based CNN classification. Our work can be used in many hand-drawn images recognition tasks like children education [3], signature matching, hand gesture recognition by plotting the coordinate points in feature vector space, sketch-based search [1], hand-written documents or bills text extraction [2].

## References

1. Cao, Y., Wang, H., Wang, C., Li, Z., Zhang, L., Zhang, L.: Mindfinder: interactive sketch-based image search on millions of images. In: Proceedings of the 18th ACM International Conference on Multimedia, pp. 1605–1608. ACM, October 2010
2. Plamondon, R., Srihari, S.N.: Online and off-line handwriting recognition: a comprehensive survey. IEEE Trans. Pattern Anal. Mach. Intell. **22**(1), 63–84 (2000)
3. Paulson, B., Eoff, B., Wolin, A., Johnston, J., Hammond, T.: Sketch-based educational games: drawing kids away from traditional interfaces. In: Proceedings of the 7th International Conference on Interaction Design and Children, pp. 133–136. ACM, June 2008
4. Davis, R.: Magic paper: sketch-understanding research. Computer **40**(9), 34–41 (2007)
5. Hammond, T., Davis, R.: LADDER, a sketching language for user interface developers. In: ACM SIGGRAPH 2007 Courses, p. 35. ACM, August 2007
6. Sun, Z., Wang, C., Zhang, L., Zhang, L.: Sketch2Tag: automatic hand-drawn sketch recognition. In: Proceedings of the 20th ACM International Conference on Multimedia, pp. 1255–1256). ACM, October 2012
7. Kokkinos, I., Maragos, P., Yuille, A.: Bottom-up & top-down object detection using primal sketch features and graphical models. In: 2006 IEEE Computer Society Conference on Computer Vision and Pattern Recognition, vol. 2, pp. 1893–1900. IEEE (2006)

8. Qi, Y., Guo, J., Li, Y., Zhang, H., Xiang, T., Song, Y.Z.: Sketching by perceptual grouping. In: 2013 20th IEEE International Conference on Image Processing (ICIP), pp. 270–274. IEEE, September 2013

9. Krizhevsky, A., Sutskever, I., Hinton, G.E.: ImageNet classification with deep convolutional neural networks. In: Advances in Neural Information Processing Systems, pp. 1097–1105 (2012)

10. Girshick, R., Donahue, J., Darrell, T., Malik, J.: Rich feature hierarchies for accurate object detection and semantic segmentation. In: Proceedings of the IEEE Conference on Computer Vision and Pattern Recognition, pp. 580–587 (2014)

11. Donahue, J., et al.: DeCAF: a deep convolutional activation feature for generic visual recognition. In: International Conference on Machine Learning, pp. 647–655, January 2014

12. Eitz, M., Hays, J., Alexa, M.: How do humans sketch objects? ACM Trans. Graph. **31**(4), 44 (2012)

13. Schneider, R.G., Tuytelaars, T.: Sketch classification and classification-driven analysis using fisher vectors. ACM Trans. Graph. (TOG) **33**(6), 174 (2014)

14. Davis, R.: Position statement and overview: sketch recognition at MIT. In: AAAI Sketch Understanding Symposium (2002)

15. Chawla, N.V., Bowyer, K.W., Hall, L.O., Kegelmeyer, W.P.: SMOTE: synthetic minority over-sampling technique. J. Artif. Intell. Res. **16**, 321–357 (2002)

16. Deepa, T., Punithavalli, M.: An E-SMOTE technique for feature selection in high-dimensional imbalanced dataset. In: 2011 3rd International Conference on Electronics Computer Technology (ICECT), vol. 2, pp. 322–324. IEEE, April 2011

17. Keller, J.M., Gray, M.R., Givens, J.A.: A fuzzy k-nearest neighbor algorithm. IEEE Trans. Syst. Man Cybern. **4**, 580–585 (1985)

18. Beyer, K., Goldstein, J., Ramakrishnan, R., Shaft, U.: When is "Nearest Neighbor" meaningful? In: Beeri, C., Buneman, P. (eds.) ICDT 1999. LNCS, vol. 1540, pp. 217–235. Springer, Heidelberg (1999). https://doi.org/10.1007/3-540-49257-7_15

19. Dudani, S.A.: The distance-weighted k-nearest-neighbor rule. IEEE Trans. Syst. Man Cybern. **4**, 325–327 (1976)

20. Suykens, J.A., Vandewalle, J.: Least squares support vector machine classifiers. Neural Process. Lett. **9**(3), 293–300 (1999)

21. Zhang, M., Zhang, D.X.: Trained SVMs based rules extraction method for text classification. In: IEEE International Symposium on IT in Medicine and Education, ITME 2008, pp. 16–19. IEEE, December 2008

22. Lau, K.W., Wu, Q.H.: Online training of support vector classifier. Pattern Recogn. **36**(8), 1913–1920 (2003)

23. Breiman, L.: Random forests. Mach. Learn. **45**(1), 5–32 (2001)

24. Jain, A.K., Mao, J., Mohiuddin, K.M.: Artificial neural networks: a tutorial. Computer **29**(3), 31–44 (1996)

25. How much data do we create every day? The mind-blowing stats everyone should know. https://www.forbes.com/sites/bernardmarr/2018/05/21/how-much-data-do-we-create-every-day-the-mind-blowing-stats-everyone-should-read/

26. Quick Draw Dataset. https://quickdraw.withgoogle.com/data

# Key Phrase Extraction System for Agricultural Documents

Swapna Johnny$^{(\boxtimes)}$ and S. Jaya Nirmala$^{(\boxtimes)}$

National Institute of Technology, Trichy, India
swapna.johnny@gmail.com, sjaya@nitt.edu

**Abstract.** Keywords play a vital role in extracting relevant and semantically related documents from a huge collection of various documents. Keywords represent the main topics covered in the document. But manual keyword extraction is a tedious and time-consuming process. Thus, there is a need for an automated keyword extraction system for easier extraction of relevant documents. In this paper, the focus is on agriculture-related documents. Agrovoc, an agriculture-based vocabulary that contains more than 35,000 concepts is used for extracting relevant keywords from the document. The proposed system extracts the relevant keywords from agricultural documents which are further used to extract relevant documents. Also, with the increasing number of documents on the Internet, the need for efficient storage for keywords with their corresponding documents is necessary. In the proposed system, a trie-based inverted index has been used for efficient storage and retrieval of keywords and the related documents.

**Keywords:** Keyword extraction · AGROVOC · Agrotags · Agricultural documents · Trie-based inverted index

## 1 Introduction

Farmers are a vital part of our society. Their queries and doubts related to agriculture need to be answered to increase crop production. There are many agricultural thesauri available, but the extraction of keywords is a manual process, thus it makes it difficult to analyze the document. Also, it is not possible for the farmers to go through the entire document. Thus, an automated query answering system based on the agriculture thesaurus is proposed. The first phase of the project deals with the extraction and storage of the keywords from various agricultural documents. The task of identifying the set of the terms that characterize a document is essential for information extraction. These terms that signify the most important information in the documents are called by various names: *key terms*, *key phrases* or just *keywords*. All these words have the same purpose – to represent the important concepts discussed in the document. With the increasing number of digital documents present on the Internet, key phrases are essential and make important metadata. Many documents still don't have keywords assigned to them and the process of manual assignment of keywords largely depends on the subject experts and authors. Thus, the need for an efficient keyword extraction is vital.

© Springer Nature Singapore Pte Ltd. 2019
A. B. Gani et al. (Eds.): ICICCT 2019, CCIS 1025, pp. 240–252, 2019.
https://doi.org/10.1007/978-981-15-1384-8_20

With agricultural documents, there are many technical terminologies that may not be commonly used and are listed only in agricultural thesaurus like AGROVOC, Centre for Agriculture and Bioscience International (CABI), National Agricultural Library (NAL) etc. Documents like these prove to be crucial for keyword extraction from various agricultural documents and are extremely useful in agriculture information systems. Linking semantically related agriculture documents can be made easier by analyzing and extracting keywords from these documents. With the use of natural language processing techniques and AGOVOC, extraction of key phrases can be done and thus the processing of these documents is made more efficient and faster.

Another issue related to information extraction is the storage of the extracted keywords and the retrieval of the corresponding documents. The searching and retrieval time should be minimized. An efficient way of doing the same is by using an inverted index. Almost all retrieval engines for full-text search today rely on an inverted index, which gives access to a list of documents related to a term when the term is given to the inverted index. An even more efficient indexing method can be achieved by combing two data structures – inverted index and trie. An inverted index helps in efficient indexing and trie helps in faster retrieval time.

The agricultural documents used in the system are taken from AGRIS (International System for Agricultural Science and Technology) which consists of more than 8 million structured bibliographical records on agricultural science and technology. The agricultural thesauri AGROVOC is an RDF file consisting of more than 35,000 concepts in multiple languages in the form of triples. It is made available by the Food and Agriculture Organization of the United Nations (FAO).

## 2 Keyword Extraction Approaches

### 2.1 Supervised Approach

This approach refers to the system learning to predict keywords from an already tagged document set. The keyword extraction system is trained with the training document set and then used to predict keywords from new documents. One example of this approach is the Keyphrase Extraction Algorithm (KEA). On the training documents, using TF-IDF and first occurrence, KEA applies Bayes theorem to extract the key phrases from new documents [6]. In KEA, TF-IDF stands for Term Frequency Inverse Document frequency. A word having a higher TF-IDF implies that the term occurs frequently only in that document and not in any other document. Thus, the word is a keyword.

### 2.2 Unsupervised Approach

This approach refers to the system learning to predict keywords without any training document set. The system learns by itself on how to extract keywords. Two examples under this approach are TextRank and Rapid Automatic Keyword Extraction (RAKE). TextRank extracts key phrases by creating a graph from the text and ranking the key phrases based on the co-occurrence links between words. The advantage of TextRank is that it is language independent as its extraction process depends on the word occurrences [11]. RAKE is also a language-independent keyword extraction process. It works on the assumption that keywords do not contain stop words and punctuation.

## 3   Existing Work

Agrotagger, an automated key phrase extractor for various agricultural documents, identifies the agrovoc terms from the document and maps them to Agrotags terms. The paper [1] describes a mapping process which consists of narrowing down the keywords. The agrovoc terms are mapped according to the broader term and non-descriptor term. All the non-descriptor terms get mapped to a single Agrotag. For example, 'paddy' the non-descriptor for 'rice' is mapped to 'rice'. Then various terms are again mapped down to the broader concept like, for example, 'garden waste' is mapped to 'organic waste'. Also, during the mapping process, all the non-descriptors, scientific names, geographical terms, and fishery-related terms are removed [1]. Agrotagger helps to link documents related to agriculture so that there is a more efficient and faster retrieval of information. The concept of using AGROVOC has helped to achieve this. But, one of the issues is that if the terms are not in AGROVOC, they are not detected.

The paper [6] deals with the study and use of social networking sites for agro-produce marketing. It helps to connect the farmers directly with the merchants. The posts on various social networking sites can be analysed to retrieve relevant information which can be propagated to the farmers and merchants as suggestions. The information extraction system extracts entities and related data by using domain-specific named entity recognizer from the tweets on Twitter. This extracted information is then exchanged between the farmers and merchants [7]. This system can be used to benefit the farmers. But in case of an unrecognizable or new entity, it is difficult to extract keyphrases. Since many of agricultural terms are not recognized, keyphrase extraction is not that efficient in such a case.

The paper [12] deals with the process of unsupervised keyword extraction by using noun clustering. In the system proposed in the paper, two extracted nouns are in the same cluster if they have at least one common word. Then keyphrase extraction is done by choosing the top k clusters, scores of which are calculated by averaging the scores of all the noun phrases in that cluster. The score of each noun phrase is the sum of the frequencies of each individual word in that phrase. The paper has shown that keyword extraction is more efficient if nouns are clubbed with their synonyms [12]. But, if the keyphrase to be extracted is not a noun then the keyphrase is not detected. Limiting the keyphrases only to nouns thus reduces the efficiency of the system. Also, if the frequency of the noun phrase is low, then keyphrase is not detected.

An inverted index is a data structure that contains a list of all the distinct words that are present in all the documents, and for each word, a list of the documents in which it appears is also present. The inverted index helps to map the words to its corresponding documents.

An inverted index consists of two parts:

- Vocabulary – set of all unique words
- Occurrences – list containing all the important information related to the word like frequency of word, documents in which the word appears.

Trie is an information retrieval tree data structure used to store multiple strings in which each node has a maximum of 26 children, one for each alphabet. The last node of a string is denoted by *EndOfWord* field in the node. All the descendants a node have a common prefix of the string associated with that node. Figure 1 shows an example of example of a trie data structure.

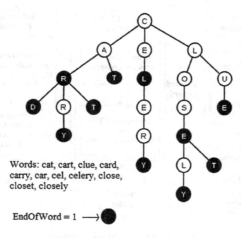

Words: cat, cart, clue, card, carry, car, cel, celery, close, closet, closely

EndOfWord = 1 $\longrightarrow$

**Fig. 1.** Trie data structure

The advantage of trie is that the time complexity of searching is reduced to $O(m)$ where $m$ is the maximum length of the string. Thus, searching and information retrieval is very efficient with this data structure. It can be used to store an inverted index. The index which contains the list of documents in which the keyword is present can be placed alongside *EndOfWord* field. The paper [4] describes the use of a trie as an inverted index for a search engine. The performance of a search engine depends on the document searching and retrieval time and thus, the proposed data structure facilitates the fast searching of relevant information over the web. The system removes the step words, stems the keywords and then creates the index. Thus, the user queries are processed much faster with the proposed system [4].

## 4  Proposed Solution

The proposed solution makes use of natural language processing techniques for the extraction of keywords and a trie data structure for the indexing and storage of the thus obtained keywords. A text file consisting of links to different agricultural documents present on AGRIS (International System for Agricultural Science and Technology) is given as input by the user for the keyword extraction process. Once the keyword extraction is done with the help of AGROVOC, the extracted keywords and its corresponding documents are indexed by using a trie data structure. Figure 2 shows the block diagram of the proposed solution.

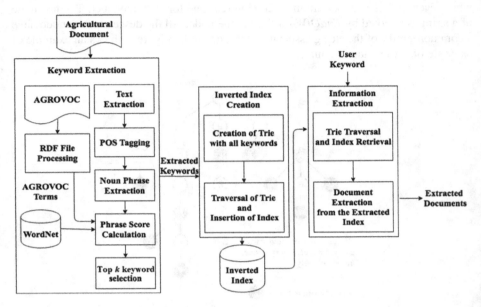

**Fig. 2.** System block diagram

## 4.1   Keyword Extraction

Keyword extraction is an important process to identify the major concepts and terms present in the document. It is easier to select which document to read if the appropriate keywords are extracted. Here, an unsupervised keyword extraction technique is proposed where different natural language processing techniques and statistical techniques have been used for the extraction process. In this module, for each of the agricultural documents, the corresponding keywords are extracted with the help of AGROVOC and WordNet. These extracted keywords help to identify what the document is about and are thus essential for information extraction.

**RDF File Processing.** AGROVOC contains all the agriculture-related terms and concepts. It is an RDF (Resource Description Framework) file and processing is required to extract the AGROVOC terms from it. The RDF file is used to represent information or resources present on the web and is made up of triples each of which consists of 3 components – subject, predicate and object. These three components are also known as resource, property and property value. The AGOVOC RDF file used in this work consists of triples that describe schemas and concepts. To access the

AGOVOC terms, only those triples corresponding to a concept are required. An example of a triple is given below:

```
<rdf:Description
rdf:about="http://aims.fao.org/aos/agrovoc/c_30790">
  <rdf:type
rdf:resource="http://www.w3.org/2004/02/skos/core#Concept
"/>
</rdf:Description>
```

In the triple,
    <rdf:type    rdf:resource="http://www.w3.org/2004/02/skos/core#Concept"/> indicates that the triple is that of a concept. <rdf:Description
rdf:about="http://aims.fao.org/aos/agrovoc/c_30790"> gives the link to the AGROVOC term.

SPARQL, a query processing language for RDF files, is used for processing each triple and selecting those triples which correspond to a concept and extracts the link to the AGROVOC term. Since AGROVOC consists of agricultural terms in multiple languages, the English name of the concept is extracted.

**Text Extraction.** Since the input is a text file consisting of links to agricultural documents taken from AGRIS, the system needs to identify and then extract the text from its corresponding web page. The extracted text is then divided into sentences and these sentences are then further processed by the next block.

**Parts of Speech Tagging.** In a sentence, each word belongs to a category under parts of speech (POS) like noun, adjective, adverb, verb etc. In this module, the POS of each word in the sentence is assigned. This step is important because the same word can have a different POS tag in different contexts. Thus, word sense disambiguation is done here for identifying the meaning of the word when the same word may have different meanings. Each word along with its POS tag is forwarded to the next block for further processing.

**Noun Phrase Extraction.** Based on the POS tagging of each word, the noun phrases are extracted in this block. This is essential to the keyword extraction process because the noun represents most of the important information in a document and thus extracting noun phrases would simplify the keyword extraction process. The various possible patterns for a noun phrase are given below:

- Noun
- Nouns
- Adjectives Noun.

To identify a noun phrase, a regular expression containing the different patterns for a noun phrase is used to extract the noun phrases. The regular expression used by the system is <NN.*|JJ>*<NN.*> where NN represents a noun and JJ represents an adjective. The system uses the regular expression and looks for the matching pattern in the sentence with the POS tags. The identified noun phrases are then forwarded to the next block for further processing.

**Phrase Score Calculation.** In this block, each phrase is given a score based on the frequency of similar words. WordNet, a database consisting of words and its synonyms, helps to calculate the similarity between two words. Higher the value more similar the words are. It's been shown that the co-occurrence distribution of words in different phrases represents how important a phrase is in a document. A phrase $w$ is a key phrase if the probability distribution of the phrase $w$ and the frequent phrases is biased to a subset of frequent words [5]. Thus, in order to calculate the phrase score of a phrase, the frequency of each word in the phrase is calculated.

Given phrase = word [1] word [2] ... word[$n$]

$$Phrase\_Score = \sum\nolimits_{i=1}^{n} (frequency[word[i]]) \tag{1}$$

where $n$ = number of words in the phrase.

**Top $k$ Keyword Selection.** The phrases are sorted according to their phrase scores and from these phrases, the top $k$ phrases are selected as the extracted keywords.

### 4.2  Inverted Index Creation

An inverted index is a data structure that is used to map the words to the documents it is present in. One of the major disadvantages of using a traditional inverted index is that there is a large storage overhead and the cost of insertion, updation, and deletion increases with the increase in the number of keywords. Trie data structure's advantage is that it can store many words in limited space and the search time is also drastically decreased to the length of the keyword. Thus, in the proposed system a trie-based inverted index is used. It combines the advantages of both an inverted index and trie for efficient and fast information retrieval system.

**Trie Node Creation.** For every keyword extracted, first, the trie is traversed to check if the keyword already exists in the inverted index. If the keyword doesn't exist, the trie is traversed until the longest matching prefix is found and then for the remaining part of the word the corresponding nodes are inserted into the trie.

**Insertion of Index.** Once the keyword is found in the trie, the document number to which the keyword corresponds to is appended to the list of documents in the leaf node of the keyword.

### 4.3  Information Extraction

Retrieval of documents related to the user entered keyword is done in this block. The trie based inverted index is traversed to locate the document numbers in which the keyword is present.

**Trie Traversal and Index Retrieval.** The trie based inverted index which stores all the keywords and its corresponding documents is traversed to locate the user entered keyword. The index consisting of the list of related documents is obtained in the last node of the user entered keyword in the trie. That index is then forwarded to the next block.

**Document Extraction from the Extracted Index.** In this module, the documents in the index are extracted and displayed to the user. These documents that are extracted are the documents that are related to the user entered keyword.

# 5  Results

## 5.1  Extraction of AGROVOC Terms

AGROVOC contains more than 35,000 concepts related to agriculture, food, nutrition, fisheries, forestry, environment etc. The AGROVOC RDF file needs to be processed in order to obtain the AGROVOC terms. A total of 35859 terms were extracted by the system. Some of the AGROVOC terms extracted after processing the AGROVOC RDF file are shown in Fig. 3.

```
swapna@swapna-Lenovo-ideapad-500-15ISK:~/Documents/NITT/Project$ python rdf.py
Total no of concepts =  35859
1     Cottocomephorus grewingkii
2     Scomberesox saurus
3     neurotoxicity
4     Edovum puttleri
5     sustainable development
6     Clarias werneri
7
7     random sampling
8     Zygophyllaceae
9     radiation
10
10    international relations
11    Thysaniezia
12    Acremonium coenophialum
13
13    Metcalfa pruinosa
14    Oxymonacanthus halli
15    Parthenium
16    Chagas' disease
17    isoelectric focusing
18    Mullus
19    Scabiosa caucasica
20    Phycis
```

**Fig. 3.** AGROVOC terms

## 5.2  Extracted Keywords

After applying Natural Language processing techniques, the top $k$ phrases with maximum phrase score are selected as the extracted keywords. The keywords extracted for Document 1 is given below:

Keywords for Document 1
['PPR VIRUS', 'PPR', 'INDIA', 'SHEEP', 'TOTAL', 'LYMPH NODE', 'DISEASE', 'BLOOD', 'RUMINANTS', 'RINDERPEST']

## 5.3    Extraction of Documents from User Entered Keyword

The system reads the user entered keyword and then the documents which contain the user entered keyword are displayed. The documents in which the keyword "RICE" exists is displayed in Fig. 4.

```
Enter user keyword rice
The docs are
[9, 11, 66]
```

**Fig. 4.** Extracted documents from user entered keyword

## 5.4    Experiment Evaluation

Table 1 denotes the evaluation metrics for the system.

**Table 1.** Evaluation Metric

| Metric | Value |
|---|---|
| Accuracy | 91.79% |
| True positive rate (Sensitivity) | 82.75% |
| Precision | 63.06% |

For evaluating the performance of the system accuracy, true positive rate and precision metrics have been used. The accuracy calculated was 91.79%, true positive rate was 82.75% and precision was 63.06%.

Since the number of keywords to be predicted is lower than the words that are not supposed to be predicted, the true positive value is always lower than the true negative value. Thus, even if the correctly predicted keywords are less, the effect on the accuracy is less because the true negative is large. This is the reason why the true positive rate is lower than the accuracy rate. The accuracy obtained for each of the documents is depicted in Fig. 5. The overall average accuracy obtained is 91.79%.

Although the overall accuracy was calculated as 91.79%, the range of accuracy is from 56% to 100%. Analysis of the system denotes that one of the reasons of lower accuracy in some documents is that the system is unable to predict chemical compounds like Nitrogen (N), Phosphorous (P) etc. since the keyword is the chemical name, but the chemical symbol is mentioned in the document. Another reason is that the keywords that are not noun phrases are not detected. Keywords that are verbs will not be detected as keywords by the system. This again lowers the accuracy rate. The value of $k$ in the top $k$ keyword selection also affects the accuracy rate. A lower value of

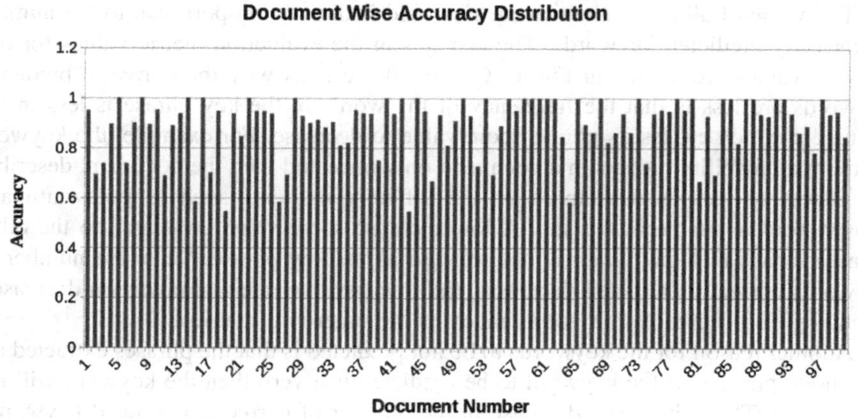

**Document Wise Accuracy Distribution**

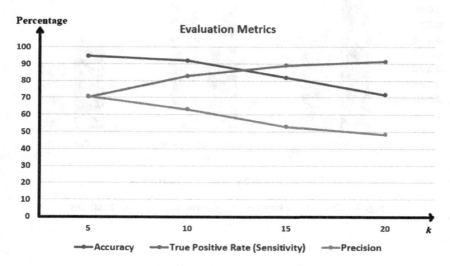

**Fig. 5.** Document accuracy

$k$ may lead to the elimination of the keyword due to a lower phrase score. A higher value of $k$ may lead to the inclusion of all keywords but will also detect words that are not keywords.

True positive rate (sensitivity) denotes the number of correctly predicted keywords out of the actual keywords. Since the number of actual keywords is usually around 5–7 keywords, a small decrease in the number of predicted keywords causes the true positive rate to decrease drastically. Also, with a larger value of $k$ in the top $k$ keyword selection, the precision increases because a larger set of keywords is extracted.

Precision denotes the number of correctly predicted keywords out of all the keywords predicted by the system. Since the value of $k$ taken by the system is usually larger than the actual number of keywords, the precision is low. This implies that a superset of keywords is extracted. As the value of $k$ increases, the precision decreases.

**Fig. 6.** Evaluation Metric Values for different $k$ values

The value of all the evaluation metrics used is directly proportional to the number of correctly predicted keywords. The changes in the evaluation metric values for different $k$ values are shown in Fig. 6. One of the reasons why the correctly predicted keywords are less is that the frequency of the words in the key phrase is less in the document which causes its phrase score value to decrease. For example, the keyword 'nutrient availability' occurs in a document only once and then the document describes the effects of various substances on the different nutrients such as potassium and phosphorous available in the soil. A lower phrase score value compared to the other phrases means that the phrase is not selected as the key phrase. Since the number of keywords correctly predicted decreases, the true positive rate and precision decreases. Accuracy also decreases but not as much as the other two metrics.

Another reason for the keywords to be not predicted is that the phrases extracted are only noun phrases. If the keyword to be predicted is a verb then the keyword will not be detected. Thus, there is a decrease in the number of correctly predicted keywords. Another reason for incorrect keyword prediction is that the document contains the abbreviation or chemical composition, but the keyword is the full form or the chemical name.

### 5.5    Comparison

The system is compared with Summa 1.1.0 which is the implementation of TextRank in python 3. TextRank extracts keywords by creating a graph from the text and ranking the key phrases based on the co-occurrence links between words. The comparison between the two techniques is given in Fig. 7.

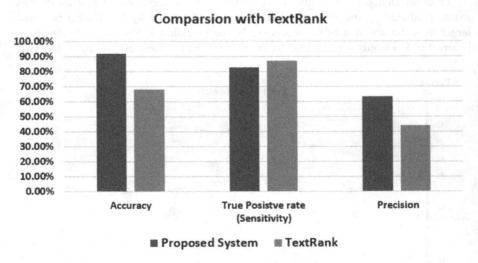

**Fig. 7.** Comparison between the proposed system and TextRank

The number of keywords extracted by TextRank is comparatively higher than that by the proposed system. TextRank extracts a superset of the keywords. One of the reasons behind the large set of keywords is that two similar terms are treated as separate keywords. WordNet does not contain the synonyms for many phrases. The system can handle the cases where the number of words in two similar phrases are the same and each corresponding word is similar. For example, 'health hazard' and 'health risk' are similar phrases but WordNet does not contain these phrases. The system thus compares the corresponding words, 'health' with 'health' and 'hazard' with 'risk'. In such a case, the system can treat both the phrases as similar. TextRank on the other hand treats 'health hazard' and 'health risk' as two different keywords. Because of the larger keyword set, the precision of TextRank is lower compared to that of the proposed system. But because of the same reason, the True positive rate is higher for TextRank in comparison with the proposed system. Since the extracted keywords are more in the case of TextRank, it also implies that there is an increase in the number of false positives and thus a decrease in the number of true negatives. Thus, the overall accuracy for TextRank is lower in comparison with the proposed system.

## 6 Future Scope

The system can be enhanced by detecting keyphrases that are not only nouns but also verbs that are related to the noun phrases. Also, the detection of abbreviations and chemical compositions can be done to improve the efficiency of the system. The proposed system can be used as an agricultural information extraction system for a farmer query answering system. The extraction of relevant documents which contain the appropriate query answers could be achieved with the help of this system.

## 7 Conclusion

This paper examined a system for extraction and indexing of keywords from agricultural documents with the help of AGROVOC. The system uses various natural language processing techniques to extract keywords from the documents and makes use of a trie-based inverted index data structure for indexing. The user entered keyword is searched in the inverted index and the corresponding documents are extracted. It was found that the accuracy of the prediction of the keywords was 91.79%, the true positive rate was 82.75% and the precision was 63.06%. The performance of the system for different values of $k$ was performed and analyzed. The proposed system was also compared with TextRank and the results were analyzed. The system, however, was not able to detect keywords that were verbs and whose chemical composition was mentioned in the document.

# References

1. Balaji, V., et al.: Agrotags – a tagging scheme for agricultural digital objects. In: Sánchez-Alonso, S., Athanasiadis, Ioannis N. (eds.) MTSR 2010. CCIS, vol. 108, pp. 36–45. Springer, Heidelberg (2010). https://doi.org/10.1007/978-3-642-16552-8_4

2. Cutting, D., Pedersen, O.: Optimizations for dynamic inverted index maintenance. In: Proceedings of the 13th Annual International ACM SIGIR Conference on Research and Development in Information Retrieval, pp. 405–411, January 1990

3. Sun, H.F., Hou, W.: Study on the improvement of TFIDF algorithm in data mining. In: Advanced Materials Research, vol. 1042, pp. 106–109 (2014)

4. Zaware, P.S., Todmal, S.R.: Inverted indexing mechanism for search engine. Int. J. Comput. Appl. 123, 15–19 (2015)

5. Matsuo, Y., Ishizuka, M.: Keyword extraction from a single document using word co-occurrence statistical information. Int. J. Artif. Intell. Tools 13, 157–169 (2003)

6. Siddiqi, S., Sharan, A.: Keyword and keyphrase extraction techniques: a literature review. Int. J. Comput. Appl. 109, 18–23 (2015)

7. Joshi, P., Chaudhary, S., Kumar, V.: Information extraction from social network for agro-produce marketing. In International Conference on Communication Systems and Network Technologies, Rajkot, pp. 941–944 (2012)

8. Luthra, S., Arora, D., Mittal, K., Chhabra, A.: A statistical approach of keyword extraction for efficient retrieval. Int. J. Comput. Appl. 168, 31–36 (2017)

9. Terrovitis, M., Passas, S., Vassiliadis, P., Sellis, T.: A combination of trie-trees and inverted files for the indexing of set-valued attributes. In: Proceedings of the 15th ACM International Conference on Information and Knowledge Management, pp. 728–737 (2006)

10. Balcerzak, B., Jaworski, W., Wierzbicki, A.: Application of TextRank algorithm for credibility assessment. In: IEEE/WIC/ACM International Joint Conferences on Web Intelligence (WI) and Intelligent Agent Technologies (IAT), Warsaw, pp. 451–454 (2014)

11. Yen, S.-F., Chen, J.-J., Tsai, Y.-H.: Efficient cloud image retrieval system using weighted-inverted index and database filtering algorithms. J. Electron. Sci. Technol. 15(2), 161–168 (2017)

12. Rezaei, M., Gali, N., Fränti, P.: ClRank.: a method for keyword extraction from web pages using clustering and distribution of nouns. In: IEEE/WIC/ACM International Conference on Web Intelligence and Intelligent Agent Technology (WI-IAT), Singapore, pp. 79–84 (2015)

13. AIMS AGROVOC. http://aims.fao.org/vest-registry/vocabularies/agrovoc. Accessed 30 Jan 2019

# Performance Analysis of Flappy Bird Playing Agent Using Neural Network and Genetic Algorithm

Yash Mishra$^{(\boxtimes)}$ ⓘ, Vijay Kumawat ⓘ, and K. Selvakumar

Vellore Institute of Technology, Vellore, TN 632014, India
yashmishra12@hotmail.com, vijaykumawat256@gmail.com

**Abstract.** The aim of this paper is to develop and study an artificial intelligence based game-playing agent using genetic algorithm and neural networks. We first create an agent which learns how to optimally play the famous "Flappy Bird" game by safely dodging all the barriers and flapping its way through them and then study the effect of changing various parameters like number of neurons on the hidden layer, gravity, speed, gap between trees has on the learning process. The gameplay was divided into two level of difficulty to facilitate study on the learning process. Phaser Framework was used to facilitate HTML5 programming for introducing real-life factors like gravity, collision and Synaptic Neural Network library was used to implement neural network so as to avoid creating a neural network from scratch. Machine Learning Algorithm which we have adopted in this project is based on the concept of Neuro-Evolution and this form of machine learning uses algorithms which can evolve and mature over time such as a genetic algorithm to train artificial neural networks.

**Keywords:** Artificial Intelligence · Neural network · Genetic algorithm · AI · Game-playing agent · Flappy bird

## 1 Introduction

The main aim of this work is to develop a Genetic Algorithm based approach for constructing an Artificial Intelligence based agent which can play the Flappy Bird game and improve upon itself, thereby, ultimately beating any person.

Flappy Bird is an ever-engaging game developed by Vietnamese video game artist and programmer Dong Nguyen, under his game development company dotGears [1]. The gameplay action in Flappy Bird can be viewed from a side-view camera angle and the on-screen bird can flap to rise against the gravity which pulls it towards the ground. The agent is supposed to learn when to flap so as to avoid collision. Passing through each obstacle-gap leads to the increment in score while collision leads to instant death.

There are various factors which can affect the difficulty of the gameplay and ultimately make it more challenging for the agent to learn and decide when to flap and when not to flap. Our work divided the gameplay into two levels: Easy and Difficult. In the Easy Level, the velocity of the obstacle coming towards the agent was kept at 200, gravity at 2000 and gap between the upper obstacle and lower obstacle, that is, the gap

© Springer Nature Singapore Pte Ltd. 2019
A. B. Gani et al. (Eds.): ICICCT 2019, CCIS 1025, pp. 253–265, 2019.
https://doi.org/10.1007/978-981-15-1384-8_21

for the agent to flap through as 130. While in the Difficult Level, the gameplay was made harder by tweaking the above parameters with the obstacle velocity increased to twice the original value, gravity increased to 2500 so as to force the agent to flap more and the gap was reduced by 10 units.

In the Easy Level, we observed that the agent apparently became immortal after a certain number of generations which forced us to make the gameplay harder and study the results. Though it is worth mentioning that the idea of immortality of the agent is an idea assumed by the authors as they observed agents surviving for as long as 15 min but the notion of their death at the 16th min cannot be denied. For simplicity, a score of 150+ made the agent fall under the category of immortal and optimal agent.

Achieving a score of more than 150 was observed to be very difficult and almost far-fetched in the Difficult level and therefore, the study was conducted up to 30 generations.

## 2 Theory

### 2.1 Genetic Algorithm

Genetic Algorithms are adaptive heuristic search algorithms which support the organic process ideas of selection and genetic science. Intrinsically they represent an intelligent exploitation of a random search space accustomed to solve improvement issues and optimization problems. Though irregular, these are by no means random, instead they exploit historical knowledge and information to direct the search into regions of higher and improved performance within the search space. The fundamental techniques are designed to mimic and simulate processes which lead to evolution in nature, they particularly follow the principles first laid down by Charles Darwin. Since in nature, competition among living beings for scanty resources leads to the fittest being dominating over the weaker ones.

Some important terms in Genetic Algorithm and our and their role in our work are discussed below:

**Search Space:** Aset of individuals is maintained within a search space. Individuals are like chromosomes and the entries in that vector represent genes. The fitness function defines the competing ability of an individual. The main idea is to create a new individual which is carry the best characteristics of the parents.

**Selection:** The most fit individual is chosen and passed on to the next generation. In our implementation, the top 4 fit values pass on to the next generation without any hinderance.

**Cross Over:** The portion of the string is exchanged from the individuals which are selected due to highest level of fitness from the rest. The string is selected randomly for this purpose. The idea behind this is to introduce randomness while maintain the characteristics of the fittest individuals.

**Mutation:** Our project has set the mutation to 0.2 to introduce randomness. Mutation is done on a very limited scale to flip the values of bits. This ensures that the generation is always changing and not stuck at some undesirable point.

## 3 Implementation

### 3.1 Artificial Intelligence Neural Network

An artificial neural network comes under the subset of machine learning algorithm. It is highly influenced by the structure and functions of the biological neural networks found in a living being and nature. These networks are created out of the many neurons that send signals to every alternative. Therefore, to make a synthetic brain we would have to replicate the working functions of a neuron and join them together so that they can form a neural network.

The most conventional artificial neural network consists of an input layer, one or more hidden layers and an output layer with each layer constituted by several neurons. Hidden neurons are connected between input and output neurons. In our work, each unit (bird) has its own neural network which is used as its Artificial Intelligence brain for playing the game. It consists of 3 layers as follows:

- **Input Layer.** An input layer two neurons shows the agent's perspective and represents what it sees:
  Horizontal Distance to the closest gap
  Height difference to the closest gap
- **Hidden Layer.** Number of neurons in the hidden layer were varied from 1, 5, 10 and 20 and the performance of each was studied.
- **Output Layer.** An output later with one neuron which provides an action of flap if output >0.5 else do nothing.

### 3.2 Genetic Algorithm

Genetic algorithm is an optimization technique which is heavily inspired by the process of evolution, natural selection and genetics. It uses the same processes discussed above which is used by evolution to evolve the initial random population.

The main steps of our genetic algorithm implementation:

1. Random neural network helps in creating the initial population of 10 birds.
2. Each unit has its own neural network which helps them to play autonomously.
3. Fitness function is calculated for each bird separately.
4. When all die, evaluate the current population to the next.
5. Go back to the second step

### 3.3 Fitness Function

Fitness function is the metric to measure desirability of an entity or an actor. Fitness function helps us select the birds which are most likely to give us the highest score and help us create a new population with their characteristics slightly changed.

We reward a bird by adding its travelled distance and penalize it by subtracting its current distance to the closest gap. Subtracting distance to the closest gap from the total distance travelled also eliminates the significance of the barrier gap, i.e. the distance between each pair of obstacles (Fig. 1).

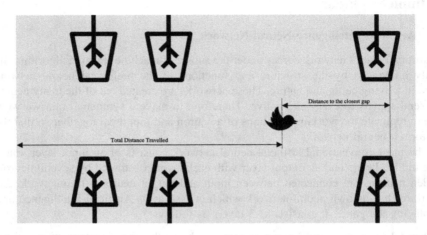

**Fig. 1.** Fitness function is total distance travelled minus distance to the closest gap

### 3.4   Replacement Strategy

In addition to the genetic algorithm, the steps mentioned below is used to create a new population. We had set the random mutation level to 0.2. Basically, the most desirable units survive and their children replace the worst units in the following way [2]:

1. Sort the units of the current population by their fitness function value in descending order.
2. Select the best four units and pass them directly to the next population.
3. Create one offspring as a crossover product of two best winners.
4. Create three offspring as a crossover products of two random winners.
5. Create two offspring as a replica of two random winners
6. Apply random mutation on each offspring to add some variation.

## 4   Results

### 4.1   Easy Level

As discussed earlier, we found the agents slowly becoming fit enough to score more than 600 points and observing each generation started taking a considerable amount of time. For the ease of study, we have made an assumption that any generation with a score of 150+ is immortal and has achieved an optimum fitness level. Once the score of 150 is achieved, we have stopped the game and carried the same birds for future generation so that graphical comparison with other variants of hidden layer can be done (Figs. 2 and 3).

In Easy Level, the velocity of trees was kept to 200, gravity was set to 2000 and gap was set to 150. Effect of the number of neurons in the hidden layer were observed (Tables 1, 2, 3 and 4).

**Table 1.** One Neuron

| One (1) Neuron in Hidden Layer | | |
|---|---|---|
| Generation | Fitness | Score |
| 1 | −177.02 | 0 |
| 2 | 421.37 | 0 |
| 3 | 510.12 | 0 |
| 4 | 1998.34 | 5 |
| 5 | 2927.18 | 8 |
| 6 | 10176.52 | 32 |
| 7 | 45586.40 | 150+ |

**Table 2.** Five Neuron. Two birds scored more than 150.

| Five (5) Neuron in Hidden Layer | | |
|---|---|---|
| Generation | Fitness | Score |
| 1 | 176.86 | 0 |
| 2 | 512.77 | 0 |
| 3 | 556.16 | 0 |
| 4 | 45890.12 | 150+ |

**Table 3.** Ten Neuron.

| Ten (10) Neuron in Hidden Layer | | |
|---|---|---|
| Generation | Fitness | Score |
| 1 | −215.14 | 0 |
| 2 | 335.24 | 0 |
| 3 | 320.21 | 0 |
| 4 | 325.95 | 0 |
| 5 | 324.74 | 0 |
| 6 | 368.33 | 0 |
| 7 | 478.38 | 0 |
| 8 | 426.55 | 0 |
| 9 | 10406.69 | 33 |
| 10 | 17699.86 | 57 |
| 11 | 45680.13 | 150+ |

**Table 4.** Twenty Neuron

| Twenty (20) Neuron in Hidden Layer | | |
| --- | --- | --- |
| Generation | Fitness | Score |
| 1 | −280.00 | 0 |
| 2 | −453.12 | 0 |
| 3 | −315.29 | 0 |
| 4 | 547.67 | 0 |
| 5 | 3655.72 | 12 |
| 6 | 5751.12 | 19 |
| 7 | 14506.78 | 48 |
| 8 | 45691.82 | 150+ |

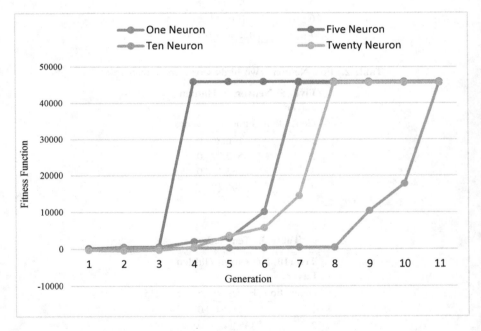

**Fig. 2.** Easy Level: Fitness Function vs Generation

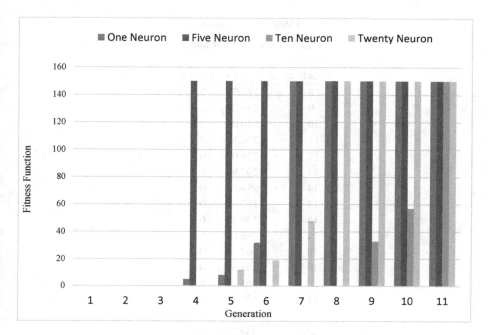

**Fig. 3.** Easy Level: Score vs Generation

## 4.2 Difficult Level

The gameplay was tweaked to make it harder to score points by decreasing gap between obstacles to 120, increasing tree velocity to 400, twice the speed at easy level and increasing gravity to 2500 so that the agent has to flap more. We have considered up to 30 generations for all the number of neurons in the hidden layer as the empirical evidence showed no improvement in score even after considering up to 60 generations (Tables 5, 6, 7 and 8) and (Figs. 4 and 5).

**Table 5.** One Neuron

| One (1) Neuron in Hidden Layer | | |
|---|---|---|
| Generation | Fitness | Score |
| 1 | 158.63 | 0 |
| 2 | 219.66 | 0 |
| 3 | 207.58 | 0 |
| 4 | 293.47 | 0 |
| 5 | 117.57 | 0 |
| 6 | 140.54 | 0 |
| 7 | 290.0 | 0 |

*(continued)*

**Table 5.** (*continued*)

**One (1) Neuron in Hidden Layer**

| Generation | Fitness | Score |
|---|---|---|
| 8 | 170.95 | 0 |
| 9 | 256.27 | 0 |
| 10 | 140.0 | 0 |
| 11 | 206.65 | 0 |
| 12 | 258.07 | 0 |
| 13 | 202.92 | 0 |
| 14 | 95.45 | 0 |
| 15 | 157.68 | 0 |
| 16 | 259.87 | 0 |
| 17 | 288.04 | 0 |
| 18 | 562.28 | 0 |
| 19 | 585.81 | 0 |
| 20 | 563.07 | 0 |
| 21 | 555.40 | 0 |
| 22 | 556.35 | 0 |
| 23 | 3432.81 | 10 |
| 24 | 17887.81 | 57 |
| 25 | 4036.65 | 12 |
| 26 | 2208.92 | 6 |
| 27 | 2519.74 | 7 |
| 28 | 13230.75 | 42 |
| 29 | 21820.14 | 70 |
| 30 | 26118.90 | 84 |

**Table 6.** Five Neuron

**Five (5) Hidden Layer**

| Generation | Fitness | Score |
|---|---|---|
| 1 | 378.90 | 0 |
| 2 | 397.25 | 0 |
| 3 | 453.63 | 0 |
| 4 | 398.25 | 0 |
| 5 | 540.18 | 0 |
| 6 | 515.95 | 0 |
| 7 | 551.02 | 0 |
| 8 | 990.43 | 2 |
| 9 | 8013.30 | 25 |

(*continued*)

**Table 6.** (*continued*)

### Five (5) Hidden Layer

| Generation | Fitness | Score |
|---|---|---|
| 10 | 14167.36 | 45 |
| 11 | 27619.94 | 89 |
| 12 | 26435.12 | 85 |
| 13 | 38383.66 | 124 |
| 14 | 23044.10 | 74 |
| 15 | 21832.11 | 70 |
| 16 | 20903.82 | 67 |
| 17 | 29448.02 | 95 |
| 18 | 8960.55 | 28 |
| 19 | 42077.80 | 136 |
| 20 | 6466.81 | 20 |
| 21 | 3123.06 | 9 |
| 22 | 7716.37 | 24 |
| 23 | 8946.76 | 28 |
| 24 | 26723.05 | 28 |
| 25 | 7730.05 | 24 |
| 26 | 39620.36 | 128 |
| 27 | 5280.06 | 16 |
| 28 | 15092.64 | 48 |
| 29 | 9873.50 | 31 |
| 30 | 17835.71 | 47 |

**Table 7.** Ten Neuron

### Ten (10) Hidden Layer

| Generation | Fitness | Score |
|---|---|---|
| 1 | 367.08 | 0 |
| 2 | 516.09 | 0 |
| 3 | 397.59 | 0 |
| 4 | 6770.94 | 21 |
| 5 | 972.40 | 2 |
| 6 | 1285.66 | 3 |
| 7 | 19064.09 | 61 |
| 8 | 13845.40 | 44 |
| 9 | 2513.14 | 7 |
| 10 | 16060.63 | 51 |
| 11 | 3123.54 | 9 |
| 12 | 33185.41 | 107 |

(*continued*)

**Table 7.**  (*continued*)

**Ten (10) Hidden Layer**

| Generation | Fitness | Score |
|---|---|---|
| 13 | 18446.62 | 59 |
| 14 | 2518.85 | 7 |
| 15 | 15079.15 | 48 |
| 16 | 35022.63 | 113 |
| 17 | 988.86 | 2 |
| 18 | 1585.55 | 4 |
| 19 | 21519.06 | 69 |
| 20 | 31339.33 | 101 |
| 21 | 28586.50 | 92 |
| 22 | 20607.40 | 66 |
| 23 | 15088.47 | 48 |
| 24 | 6495.66 | 20 |
| 25 | 11408.49 | 36 |
| 26 | 31341.45 | 101 |
| 27 | 7716.02 | 24 |
| 28 | 3433.39 | 10 |
| 29 | 16313.71 | 52 |
| 30 | 18446.62 | 59 |

**Table 8.**  Twenty Neuron

**Twenty (20) Hidden Layer**

| Generation | Fitness | Score |
|---|---|---|
| 1 | 473.86 | 0 |
| 2 | 553.23 | 0 |
| 3 | 539.76 | 0 |
| 4 | 553.35 | 0 |
| 5 | 1222.22 | 3 |
| 6 | 611.32 | 1 |
| 7 | 1198.05 | 3 |
| 8 | 1506.79 | 4 |
| 9 | 1209.45 | 3 |
| 10 | 641.51 | 1 |
| 11 | 2438.69 | 7 |
| 12 | 1660.71 | 4 |
| 13 | 5276.07 | 16 |
| 14 | 12318.30 | 39 |
| 15 | 9874.10 | 31 |
| 16 | 10792.50 | 34 |

(*continued*)

**Table 8.** (*continued*)

| Twenty (20) Hidden Layer | | |
|---|---|---|
| Generation | Fitness | Score |
| 17 | 8654.58 | 27 |
| 18 | 2859.48 | 8 |
| 19 | 4358.95 | 13 |
| 20 | 3745.56 | 11 |
| 21 | 10796.09 | 34 |
| 22 | 13887.94 | 44 |
| 23 | 3768.68 | 11 |
| 24 | 5620.93 | 17 |
| 25 | 6813.21 | 21 |
| 26 | 4086.47 | 12 |
| 27 | 8675.23 | 27 |
| 28 | 16634.58 | 53 |
| 29 | 15086.20 | 48 |
| 30 | 12023.61 | 38 |

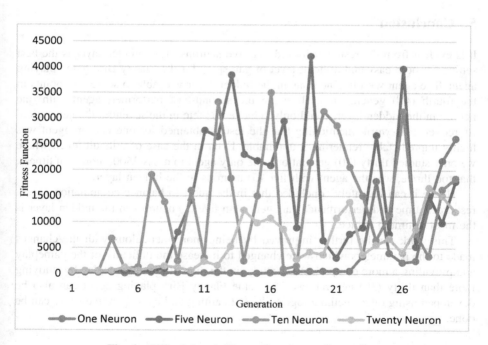

**Fig. 4.** Difficult Level: Fitness Function vs Generation

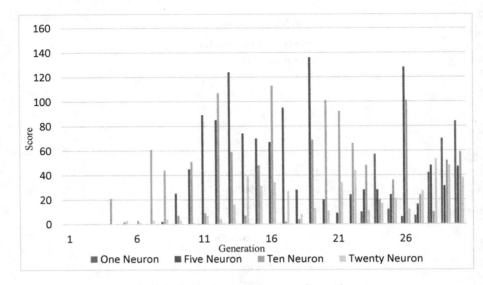

**Fig. 5.** Difficult Level: Score vs Generation

## 5   Conclusion

It is evident from the results achieved that five neurons in the hidden layer is the best setting for both easy and difficult level of gameplay for the Flappy Bird game playing agent. The agent with five neurons in the hidden layer was able to score 150 points in the fourth (4th) generation itself while the second-best performer, agent with one neuron in the hidden layer, could only achieve this fete in the seventh (7th) generation. Moreover, it is worth mentioning that the results obtained for one neuron agent was highly unpredictable rendering it unreliable. Even in the case of difficult level, where we only studied thirty (30) generations, the only agent to cross 35000 points of fitness, that too thrice, was the agent with five (5) neurons in the hidden layer.

Hence it can be safely concluded that in the quest to balance computation speed, resources, time and performance, an agent with five (5) neurons in the hidden layer is the most optimum choice.

This work can be further improved by being more meticulous with the changes made to the parameters which were changed to increase the difficulty of the gameplay and providing a more comprehensive study on the more difficult gameplay by studying more than thirty (30) generations. The same Flappy Bird playing agent can also be developed using other methodologies like Q-Learning and a comparative study can be done.

## References

1. Rhiannon, W.: What is Flappy Bird? The game taking the App Store by storm. The Daily Telegraph. January 30 (2014)

2. ASK for Game Task. https://www.askforgametask.com/tutorial/machine-learning-algorithm-flappy-bird/. Accessed 10 Feb 2019
3. Kumar, A., Garg, S., Garg, S., Chaudhry, R.: Flappy Bird Game AI using Neural Networks, 1–8 (2016)
4. Chen, K.: Deep Reinforcement Learning for Flappy Bird. Cs229.Stanford. Edu, 6 (2015). https://doi.org/10.1016/0166-445X(88)90132-4
5. Sahin, A., Atici, E., Kumbasar, T.: Type-2 fuzzified flappy bird control system. In: 2016 IEEE International Conference on Fuzzy Systems, FUZZ-IEEE 2016, pp. 1578–1583 (2016). https://doi.org/10.1109/FUZZ-IEEE.2016.7737878
6. Lucas, S.M., Kendall, G.: Evolutionary computation and games. IEEE Comput. Intell. Mag. 1 (1), 10–18 (2006). https://doi.org/10.1109/MCI.2006.1597057
7. Sipper, M., Giacobini, M.: Introduction to special section on evolutionary computation in games. Genet. Program Evolvable Mach. 9(4), 279–280 (2008). https://doi.org/10.1007/s10710-008-9066-x
8. Nie, Y.: How does a Flappy Bird Learn to Fly Itself How does a Flappy Bird Learn to Fly Itself, August (2017)

# Open Domain Conversational Chatbot

Vibhashree Deshmukh$^{(\boxtimes)}$ (iD) and S. Jaya Nirmala (iD)

National Institute of Technology, Trichy, India
vibhadeshmukh09@gmail.com, sjaya@nitt.edu

**Abstract.** The main medium for communication between human beings is natural language in oral or textual form. Artificial Intelligent systems are supposed to be of use to humans such that they help in solving queries related to health, education, social and various other domains. Chatbots are faster and always available compared to traditional techniques of communication, ensuring quick and easy answers related to problems in different domains. The proposed solution implements a conversational chatbot using the *seq2seq* encoder-decoder model. The model is trained on a Twitter corpus containing conversational exchanges. The recurrent encoder-decoder performs the encoding of text conversations. RNN deals with sequential data which ideally captures the semantic summary of the input sequence and then based on context, the decoder generates output one word at a time step.

**Keywords:** Chatbot · Open domain · Conversational · Generative · Deep learning · Twitter based

## 1 Introduction

For humans to communicate with machines, machines should understand natural language. The need of the hour is to make conversational agents which have the capability of solving users queries in various domains. There is a need for a transition from traditional interfaces to conversational interfaces. Easy accessibility, handling capacity, flexibility, work automation, and personal assistant are some of the attributes that encourage the idea of developing chatbots.

One of the toughest challenges in the trending field of AI is modeling conversations. The machine should be able to understand natural language. Chatbot should be able to provide human-like responses. There is a tremendous amount of conversational data available on the Internet with the development of society. Open domain chatbot makes it possible for the users to ask queries from any domain irrespective of any constraints. The vast availability of conversational data enables the generation of proper responses with regard to user's messages. An open domain setting allows the user to take a conversation anywhere. A well-defined goal or intention is not mandatory. The proposed solution implements an open domain conversational chatbot based on Twitter dataset that generates responses to queries fired by users. Retrieval based chatbots use specification languages like AIML to manually define the predefined interactions and generate predefined responses. On the other hand, the generative

© Springer Nature Singapore Pte Ltd. 2019
A. B. Gani et al. (Eds.): ICICCT 2019, CCIS 1025, pp. 266–278, 2019.
https://doi.org/10.1007/978-981-15-1384-8_22

models have the ability to generate new responses that previously did not exist in the knowledge base of the chatbot.

Conversations on social media like Twitter and Reddit are typically open domain. They can be carried in all kinds of direction. There are infinite numbers of topics available to create reasonable responses. The conversation systems try to imitate human dialog in all respects. They are required to have a certain amount of world knowledge and commonsense reasoning capabilities so that they are capable of conducting conversations about any topic.

## 1.1 Chatbot Approaches

A Chatbot presents the capability of impersonating a human being by generating meaningful responses while having a conversation. Eliza is the first ever chatbot program written. Handwritten templates are used to generate the replies that resemble the user's input query. This model captures the input and tries to match the keywords with predefined responses. Jabberwacky was the first to learn meaningful human conversations. It learns from the real-time human conversation and not being driven by some static knowledge base. ALICE is a chatterbot based on natural language processing that converses with a human by applying pattern matching rules on the input. It was first implemented in a language based on set theory and mathematical logic. It is purely based on pattern matching techniques.

## 1.2 Chatbot Strategies

Chatbot assists in the interaction of humans with a computer. Chatbots need a good knowledge base in order to generate appropriate responses. Knowledge base creation can be done by some predefined handwritten rules. A chatbot is divided into three parts namely responder, classifier and graph master.

- Responder: Its main role is to provide the interfacing between the user and the classifier.
- Classifier: It acts as an intermediate agent between the responder and graph master. It performs normalization of input and segments input into logical components. It transfers normalized sentence into graph master and performs handling of database instructions.
- Graph master: It stores the pattern matching algorithm.

## 1.3 Chatbot Fundamental Design Techniques

- Parsing: It includes the analysis of input and manipulation by NLP functions.
- Pattern Matching: It is commonly used in question-answering on matching types.
- AIML: It consists of AIML objects consisting of topics and categories. The top-level element is a topic and it has several categories.
- Chat Script: It generates a default answer when no pattern matches in the knowledge base.
- SQL and Relational database: It basically helps a chatbot in remembering previous conversations.

### 1.4  Artificial Intelligent Markup Language

Artificial Intelligent Markup Language, a derivative of XML represents the knowledge base put into a chatbot. AIML objects consist of topics and categories. It basically does conversational modeling. The category is the unit of conversation. The pattern is for the identification of the input from the user. The template is the response for a particular input.

### 1.5  Taxonomy of Chatbot Models

Several experiments have been done in AI in creating intelligent chatbots that can assist humans in interacting with computers in natural language. The two approaches of modelling chatbot are given below:

- Retrieval Based Model: These models have a repository of predefined responses and appropriate responses are picked on the basis of input and context using some sort of heuristics. The heuristics can be simple or complex from expression matching using rules to complicated machine learning classifiers.
- Generative Model: These models do not work on a pre-defined set of responses, rather generate new responses. They are a sort of machine translation technique in which instead of translating languages, translation occurs from input to output sequence (response).
- Vector space model: This model performs data preprocessing and creation of term vectors for the user query using term frequency-inverse document frequency. The dataset is then searched for answer for the query using cosine similarity.
- Machine learning model: XGBoost, neural networks, random forest and SVM are the machine learning models used. The dataset question is retrieved based on the confidence level and the answer to the corresponding matching query is displayed to the user.

### 1.6  Chatbot Domains

- Open Domain Chatbot: In open domain there is no such predefined goal of conversation rather it can go anywhere like the conversational exchanges on social media sites. There is a huge range of topics on which conversation can take place. Hence, such chatbots are able to solve user queries in a lot more dimensions rather constraining to a particular field. For the same reason, such chatbots need to have a lot of knowledge base to create apt responses for the users.
- Closed Domain Chatbot: These chatbots have limited input and output. The system has a specific goal rather than multiple. The user can take the conversation in any direction but then the chatbot may not generate proper responses because it is not trained to handle such cases.

## 2  Existing Work

Research on a large scale has been actively done with the evolution of chatbots. Various chatbots have been developed for the purpose of seeking information, FAQ answering in different domains like education, health, entertainment etc. With the development of technology, there is a transformation of traditional interfaces into conversational interfaces. There is a huge range of chatbots identified based on the learning capacity and the manner of interaction with the users. Some of the related works are listed below.

Joseph Weizenbaum et al. proposed ELIZA in which the user input is identified for keywords and then the sentences are transformed according to the defined rules. After the context is discovered appropriate transformation is applied and the response is generated [1].

Wallace et al. proposed A.L.I.C.E (Artificial Linguistic Internet Computer Entity) which uses a dialogue corpora to train the chatbot. It partially describes the behavior of the computer program into parsed and unparsed data. In the normalization process, pattern matching is performed and the input path is generated [2].

Lokman et al. proposed virtual dietitian (ViDi), a chatbot for Diabetic patients. It has Vpathas a parameter which remembers the conversation path and the patient is diagnosed by analysis of the Vpath parameter. It basically remembers the conversation flow and if *ViDi* did not know any keyword then it will further ask several questions [3].

Yan et al. proposed *DocChat*. From a document set, the appropriate response is retrieved based on the given utterance. The retrieved responses are ranked on basis of certain algorithm and response is triggered based on whether it is confident enough. Highest probability sentences are shortlisted from the set of documents to contain the query's response by doing indexing such that the next and previous sentence of the selected sentence should be relevant to the utterance. Response ranking is done using word level, phrase level & sentence level feature extraction and the suitable response is triggered [4].

Harris et al. proposed the Bag of Words model which ignores the sentence structure order and syntax and keeps the count of the occurrence of various words. After the removal of stop words, the morphological variants are passed through the process of lemmatization and stored as instances of the basic lemma. Then in the second phase, the bot's knowledge base is tested against the resulting words and the documents with the most similar keywords to the query are listed [5].

Deerwester et al. proposed Latent Semantic Analysis which creates a matrix where each column represents a document and each row represents the words in the vocabulary whereas the cell at the intersection of row and column represents the frequency of the occurrence of a word in the sentence. The distance between the vector representing the utterance and the document is calculated using the single value decomposition [6].

Breiman et al. used Decision Tree the supervised learning method. It does not rely on data normalization rather requires little data preparation. It depends on decision thresholds to operate directly on both numerical data. Complex trees are generated as the tree grows [7].

Amit and Geman et al. used Random Forests in language modeling, it does fitting on various subsamples of the dataset for a number of decision trees. RF improves predictive accuracy using averaging and also combats the overfitting of individual decision trees [8].

Porter used the Naïve Bayes supervised learning approach. Gaussian assumes the distribution as continuous and multinomial used in text classification. The input data represented as word vector counts [9].

# 3  Proposed Solution

The proposed solution implements a generative conversational chatbot using deep learning. The interface supports processing of user queries and generation of a concise answer using an encoder-decoder structure based *seq2seq* model.

## 3.1  Preprocessing

Preprocessing involves the tokenization of sentences into words, the removal of infrequent words, the addition of start and end token, the filtering of very long and very short sentences and the creation of a dictionary which contains the mapping from word to index and index to the word. Every word is transferred to its numerical representation by word embedding. This is done to basically capture semantic similarity between words. The frequency distribution of various words is evaluated and zero padding performed.

## 3.2  Splitting the Dataset

The corpus is split into training, testing and validation data in the ratio of 70%, 15%, 15% for encoder and decoder network in the *seq2seq* model. The word to index embedding is fed into the encoder for training and the pad sequences are removed. The end-id and start-id are added to the target sequence and decode sequence and place-holders are assigned for training the model (Fig. 1).

## 3.3  Seq2Seq Model

The Seq2Seq Model uses Encoder-Decoder framework. It consists of two RNNs (Recurrent Neural Networks). The input is fed to the encoder as a sequence and processed one word at a time. Encoder does the conversion of words into feature vector of fixed size. Each hidden state of the encoder is influenced by the previous hidden state and the final hidden state is called the thought vector. From the context, decoder generates the final output sequence one word at a time. Output sequence is then expressed as a conditional probability of

$$p(y_1,\ldots\ldots,y_{N'}|x_1,\ldots\ldots\ldots,x_N) \tag{1}$$

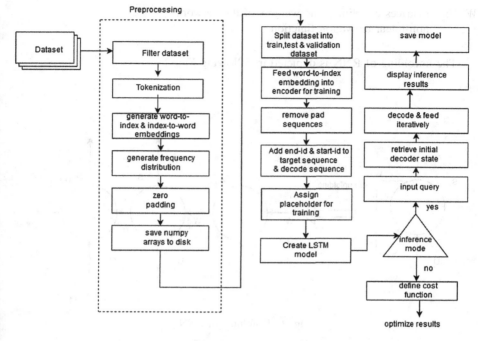

**Fig. 1.** Block Diagram

where input sequence is denoted by $x_1,\ldots\ldots,x_n$ and output sequence by $y_1,\ldots\ldots,y_n$. The length of sequences $N$ and $N'$ can be different since different RNNs are used.

### 3.4 Encoder-Decoder Framework

Encode & decode embedding input layer is created by specifying the encoding sequence, decoding sequence & vocabulary size. When the neural network is applied to natural language processing transformation of each word to its numerical representation is needed. This task is accomplished using word embeddings, where each word is a fixed size vector of real numbers. The advantage of word embeddings lies in the fact that instead of handling huge vectors of vocabulary size, the representation is done in lower dimensions. The training of word embeddings are done on a large amount of natural data and vector representations are build to capture the semantic similarity between the words. Deep learning models do not rely on handwritten rules rather they rely on matrix multiplication and nonlinear functions.

A recurrent neural network takes as input a sequence $x_1,\ldots\ldots,x_n$ and produces a sequence $h = (h_1,\ldots\ldots,h_n)$ by using recurrence. In unrolling of the network, at every step input $x_i$ and previously hidden state $h_{i-1}$ are taken and a hidden state $h_i$ is generated

$$h_i = f(Wh_{i-1} + Ux_i) \tag{2}$$

*W, U*:     matrices containing the weights of the network
*f*:         non-linear activation function

The unfolding of RNN is depicted in following Fig. 2.

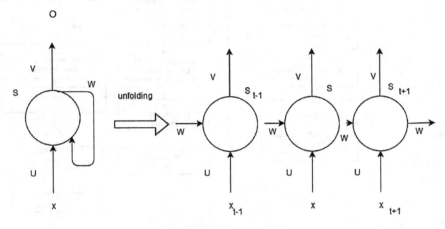

**Fig. 2.** Unfolding of RNN

LSTM networks use a gating mechanism to manage information flow. In RNN, the prediction of the next word in the sequence is done by generating the probability distribution from the hidden state in the last state. The predicted word is then fed into the softmax activation function. The probability distribution is represented in the following Eq. (3):

$$p\left(x_{i,j}|x_{i-1},\ldots\ldots.x_1\right) = \sum_{j=1}^{K} \exp\left(v_j h_i\right) \tag{3}$$

$v_j$:                        rows in the V weight matrix of the softmax function.
$x_{i-1},\ldots\ldots,x_1$:     input sequence
$h_i$:                        hidden state at step i

The repeating module in the chain structure of LSTM has a different structure. There are four neural network layers instead of one interacting in a very special manner depicted in following Fig. 3.

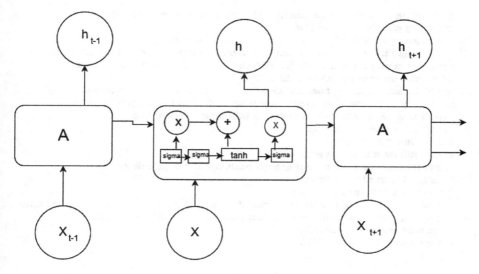

**Fig. 3.** LSTM cell structure

Training of the network is done by generalized backpropagation algorithm. Hence, in every time step the error is backpropagated. The cross entropy loss function calculates the difference in the predictions made by the model and the true labels.

## 4   Results

Data is collected by constructing Tweet Scrapper where Twitter's public stream tweets are mined. We used Twitter dataset with 377265 conversational exchanges. The dataset is divided into training, testing and validation dataset for encoder and decoder in the ratio of 70%, 15%, 15%. Some sample tweets from twitter dataset for training the model are shown in Fig. 4.

The model is trained for 50 epochs and loss and perplexity scores are computed for every epoch. During training, the model learns new capabilities from existing data. The training of the model is depicted in Fig. 5.

The generative model generates new responses based on the query. Hence, it can handle previously unseen queries. During inference, the same capability is applied to new data. Fig. 6 depicts the testing of the model.

79  if you haven't yet then do not update your phones with ios 10 update. it drains out battery like anything
80  but most of the guys are enjoying the ios 10 vg
81  - spent many sunday's there when i lived in cali!! ricky always too great care of me!!
82  love hearing that!! ricky still walks the floor during events to make sure everyone is having a good time. great host!
83  trump attracts the best people. amazing people. you wouldn't believe the caliber of people he brings to the party.
84  wow! you really bring out the best don't you girlfriend! ŏϊↀ
85  rep marsha blackburn you lying sack of camel dung! answer the questions asked instead of ducking and redirecting!
86  ps...at your age, cute act does not work any more!...maybe in high school, but you are like 80 years old now!
87  you people are tedious. he was told to get on the ground. he walked away and reached into his car.
88  i bet they don't knowxthe word feral, either
89  refer to my latest tweet for updates.
90  whatever i still want an invite to the wedding
91  woman who lied about pneumonia eager to release her medical records, right after she purges the ones about her yoga.
92  looking for independent medical exam for both!
93  when i talk to you, my day gets a whole lot better
94  i seriously, again wish you would contact me anyway possible. i'm not dumb. just in love. any connection to my life feels vryfine
95  sorry to bother, i need feature. hopefully you can help me by provide premium course video or tutorial, thx
96  saw your tweets, been preoccupied with an event
97  who's live tweeting on the west coast? ŏϊↀŏϊↀ
98  right here.

**Fig. 4.**  Sample Tweets for training the model

```
Epoch [2/5]: loss 3.6922 perplexity12.9261
Query > happy birthday have a nice day
Seconds:  0.03
 > thanks so much
Seconds:  0.04
 - thanks a lot much appreciate it
Seconds:  0.04
 > thanks a lot much appreciate it
Seconds:  0.04
 > thanks so much
Seconds:  0.04
 > thank you
Query > donald trump won last nights presidential debate according to snap online polls
Seconds:  0.06
 > end_id security is a lie
Seconds:  0.04
 >
Seconds:  0.04
 > end_id kaine
Seconds:  0.05
```

**Fig. 5.**  Training of Seq2seq Model

```
[TL]         batch_size (concurrent processes): 1
[TL] DynamicRNNLayer model/seq2seq/decode: n_hidden: 1024, in_dim: 3 in_shape: (1, ?, 1024) cell_fn: LSTMCell drc
: None n_layer: 3
[TL]         batch_size (concurrent processes): 1
[TL] DenseLayer  model/output: 8004 No Activation
[TL] [*] Load model.npz SUCCESS!
Inference Mode
--------------
Enter Query: happy new year
Seconds:  0.62
 > thanks man
Enter Query: how are you
Seconds:  0.11
 > doing quite well i will have to wait for this one
Enter Query: c programming is tough
Seconds:  0.52
 > you should be the first one
Enter Query: tell me about science
Seconds:  0.41
 > i was thinking the same thing
Enter Query: good
Seconds:  0.07
 > yeah i have to wait until next week
Enter Query: health is important
Seconds:  0.06
 > i think its a good thing
Enter Query: twitter is trending
Seconds:  0.07
 > i think its a great thing

Enter Query: |                                              |
```

**Fig. 6.** Testing the model

## 4.1 Performance Evaluation

The model is evaluated on Twitter datasets and loss and perplexity measures are evaluated for qualitative judgment.

**Loss.** The Cross entropy loss function calculates how different the predictions are from the true labels. As the predicted probability converges to true label there is a decrease in the value of the loss. The proposed algorithm is using the cross_entropy_seq_with_mask function for the prediction of loss in the model. It returns the cross-entropy of two sequences with softmax implemented internally. The variation of loss with various epochs during the training of model is depicted in the following Fig. 7.

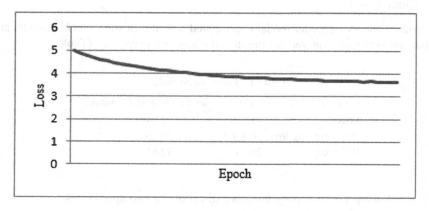

**Fig. 7.** Cross Entropy Loss

**Perplexity.** Perplexity is the measure of how well a probability distribution predicts. The formula for calculating is given in equation:

$$Perplexity = 2^{\left(\frac{-1}{N}\sum_1^N \log_2 q(x_i)\right)} \tag{4}$$

$x_i$:    test samples or the words in the sentence.
N:    length of the sentence.

The goal of the model is assigning a high probability to the test samples. Thus, lower the perplexity score the better. Models with lower perplexity have highly varied probability values and hence the model makes stronger predictions. The following Fig. 8 shows the perplexity over the entire training of the model.

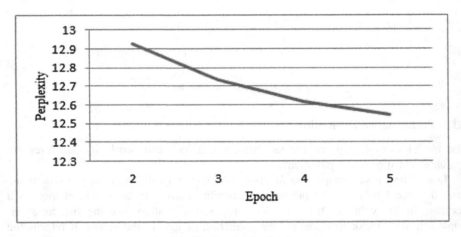

**Fig. 8.** Perplexity Graph

## 4.2    Comparison

The *seq2seq* encoder-decoder model is compared with the existing *seqGAN* model and comparison performed on various metrics like loss and perplexity (Table 1).

**Table 1.** Comparison table

| Parameter | Existing solution | Proposed solution |
|---|---|---|
| Loss | 4.3261 | 3.6491 |
| Cross entropy loss | 0.6360 | 0.5621 |
| Perplexity | 29.09 | 12.61 |

The following Fig. 9 depicts the comparison of the two approaches.

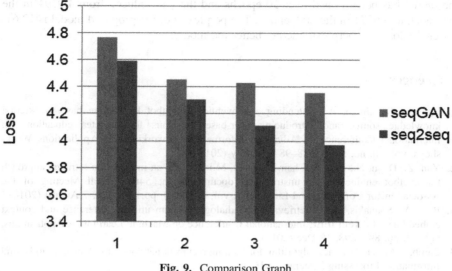

**Fig. 9.** Comparison Graph

Following Table 2 depicts the perplexity scores obtained based on the performance of various GAN models.

**Table 2.** Perplexity result comparison on different models

| Model | Test perplexity |
|---|---|
| GAN with policy gradient on small channel | 8560 |
| RNN on small channel | 24.8 |
| GAN with policy gradient and cross entropy on small channel | Infinite |
| RNN on large channel | 17.8 |

## 5   Conclusion

For humans to communicate with machines, machines should understand natural language. Conversational agents provide the capability of solving user queries. The generative model generates new responses based on the query. Hence, it can handle previously unseen queries. The proposed solution implements a generative conversational model. The encoder-decoder model makes the conversational chatbot more natural and human-like. The chatbot works in two modes, inference, and training mode. The training mode needs the specification of data corpus, batch size, number of epochs and learning rate. Training of the network is done using a generalized backpropagation algorithm and the error is computed using cross entropy loss function. In inference mode, the user enters his query and the suitable response is generated by the model. The model is evaluated on cross-entropy loss and perplexity measures during training.

The model has been trained over 50 epochs and the loss reduced from 4.9791 in the first epoch to 3.6271 in the 50th epoch. The perplexity of the proposed model is 12.61. Lower the loss and perplexity score, better the model.

# References

1. Hussain, S., Ginge, A.: Extending a conventional chatbot knowledge base to external knowledge source and introducing user-based sessions for diabetes education. In: International Conference on Advanced Information Networking and Applications Workshops, vol. 30, no. 5, pp. 978–982, 23 July (2018)
2. Yan, Z., Duan, N., Bao, J., Chen, P., Zhou, M.: DocBot: an information retrieval approach for chatbot engines using unstructured documents. In: 54th Annual Meeting of the Association for Computational Linguistics, vol. 43, no. 8, pp. 516–525, 7 August (2016)
3. Bart, A., Spanakis, G.: A retrieval-based dialogue system utilizing utterance and context embeddings. In: 16th IEEE International Conference on Machine Learning & Applications, vol. 13, pp. 890–898, 18 Dec (2017)
4. Lamb, A.M., et al.: A new algorithm for training recurrent networks. In: Advances in Neural Information Processing Systems (2016)
5. Varghese, E., Rajappan Pillai, M.T.: A standalone generative conversational interface using deep learning. In: 2nd International Conference on Inventive Communication and Computational Technologies, pp. 1915–1920 (2018)
6. Sun, X., Chen, X., Pei, Z.: Emotional human-machine conversation generation based on seqGAN. In: 1st Asian conference on affective computing and intelligent interaction (2018)
7. Hingston, P.: A turing test for computer game bots. In: IEEE Transactions on, Computational Intelligence and AI in games, vol. 1, no 3, pp. 169–186 (2009)
8. Higashinaka, R., Imamura, K., Meguro, T., Miyazaki, C., Kobayashi, N.: Towards an open domain conversational system fully based on natural language processing. In: Computational Linguistics and Chinese Language Processing (2014)
9. Ryu, P.-M., Jang, M.-G., Kim, H.-K.: Open domain question answering using Wikipedia-based knowledge model. In: Information Processing and Management, Elsevier, vol. 50, no. 2014, pp. 683–692 (2014)
10. Ly, P., Kim, J.-H., Choi, C.-H., Lee, K.-H., Cho, W.-S.: Smart answering chatbot based on OCR and over generating transformations and ranking. In: 2016 Eighth International Conference on Ubiquitous and Future Networks (ICUFN), pp. 1002–1005 (2016)
11. John, A.K., et al.: Legalbot: a deep learning-based conversational agent in the legal domain. In: Natural Language Processing and Information Systems. NLDB 2017, vol. 10260 (2017)
12. Xu, A., Liu, Z., Guo, Y., Sinha, V., Akkiraju, R.: A new chatbot for customer service on social media. In: Proceedings of the 2017 CHI Conference on Human Factors in Computing Systems (2017)
13. Ghose, S., Barua, J.J.: Toward the implementation of a topic-specific dialogue based natural language chatbot as an undergraduate advisor. In: 2013 International Conference on Informatics, Electronics, and Vision, ICIEV (2013)

# Human Protein Function Prediction Enhancement Using Decision Tree Based Machine Learning Approach

Sunny Sharma[1]([⊠]) [ID], Gurvinder Singh[2], and Rajinder Singh[2]

[1] Hindu College, Guru Nanak Dev University, Amritsar, India
sunnysharma05@yahoo.co.in
[2] Guru Nanak Dev University, Amritsar, India
gsbawa7l@yahoo.com

**Abstract.** The interrelated complex protein sequence databases are brimming day by day with the rapid advancement in technology. The sophisticated computational techniques are required for the extraction of data from these huge loads, so that refined extracted information can be easily deployable for the progress of mankind. The human protein function prediction (HPFP) is the relevant research area whose identification or function prediction leads to the discovery of drugs, detection of disease, crop hybridization, etc. Numerous approaches are present these days for HPFP because of its wide and versatile nature of this domain. The Decision tree (DT) based white box Machine Learning (ML) approaches is enriched with computational techniques to grab the information from this important research area. This study uses the decision tree based machine learning approach together with a sequence derived features (SDF's) extraction from the human protein sequence in order to predict the protein function. The experiment has been performed by manually extracting the human protein classes and sequences from HPRD (human protein reference database) [1]. Thereafter extract the SDF's from the sequences with the help of proposed HP-SDFE server as well as with the help of web servers and the DT based different classifiers such as boosting, winnowing, pruning etc. has been used for HPF prediction. The efficacies of different DT classifiers are examined and compared with the existing benchmark. The importance of input configurations together with enhanced SDF's has been thoroughly examined, which leads the individual molecular class prediction accuracy to 97%. The proposed methodology is also applicable in other similar research areas.

**Keywords:** Bagging · C5 · See5 · Decision tree · HPF · Machine learning · HPRD

## 1 Introduction to Human Protein Function Prediction

The protein function prediction, crop hybridization, discovery of drugs, disease detection and their recommender systems are the important areas with massive information, which influenced the young researchers to carry forward their research. Especially, the protein function prediction under the huge loads gaining the popularity

© Springer Nature Singapore Pte Ltd. 2019
A. B. Gani et al. (Eds.): ICICCT 2019, CCIS 1025, pp. 279–293, 2019.
https://doi.org/10.1007/978-981-15-1384-8_23

and attracting the young researchers, but the learning about right perception is quite low because of its versatile nature of this domain. The identification of right target protein leads discovery of drugs as well as detection of disease. The testing of drugs is depending upon clinical as well as pre-clinical trials first on animals, thereafter it is delivered to the public for the progress of mankind. If the discovered drug does not show any fruitful results, then entire process must be repeated which could be a year's effort and this is a very cumbersome task as well as the wastage of time. Keeping in view the importance of this task the computational techniques should be used in order to get rid from wastage of time together with the progress of mankind. The computational approaches are fueled by burgeon of machine learning approaches. This research study uses decision tree based white box machine learning approach to predict the human protein function. This work focused on the development of a model for HPFP using decision tree based machine learning approach. The architecture of the proposed methodology using aforementioned machine learning classifier is expressed in the forthcoming sections. The implementation has been done with the help of See5 tool [2], which is the working model of C5.0 [3] as well as See5 classifier. The actual C5.0 algorithm works on the unix platform and the See5 works on the windows platform. The related work for HPFP together with the contribution of the ML is expressed in the "Related work" section, which guides this whole research task. The preprocessing of the dataset together with the extraction of features and input configurations is mentioned in the "Dataset taken". The efficacies of the results and the contribution of this research task is discussed in the "Result and discussion section", followed by the concluding remarks and future direction, which are discussed in the "Conclusion" Section.

## 2 Machine Learning Techniques for Human Protein Function Prediction

The protein function prediction (PFP), drug discovery, disease detection and their recommender systems are the immense areas of huge loads with the colossal amount of information, which is increasing day by day. This thing brimming the web databases at a very rapid pace. The living creatures on the earth, having a complex foundation of proteins and under various biological conditions human body genes performs different activities. The huge research activities for PFP as well as for protein classification are happening these days, but the learning about its accurate insight is quite low which necessitates to explore this vast and versatile domain. The machine learning (ML) approaches expedite automatic evaluation of PFP with improved accuracies and gives promising results to this not so clearly defined area of research, this gives confidence to make use of ML in order to enhance the present understanding about the human protein [4]. Freitas et al. analyzed the conventional black box protein functional approaches and identified that they are not easily interpreted and understandable to biologists. The authors suggested the importance of decision tree (DT) based ML white box approaches [5]. This research task uses the decision tree based machine learning approaches together with a human protein sequence derived features (SDF's) as depicted in Fig. 1, where Absorbance, Acidic, and ExpAA are the SDF's and the Defensin, DNARepairProtein are the human protein classes. The unknown human protein identification is possible if the right SDF's from the unknown protein are extracted accurately and the decision-paths of the DT are followed.

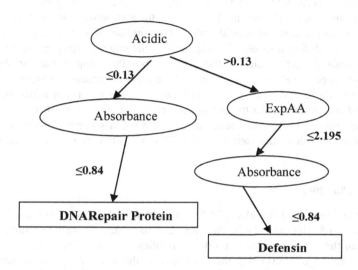

**Fig. 1.** Decision tree based machine learning Classifier

Shehu et al. surveyed the computational approaches and illustrated that for protein function prediction as well as for other genome, the machine learning approaches have been extensively used and this is the excellent area where machine learning getting some challenges to prove its dominance. In comparison to other prediction methods the machine learning together with sequence based indirect methods is doing a wonderful task [6]. The base model for the human protein function prediction classifier through SDF's using machine learning as an outcome of this research task is depicted in Fig. 2, which shows that the known functional class's and SDF's database is created and based on this database the ML classifier train itself and after that for any unknown protein sequence it is possible to predict its molecular class. Basically, this shows the supremacy of ML in the field of Bioinformatics.

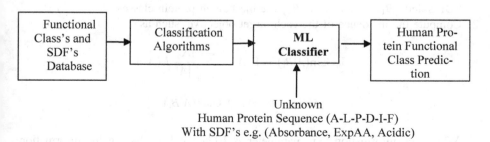

**Fig. 2.** Human protein functional class prediction using the indirect method

The decision tree based approaches are deployable to both classification and regression problems because of its simplicity and understandability for supervised learning approaches, being a white box approach it is more reliable. This guides the computational experts with step by step approach even when they are not much familiar with the concerned domain. This bridges the gap between the technical domain experts as well as the concerned field expert. The foundation stones on DT's are edges and nodes, at the different levels of computations the edges portray the functionalities so that the desired outputs among the different possible outputs can be illustrated. The DT based approaches have the ability to take various features/attributes as an input and achieve goals in a very easy way. The decision tree algorithm takes the help of $Gini_{Index}$ and information gain to find out the nodes of trees as well as in the identification of the rules. This research study uses the C5.0 Classifier in order to identify the human protein function, the detail of the C5.0 classifier is described in the next section [7].

## 2.1   C5.0 Classifier

Ross Quinlan formulated the idea of C5.0 algorithm in order to overcome the limitations of the C4.5 algorithm, whereas the C5.0 is faster as well as consumes less memory than the C4.5 in order to produce an efficient result. This research study did the implementation of C5.0 using the See5 tool on the windows platform, the See5 is enriched with the facilities of winnowing, boosting and pruning. The winnowing capability removes the useless attributes from the input data set, the facility of pruning removes the useless nodes from the decision tree, which reflects the results in a more efficient way, and the other facility of boosting under See5 builds a decision tree classifier in the most accurate way, Moreover all the above said facilities generated smaller depth decision trees as compare to C4.5. With the selection of best attribute using information gain the C5.0 shows its supremacy over the other algorithms, the selected feature/attribute becomes the root node and the process to choose the remaining child root nodes remains continue until classifier outcomes is achieved. Basically, this process splits the decision tree based on the maximum information gain.

The C5.0 algorithm pseudocode to develop the decision tree based on the human protein SDF's (sequence derived features) dataset which are obtained from the in house developed protein serve is elaborated as follows:

1. Choose the observations $\{A_1, B_1\}\ldots\{A_n, B_n\}$ Where as $\{A_1\ldots\ldots A_n\}$ are the SDF's and $\{B_1\ldots\ldots\ldots\ldots B_n\}$ are the human protein classes
2. Compute the incidence of the each target objective, such as

$$I_n\text{Gain}(A|B_i) = E(A) - \sum_{\alpha \in B_i} \frac{|A_\alpha|}{|A|} E(A_\alpha)$$

$$\text{Gain} = E(A) - I_n\text{Gain}(A|B_i)$$

Where the information gain parameter is $I_n\text{Gain}(A|B_i)$ which gains information over the SDF's and the Entropy generation is $(E(A))$ on the sampled SDF's.

3. Identify the root node using maximum information gain on SDF.
4. In order to develop the decision tree choose the splitting criteria depending upon the selected SDF information.
5. Repeat process for bifurcation of tree to get Gain on all SDF until target achieved.
6. Explore the decision tree through remaining SDF's
7. Apply the testing process to cross verified the above generated decision tree classifier.

The different decision tree classifiers have been obtained through the above said procedure on the protein dataset with the help of See5 interface and the results are carefully checked and cross-verified. The methodology to implement this research study is elaborated in the next section.

## 3  Related Work

These days the study of human protein through computational approaches is an active research area for the researchers because the proteins are the workhorse or guiding force in the human body. The varieties of computational approaches for protein function prediction have been proposed in last few decades and achieved the considerable results in terms of prediction accuracies.

King et al. proposed the methodology based on decision trees using a C4.5 algorithm to predict protein obtained from Coli and Tuberculosis and achieved 65% classification accuracy. They followed the binary feature vector consequent to common characteristics between sequences [8]. Jensen et al. revealed the ProtFun neural network classifier to predict the human protein function through PTM (Post Translational modifications) and LF (local features). The author took the help of EXPASY, PSORT and PROTPARAM web tools for feature extraction [9]. Friedberg et al. explored his valuable views for the quality of protein function prediction, according to him the sequence and the structure related data are increasing day by day, which reflecting unequal enhancement in un-portrayed gene-products and leading erroneous-annotations. The author suggested that the sequence based methods with the latest ML approaches can produce better results [10]. Singh et al. predicted the protein function based on the SDF packages on the basis of dominant SDF properties, but that was not suitable for a huge functional database's [11]. Clare et al. predicted the plant protein using data mining approaches through C4.5 algorithm and achieved the significant accuracy in prediction [12]. Singh et al. predicted the human protein function through DT based techniques and achieved 72% classification accuracy [13]. Singh et al. predicted the protein functionalities using cluster analysis techniques, which are an unsupervised learning technique. The author took the data set from HPRD and extracted features from the sequences through web based servers they considered five AA sequences together with five molecular classes. Thereafter, the author did the enhancement for PFP using C5.0 algorithm [14]. Yang et al. predicted the Yeast Prediction using KNN and extracted the fifteen SDF's from the Yeast and compared the results with existing benchmarks for improvement [15]. Singh et al. followed the C5 algorithm for HPF prediction and considered the 05 protein sequences of each 05 protein classes and achieved the accuracy between 73% to 82% [16]. Ofer et al. followed the features based technique for prediction of protein localization-structure by focusing on a specific class prediction [17]. Singh et al. followed the SVM approach for protein classification and extracted the features from Protparam at Expasy, which is a web server [18]. Shehu et al. surveyed the 397 research articles of computational approaches and illustrated that for PFP as well as for other genome, the ML approaches have been extensively used and this is the excellent area where ML getting some challenges to prove its dominance. In comparison to other prediction methods the ML together with sequence based indirect methods is doing wonderful task [6]. Sharma et al. proposed the disease-detection recommender system (DDRS) through target protein identification by extracting sequence based features. They followed the ML approaches for protein functional annotation [19].

The literature shows that the lots of protein function prediction approaches have been proposed in the last few decades, considerable results in terms of prediction accuracies have been achieved but the challenges remain. The sequence based feature extraction methodologies are performing very well together with decision tree based white box machine learning approaches and these are preferred by the researchers.

## 4   Proposed Methodology for Human Protein Function Prediction Using See5

The protein function prediction methodology of this proposed work is depicted in Fig. 3. The dataset has been taken from the HPRD database and the preprocessing is done on the data set to extract the human protein sequences together with molecular classes because the database contains a huge amount of information related to the protein-protein interaction, human disease-gene interaction and linear protein classes datasets etc. After obtaining protein classes with sequences, the class tagged/labeled sequence datasets are loaded to the in house developed HP-SDFE server as well as to some other web servers in order to extract the SDF's. The 34 SDF's have been obtained from each protein sequence in each epoch, among these 34 SDF's the 23 SDF's are obtained from the HP-SDFE server and remaining 11 features are obtained from the online web servers. After obtaining the SDF's they are feeded to the database for permanent storage. The filtration process is applied to the SDF's in order to remove the useless information. Thereafter the dataset is loaded to the machine learning C5.0 classifier to train classifers as well as for human protein function prediction. The training and testing datasets are divided in the ratio of 80:20 i.e. 80% dataset is used to train the classifer and remaining 20% is used to test the classifier. The different input configurations have been opted for training and testing of the C5.0 classifier, So that the classifier predict the functional class from the unknown protein sequence. The decision tree based machine learning classifier is performing the task of protein function prediction in a much better way as compared to other ML classifiers.

The human function prediction complexity is measured in terms of prediction accuracy and the same has been enhanced in this research study and the research task is done on intel core i3, 1.70 GHz processor using microsoft windows 10 (64-bit operating system). The See5 (32 bit classifier) consumes 8 MB memory along with <0.1% MB/s hard disk in order to produce the accurate results within 0.2 s. The decision of attributes/SDF's (sequence derived features) selection has been done manually. The SDF's which has been used in this research task are inter-correlated with each other because the human protein performs multiple functionalities only due to amino acids, which are the building blocks of human protein and the SDF's reflects these associations or functionalities. The "SDF's with Protein Classes" are stored in the database in order to choose the preset configurations, which enhance the individual classification accuracy when implemented along with the selected attributes those are obtained from the attribute filtration process.

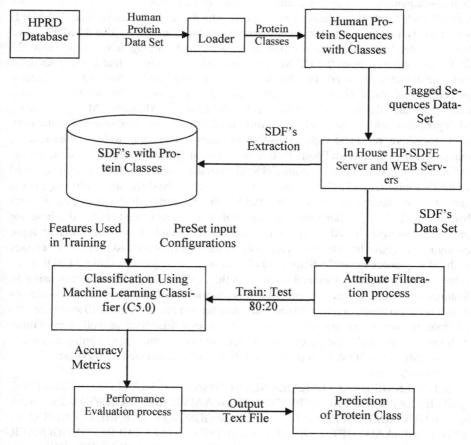

**Fig. 3.** Proposed methodology for human protein function prediction

## 5  Data Set Taken

The dataset has been taken from the HPRD database and the preprocessing is done on the data set to extract the human protein sequences together with molecular classes because the database contains a huge amount of information related to the protein-protein interaction, human disease-gene interaction and linear protein classes datasets etc. The HPRD is the platform established by john hopkins university biologists, software engineers and bioinformaticists. It contains fifteen thousand proteins related entries with unique identification along with 165 protein classes. The dataset is manually extracted from the HPRD. The human protein builds from amino acids and the proteins are soluble, non-soluble, acidic, non acidic, charged non charged etc. In this study 12 protein classes have been considered such as "DNARepairProtein, CellSurfaceReceptor, Decarboxylase, Aminopeptidase, WaterChannel, Defensin, HeatShockProtein, G-Protein, NucleotidylTransferase, Acid Phosphatase, BCellAntigenReceptor, VoltageGatedChannel". After obtaining protein classes with sequences, the class tagged/labeled sequence datasets are loaded into the in house developed HP-SDFE server as well as to some other web servers to extract the SDF's. The 34 SDF's have been obtained from each protein sequence in each epoch, among these 34 SDF's the 23 SDF's are obtained from the HP-SDFE server and remaining 11 features are obtained from the online web servers. the obtained 23 features are listed as "individual molecular percentage of each amino acid with molecular presence, MolepercPolar, MolepercNonPolar, MolepercTiny, MolepercSmall, MolepercAliphatic, MolepercAcidic, MolepercAromatic, MolepercBasic, MolepercCharged, exc(Extinction Coefficient), nneg (negative residues), npos (positive residues), molecular weight, Absorbance, expaa, predhel, IsoelectricPoint, PI (computation of IP/MW), aliphatic index, gravy, instability index, Volume, AtomsCHNOS together with Nucleotide density, and CODONS information". The remaining 12 sequence derived features "solubility, exc1, exc2, t, s, ser, thr, tyr, mean, d, prob, ProbN" are received with the assistance of web-based tools [20–23]. After obtaining the SDF's they are feeded to the database for permanent storage. Based on the above said features and classes the different input configurations have been proposed and used in the experimental task, such as 12 classes with 70 sequences using 24 SDF's, 10 classes with 58 sequences using 34 SDF's, 12 classes with 70 sequences using 34 SDF's. In the 12 classes with 70 sequences using 34 features, configuration six sequences have been considered from the 12 protein classes which reflected approximately 70 sequence data set and based on these 70 sequences the 34 features have been extracted with the help of HP-SDFE server as well as from other web servers. Similarly, the procedure has been opted for other input configurations.

The extracted output from the HP-SDFE Server is expressed as follows:

Protein-Sequence:-

MGCTLSAEERAALERSKAIEKNLKEDGISAAKDVKLLLLGAGESGKSTIVK
QMKIIHEDGFSGEDVKQYKPVVYSNTIQSLAAIVRAMDTLGIEYGDKERKADA
KMVCDVVSRMEDTEPFSAELLSAMMRLWGDSGIQECFNRSREYQLNDSAKY
YLDSLDRIGAADYQPTEQDILRTRVKTTGIVETHFTFKNLHFRLFDVGGQRSER
KKWIHCFEDVTAIIFCVALSGYDQVLHEDETTNRMHESLKLFDSICNNKWFTD

TSIILFLNKKDIFEEKIKKSPLTICFPEYTGPSAFTEAVAYIQAQYESKNKSAHKE
IYTHVTCATDTNNIQFVFDAVTDVIIAKNLRGCGLY

Total No. of AA in the Sequence = 354

Table 1 describes the detail of all the amino acids present in the protein sequence along with the presence of a molecular percentage. The detail of Table 1 is as follows:

**Table 1.** Amino Acid concentration with molecular presence

| Residue | No. of AA's | Molecular % age |
|---------|-------------|-----------------|
| A-(Ala) | 27 | 7.63 |
| B-(Asx) | 0 | 0.00 |
| C-(Cys) | 09 | 2.54 |
| D-(Asp) | 25 | 7.06 |
| E-(Glu) | 28 | 7.91 |
| F-(Phe) | 17 | 4.80 |
| G-(Gly) | 19 | 5.37 |
| H-(His) | 8 | 2.26 |
| I-(Ile) | 26 | 7.34 |
| K-(Lys) | 29 | 8.19 |
| L-(Leu) | 27 | 7.63 |
| M-(Met) | 8 | 2.26 |
| N-(Asn) | 13 | 3.67 |
| P-(Pro) | 06 | 1.69 |
| Q-(Gln) | 12 | 3.39 |
| R-(Arg) | 16 | 4.52 |
| S-(Ser) | 24 | 6.78 |
| T-(Thr) | 24 | 6.78 |
| V-(Val) | 20 | 5.65 |
| W-(Trp) | 3 | 0.84 |
| X-(Xaa) | 0 | 0.00 |
| Y-(Tyr) | 13 | 3.67 |
| Z-(Glx) | 0 | 0.00 |

Table 2 describes the extracted features detail from HP-SDFE together with its molecular percentage.

Figure 4 represents the nominal mass and mono-isotopic feature with C, H, N, O, S atoms.

**Table 2.** Sequence extracted features with molecular presence

| Extracted-features | Value | Mole % age |
|---|---|---|
| Absorbance | 0.923545 | – |
| Acidic | 53 | 14.97 |
| Aliphatic | 73 | 20.62 |
| Aliphatic-Index | 82.40 | – |
| Aromatic | 41 | 11.58 |
| Basic | 53 | 14.97 |
| Charged | 106 | 29.94 |
| Extinction-Coefficient | 36995 | – |
| GRAVY | −0.3361 | – |
| Instability-Index | 38.84 | – |
| Isoelectric-Point | 5.69 | – |
| Molecular-Weight | 40057.62 | – |
| Negatively-Charged-Residues | 53 | – |
| Non-polar | 175 | 49.44 |
| Polar | 179 | 50.64 |
| Positively-Charged-Residues | 45 | – |
| Small | 167 | 47.18 |
| IP/MW Computation | 0.000142 | – |
| Tiny | 103 | 29.09 |
| Volume | 48469.72 | – |
| C- | 1775 | – |
| H- | 2795 | – |
| N- | 475 | – |
| O- | 547 | – |
| S- | 17 | – |
| Total ATOMS | 5609 | – |
| A- | 27 | – |
| C- | 9 | – |
| G- | 19 | – |
| T- | 24 | – |

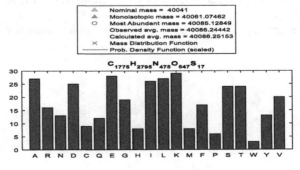

Fig. 4. Mono-isotopic and Nominal Mass

The results of the HP-SDFE server are cross verified with the results of the web servers outputs in terms of accuracy of the server and this task was handled manually and the server achieved the accurate results. The obtained 20, 24 and 34 SDF's with different molecular protein classes and sequences configurations are supplied to the decision tree based machine learning classifier with the help of See5 tool. The obtained results are listed in Tables 3 to 4 and the obtained results are compared with tree-based other classifiers as well as cross verified with the existing benchmarks. The obtained results are expressed in the next section.

# 6 Results and Discussions

The results obtained from the different DT based classifiers are are expressed as follows:

The obtained results as shown in Table 3 describes the training accuracies for the configuration such as '12 classes with 70 sequences using 24 SDF's', '10 classes with 58 sequences using 34 SDF's' and '12 classes with 70 sequences using 34 SDF's'. The results of the '12 classes with 70 sequences using 24 SDF's' input configuration obtained during the training of the classifier expressed that the winnowing classifier achieved the accuracy of 87.1% and leads during the training of the classifier, the rest of the classifiers achieved the accuracies as rule set 88.6%, ruleset sort by utility 88.6%, advance pruning 87.1% and 10 fold cross validation achieved average accuracy of 52.9% and in case of boosting maximum accuracy achieved is 87.1% during the Trial0, rest of the accuracies detail about '12 classes with 70 sequences using 24 SDF's' input configuration is shown in Table 3 along with the accuracies achieved on different classifiers construction during the training process. The test data which is considered in this study was not used during the training of the classifier. The separate testing has been performed over this configuration and the maximum accuracy has been achieved during cross validation is 47.3%, the remaining accuracies achieved by the other classifiers such as boosting 38.8%, winnowing 36.7%, cross-validation is 47.3% and in case of advance pruning 40.8%, rule set and rule sets by the utility is 44.9%, The test results demonstrate that the classifier performance is outshined and they has the ability to predict the protein function with the help of decision tree based machine learning approach. The results for the other 10 classes with 58 sequences using the 34 SDF's configuration and 12 classes with 70 sequences using the 34 SDF's input configuration are also described in Table 3. The obtained results expressed that boosting performed very well and have the ability to predict the protein classes. The detail of the Table 3 is as follows:

**Table 3.** Training accuracies on different SDF's configurations

| | (Singh, 2013) [24] | | 24 SDF's, 12 Classes, 70 Sequences | 34 SDF's, 10 Classes, 58 Sequences | 34 SDF's, 12 Classes, 70 Sequences |
|---|---|---|---|---|---|
| | 17 SDF's, 5 Classes, 25 Sequences | 21 SDF's, 5 Classes, 25 Sequences | | | |
| *Classifiers Training Accuracies in % ages* | | | | | |
| Classifier | 92 | 91.5 | 87.1 | 91.4 | 88.6 |
| Winnowing | 92 | 91.5 | 87.1 | 86.2 | 91.4 |
| Rule set | 92 | 91.5 | 88.6 | 91.4 | 88.6 |
| Sort by utility | 92 | 91.5 | 88.6 | 91.4 | 88.6 |
| *Boosting Trials* | | | | | |
| T-0 | 92 | 92 | 87.1 | 91.4 | 88.6 |
| T-1 | 88 | 85 | 85.7 | 82.8 | 84.3 |
| T-2 | 72 | 68 | 70 | 87.9 | 80 |
| T-3 | 76 | 72 | 78.6 | 72.4 | 74.3 |
| T-4 | 68 | 67 | 75.7 | 86.2 | 75.7 |
| T-5 | 96 | 95 | 75.7 | 82.8 | 75.7 |
| T-6 | 96 | 95 | 77.1 | 75.9 | 74.3 |
| T-7 | 92 | 88.9 | 71.4 | 100 | 78.6 |
| T-8 | 68 | 70.2 | 84.3 | 73.6 | 81.4 |
| T-9 | 72 | 91.5 | 74.3 | 79.3 | 68.6 |
| Advance Pruning | 92 | 91 | 87.1 | 87.9 | 88.6 |
| *10 Fold-Cross-Validation* | | | | | |
| F-1 | 50 | 50 | 57.1 | 60 | 57.1 |
| F-2 | 0 | 50 | 71.4 | 60 | 57.1 |
| F-3 | 50 | 50 | 57.1 | 33.3 | 28.6 |
| F-4 | 50 | 100 | 42.9 | 33.3 | 28.6 |
| F-5 | 50 | 50 | 42.9 | 33.3 | 42.9 |
| F-6 | 33.7 | 66.7 | 57.1 | 33.3 | 57.1 |
| F-7 | 66.7 | 33.3 | 42.9 | 33.3 | 42.9 |
| F-8 | 100 | 33.3 | 42.9 | 66.7 | 71.4 |
| F-9 | 100 | 100 | 71.4 | 50 | 42.9 |
| F-10 | 33.7 | 56.7 | 42.9 | 83.3 | 28.6 |
| Mean | 53.3 | 59 | 52.9 | 48.7 | 50 |
| SE | 90.4 | 92.5 | 96.3 | 94.3 | 95.1 |

After performing the training of the classifier the testing is performed on the input configurations. The obtained results are compared with the existing benchmarks based on the different input configuration as mentioned before such as '12 classes with 70 sequences using 24 SDF's', '10 classes with 58 sequences using 34 SDF's' and '12 classes with 70 sequences using 34 SDF's'. These input configurations are compared with the existing benchmark configuration, such as '05 classes with 25 sequences using 17 SDF's' and '05 classes with 25 sequence using 21 SDF's', which were proposed by the Singh, in 2013 [24]. Table 4 describes the comparative analysis of the obtained accuracies on the test datasets achieved during the testing session different classifiers using different input configuration. The obtained results pointed that when the dataset is increased in terms of sequences, then the accuracies got declined in a few cases, but simultaneously with enhancement of SDF's together with the sequences there is an improvement in accuracies. The results of input configuration such as '12 classes with 70 sequences using 24 SDF's' shows that when the data set grows in terms of classes, the accuracies of the classifiers decreases, on the side by side the accuracies are brimming with the enhancement in SDF's, which is expressed in the cases of '34 SDF's sequences' input configuration.

**Table 4.** Comparative analysis of testing accuracies with existing benchmark

%ages of accuracies on Test Data Set

| Classifiers | (Singh, 2013) [24] | | 12 Classes, 70 Sequences, 24 SDF's | 10 Classes, 58 Sequences, 34 SDF's | 12 Classes, 70 Sequences, 34 SDF's |
|---|---|---|---|---|---|
| | 5 Classes, 25 Sequences, 17 SDF's | 5 Classes, 25 Sequences, 21 SDF's | | | |
| Classifier | 44 | 40 | 40.1 | 61 | 32.7 |
| Rule sets | 44 | 40 | 44.9 | 58.5 | 38.8 |
| Rule sets by utility | 44 | 40 | 44.9 | 58.5 | 38.8 |
| Boosting | 64 | 64 | 38.8 | 56.1 | 64.5 |
| Winnowing | 40 | 40 | 36.7 | 53.7 | 48.6 |
| Advance pruning | 48 | 44 | 40.8 | 61.5 | 32.7 |
| Cross validation classification | 53.3 | 56.7 | 47.3 | 55.3 | 51.4 |

The improvement in the accuracies boosted the individual class prediction performance i.e. the accuracies of the individual class prediction rose to 97% especially in the case of 'defensin' molecular class and the results of the other classes also got improved. This significant enhancement reveals the contribution of input configuration because with change in input configuration the accuracies of individual class changes or we can say improved. The aforementioned results depicted that the decision tree based machine learning approach, precisely performed the experimental task of HPFP (human protein function prediction), this amelioration for prediction directly shows the contribution of enhanced SDF's and the capabilities of the machine learning white box approaches.

# 7  Conclusion

The research gaps studied during the literature survey revealed the capabilities of decision tree based machine learning approaches for human protein function prediction and shows the vast and versatile nature of this domain. The DT approaches are the best approaches because they are simple to use and easy to interpret. The critical analysis indicates that with enhancement in the SDF's from 21 to 34, brims the accuracies and ameliorates the contribution of the HP-SDFE server for protein function prediction. The state-of-the-art C5.0/See5 classifier with boosting generates outshining results with the achievement of 64.5% accuracy. The accuracies of the individual class prediction rose to 97% in the case of 'defensin' molecular class. This remarkable enhancement reflects the contribution of input configuration because with change in input configuration the accuracies of individual class changes. This shows the importance of input configurations applicability at early stages of machine learning implementation, which improve the results and does not bias the upcoming research results. The results of the boosting shows that still there is chance of improvement in prediction because the boosting belongs to the ensemble machine learning classifier. The power of input configurations together with machine learning shows that this methodology is applicable in other similar research areas for the scope of improvement in their results for classification.

# References

1. Johns-Hopkins-University and Institute-of-Bioinformatics, Human Protein Reference Database-2009 Update, January (2009). http://www.hprd.org/. Accessed 21 January, 2014
2. I. o. See5/C5.0, Information on See5/C5.0. http://rulequest.com/see5-info.html. Accessed 05 January 2016
3. Jansson, J.: Decision Tree Classification of Products Using C5.0 and Prediction of Workload Using Time Series Analysis, 30 HP Stockholm, Sverige (2016)
4. Mitchell, T.M.: Machine Learning. McGraw-Hill, New York (2005)
5. Freitas, A.A., Wieser, D.C., Apweiler, R.: On the importance of comprehensible classification model for protein function prediction. In: IEEE/ACM Transporation Computer Biology Bioinformatics, vol. 7, no. 1, pp. 172–182, January–March 2010
6. Shehu, A., Barbará, D., Molloy, K.: A survey of computational methods for protein function prediction. In: Wong, K.-C. (ed.) Big Data Analytics in Genomics, pp. 225–298. Springer, Cham (2016). https://doi.org/10.1007/978-3-319-41279-5_7
7. Quinlan, J.: Induction of decision trees. Mach. Learn. **1**(1), 81–106 (1986)
8. King, R.D., Karwath, A., Clare, A., Dehaspe, L.: Accurate prediction of protein functional class from sequence in the mycobacterium tuberculosis and escherichia coli genomes using data Mining. Yeast **17**(4), 283–293 (2000)
9. Jensen, L.J., et al.: Prediction of human protein function from post-translational modifications and localization Features. J. Mol. Biol. **319**(5), 1257–1265 (2002)
10. Friedberg, I.: Automated protein function prediction-the genomic challenge. Brief. Bioinform. **7**(3), 225–242 (2006)

11. Singh, M., Sandhu, P., Singh, H.: Decision tree classifier for human protein function prediction. In: International Conference on Advanced Computing and Communications, National Institute of Technology, Surathkal, Karnatka, India (2006)
12. Clare, A., Karwath, A., Ougham, H., King, R.D.: Functional bioinformatics for Arabidopsis thaliana. Bioinformatics (Data and text mining) 22(9), 1130–1136 (2006)
13. Singh, M., Singh, P., Wadhwa, P.K.: Human protein function prediction using decision tree induction. Int. J. Comput. Sci. Netw. Secur. 7(4), 92–98 (2007)
14. Singh, M., Singh, G.: Cluster analysis technique based on bipartite graph for human protein class prediction. Int. J. Comput. Appl. 20(3), 22–27 (2011). (0975–8887)
15. Yang, A., Li, R., Zhu, W., Yue, G.: A novel method for protein function prediction based on sequence numerical features. MATCH Commun. Math. Comput. Chem. 67, 833–843 (2012). (ISSN: 0340-6253)
16. Singh, M., Singh, D.G., Kahlon, D.K.S.: Machine learning classifiers for human protein function prediction. Int. J. Comput. Sci. Telecommun. 3(10), 21–25 (2012)
17. Ofer, D., Linial, M.: ProFET: feature engineering captures high-level protein functions. Bioinformatics 31(21), 3429–3436 (2015)
18. Singh, R., Singh, R., Kaur, D.P.: Improved protein function classification using support vector machine. Int. J. Comput. Sci. Inf. Technol. 6(2), 964–968 (2015)
19. Singh, A., Sharma, S., Singh, R., Singh, G., Kaur, A.: Quality of service enhanced framework for disease detection and drug discovery. Int. J. Comput. Sci. Eng. 6(9), 130–136 (2018)
20. PROFEAT-Server, "PROFEAT 2015 HOME,". http://bidd2.nus.edu.sg/cgi-bin/prof2015/prof_home.cgi. Accessed 15 June 2017
21. PSORT-WWW-Server, "PSORT WWW Server,". http://psort.hgc.jp/. Accessed 12 July 2017
22. TMHMM-Server, "TMHMM Server, v. 2.0,". http://www.cbs.dtu.dk/services/TMHMM/. Accessed 10 September 2017
23. NetNGlyc-Server, "NetNGlyc 1.0 Server,". http://www.cbs.dtu.dk/services/NetNGlyc/. Accessed 03 December 2017
24. Singh, M.: Machine learning classifiers for human protein function prediction. Department of Computer Science and Engineering, Guru Nanak Dev University, Amritsar, Punjab, India (2013)

# Efficient Pruning Methods for Obtaining Compact Associative Classifiers with Enhanced Classification Accuracy Rate

Kavita Mittal[1]([⊠]) [ID], Gaurav Aggarwal[1], and Prerna Mahajan[2]

[1] JaganNath University, Bahadurgarh, Haryana, India
kavitamittal.it@gmail.com,
gaurav.aggarwal@jagannathuniversityncr.ac.in
[2] Institute of Information Technology and Management,
Janak Puri, New Delhi, India
prerna.mahajan00@gmail.com

**Abstract.** The integrated approach involving supervised classification and association rule mining for development of classification model is becoming a promising strategy for building compact and accurate classifiers. As discussed in literature, in large databases the association rule mining techniques may produce large rule sets, this paper attempts to propose and implement several pruning methods to overcome the problems underlying the Associative Classification approach. The aim is to efficiently utilize the rules produced for classification in compact form and represent a rule set to be the part of classifier model with maximum data coverage and enhanced accuracy. This paper also presents experimental evaluation of the pruning methods on various datasets taken from UCI machine learning repository with the consideration of earlier approaches.

**Keywords:** Association Classification (AC) · Pruning · Full rule matching · Partial rule matching · Class correctness · Prediction · Classifier · Classification accuracy

## 1 Introduction

The Association Classification (AC) approach integrates the most common data mining technique i.e. classification and association rule mining aiming to raise the efficiency of the classification models. Many researches including (Liu et al. 1998; Thabtah et al. 2006) discussed that classification based on association rule mining concept is more capable of building the classifier models with high accuracy rates as compared to other classification methods like KNN (Mittal et al. 2018; Sahu et al. 2018), decision tree based C4.5 etc. Inspite of many improvements, AC approach suffers from few drawbacks such as generation of large number of rules based on correlation between rule items and the classes. The large number of rules including in significant rules have negative impact on the efficiency of the classifier. Thus, in order to discard redundant and insignificant rules applying pruning procedures is now becoming an important task.

Many attempts have been done to minimize the number of rules to avoid redundant and insignificant rules from participating in the prediction process in AC algorithm

© Springer Nature Singapore Pte Ltd. 2019
A. B. Gani et al. (Eds.): ICICCT 2019, CCIS 1025, pp. 294–311, 2019.
https://doi.org/10.1007/978-981-15-1384-8_24

thereby improving the classification accuracy. Second problem lies in predicting the accurate class for a test case. This chapter elaborates the rule pruning techniques and class prediction step in AC approach with an aim to:

(a) Discard insignificant rules in the evaluation process on the training data while developing the classifier.
(b) Improve the accuracy rate of class prediction by applying high confidence rules.

For pruning process different rule pruning procedures are developed and implemented with an AC algorithm called **CBA_Optimized.**

These techniques are then tested on various datasets taken from the UCI Machine Learning data repository (Merz and Murphy 1996). Their performance is then analyzed in comparison to the other pruning methods discussed in literature. The comparisons are made based on two factors: the number of rules generated and the classification accuracy rate.

## 2   The Effective Pruning Criteria

The proposed pruning methods consider the rule significant to be included in the classifier rule based on the concepts of partial rule matching and full rule matching.

The aim of pruning is to reduce the number of rules produced in the classifier and secondly to analyze the impact of pruning on classification accuracy rate. Evaluating a rule during classifier construction is based on two criterions:

Firstly the type of rule matching with the test case, i.e. whether the rule matches the test case partially or fully. Secondly, the rule's class correctness with training case class. Both the criterions play significant role in identifying the effect of pruning on the classifiers.

A rule partially matches the training case T if value of any of rule item's attribute is similar to one or more items of the training data case. Besides matching of the attribute value the rule's class must also match with the class of the training data case, i.e. the correctness of the class. Some pruning methods consider correctness of the class while others do not (Thabtah 2006).

This paper investigates partial and full matching cases with and without the correctness of the class.

This study presents the role of pruning based on partial matching criteria and class correctness in enhancing the accuracy rate. To accomplish this task several research questions need to be answered in this chapter:

(1) whether the proposed pruning method able to produce the minimum no of rules.
(2) whether the rules produced able to cover maximum no of records in the dataset.

The study proposes different pruning procedure employing partial and full matching criteria and class correctness criteria as follows:

A. Partial Matching Pruning Method (PM)
B. Full Matching Pruning Method (FM)
C. Full and Partial Matching Pruning Method (FPM)

D. Partial Matching with Correct Class (PMCC)

E. Full and Partial Matching with Correct Class (FPMCC)

Partial Matching procedure takes care of partial matching between the rule items and attribute values of the training data case. Partial Matching with correct class (PMCC) take care of Partial Matching of rule items along with the rule class and training data case to mark the rule significant. Full matching (FM) procedure ensure that the rule items must be fully matching with that of training case without caring for correct class. Full and Partial matching (FPM) concept is a hybrid approach for considering a rule potential if it fully matches the training or otherwise it cover the training the case partially with no attention towards class correctness. Lastly Full and Partial Correct Class (FPCC) works similar to FPM but takes care of class correctness of the rule with that of the training data case.

This section explains the suggested pruning methods mentioned above with an example explaining each method.

The Table 1 shows the sample data taken from car evaluation dataset from UCI data repository to implement the rule based classifier model called CBA_Optimized. The dataset contains 1728 records with 6 attributes as shown below:

**Table 1.** Example data from car evaluation dataset

| S.No | Buying | Maintenance | Doors | Persons | Luggage boot | Safety | Class |
|------|--------|-------------|-------|---------|--------------|--------|-------|
| 1 | 1 (vhigh) | 4 (med) | 1 (2) | 3 (4) | 1 (small) | high | acc |
| 2 | 1 (vhigh) | 4 (med) | 1 (2) | 3 (4) | 2 (med) | low | unacc |
| 3 | 1 (vhigh) | 4 (med) | 1 (2) | 3 (4) | 2 (med) | med | unacc |
| 4 | 1 (vhigh) | 4 (med) | 1 (2) | 3 (4) | 2 (med) | high | acc |
| 5 | 2 (med) | 3 (low) | 1 (2) | 3 (4) | 2 (med) | high | good |
| 6 | 2 (med) | 3 (low) | 1 (2) | 3 (4) | 3 (big) | low | unacc |
| 7 | 2 (med) | 3 (low) | 1 (2) | 3 (4) | 3 (big) | med | good |
| 8 | 2 (med) | 3 (low) | 1 (2) | 3 (4) | 3 (big) | high | vgood |
| 9 | 2 (med) | 3 (low) | 1 (2) | 2 (more) | 1 (small) | low | unacc |
| 10 | 2 (med) | 3 (low) | 1 (2) | 2 (more) | 1 (small) | med | unacc |
| 11 | 2 (med) | 3 (low) | 1 (2) | 2 (more) | 1 (small) | high | unacc |
| 12 | 2 (med) | 3 (low) | 1 (2) | 2 (more) | 2 (med) | low | unacc |
| 13 | 2 (med) | 3 (low) | 1 (2) | 2 (more) | 2 (med) | med | acc |
| 14 | 2 (med) | 3 (low) | 1 (2) | 2 (more) | 2 (med) | high | good |
| 15 | 2 (med) | 3 (low) | 1 (2) | 2 (more) | 3 (big) | low | unacc |
| 16 | 2 (med) | 3 (low) | 1 (2) | 2 (more) | 3 (big) | med | good |
| 17 | 4 (low) | 4 (med) | 3 (5more) | 1 (2) | 3 (big) | med | unacc |
| 18 | 4 (low) | 4 (med) | 3 (5more) | 1 (2) | 3 (big) | high | unacc |
| 19 | 4 (low) | 4 (med) | 3 (5more) | 3 (4) | 1 (small) | low | unacc |
| 20 | 4 (low) | 4 (med) | 3 (5more) | 3 (4) | 1 (small) | med | acc |

Table 2 represents the attributes and its values:

**Table 2.** Attributes names and values

| Index | Attribute name | Attribute value |
|-------|----------------|-----------------|
| 0 | Buying | low, med, high, v-high, |
| 1 | Maintenance | low, med, high, v-high |
| 2 | Doors | 2, 3, 4, 5more |
| 3 | Persons | 2, 4, 5more |
| 4 | Luggage boot | small, med, big |
| 5 | Safety | low, med, high |

Table 3 represents the codes used against attribute values used during rule generation:

**Table 3.** Attribute values with codes

| | |
|---|---|
| Buying | 'vhigh': 1, 'med': 2, 'high': 3, 'low': 4 |
| Maintenance | vhigh': 1, 'med': 4, 'high': 2, 'low': 3 |
| Doors | '2': 1, '3': 2, '5more': 3, '4': 4 |
| Persons | '2': 1, 'more': 2, '4': 3 |
| Luggage boot | 'small': 1, 'med': 2, 'big': 3 |

The data is categorized under four classes namely Unacceptability, Acceptability, Good, V-Good. The Class labels are utilized for building the classifier and testing the unseen data.

Table 4 represents the class Distribution of the dataset.

**Table 4.** Class distribution

| Class name | Number of samples (S) | S% |
|------------|----------------------|------|
| V-good | 65 | 3.7620% |
| Good | 69 | 3.9930% |
| Acc | 384 | 22.222% |
| Unacc | 1210 | 70.023% |

The text files are tested against the classifier model and predicted in any one of the category- Unacceptability, Acceptability, Good, V-Good.

The Table 5 below indicates the example of classifier model based on association rules that consists of 20 rules produced by CBA on AC algorithm with thresholds minsupp of 2% and minconf 50% from the training dataset.

**Table 5.** Classifier rules with rule ranking

| Rule ID | Rule description | Rule confidence | Rule support | Rule rank |
|---|---|---|---|---|
| 1 | {(5, low)} - > (class, unacc) | 1 | 0.33419 | 1 |
| 2 | {(3, 1)} - > (class, unacc) | 1 | 0.327763 | 2 |
| 3 | {(0, 1), (1, 1)} - > (class, unacc) | 1 | 0.062339 | 3 |
| 4 | {(0, 3), (1, 1)} - > (class, unacc) | 1 | 0.061697 | 4 |
| 5 | {(0, 1), (1, 2)} - > (class, unacc) | 1 | 0.059126 | 5 |
| 6 | {(2, 1), (4, 1), (5, med)} - > (class, unacc) | 0.909091 | 0.025707 | 6 |
| 7 | {(0, 1), (4, 1)} - > (class, unacc) | 0.907692 | 0.075835 | 7 |
| 8 | {(1, 1), (4, 1)} - > (class, unacc) | 0.890625 | 0.073265 | 8 |
| 9 | {(0, 1), (2, 1)} - > (class, unacc) | 0.873684 | 0.053342 | 9 |
| 10 | {(2, 1), (4, 1)} - > (class, unacc) | 0.87218 | 0.07455 | 10 |
| 11 | {(0, 3), (4, 1)} - > (class, unacc) | 0.859259 | 0.07455 | 11 |
| 12 | {(1, 1), (2, 1)} - > (class, unacc) | 0.852632 | 0.052057 | 12 |
| 13 | {(1, 2), (4, 1), (5, med)} - > (class, unacc) | 0.847826 | 0.025064 | 13 |
| 14 | {(0, 1)} - > (class, unacc) | 0.833773 | 0.203085 | 14 |
| 15 | {(1, 1)} - > (class, unacc) | 0.829457 | 0.206298 | 15 |
| 16 | {(0, 3), (2, 1)} - > (class, unacc) | 0.795918 | 0.050129 | 16 |
| 17 | {(2, 1), (4, 2), (5, med)} - > (class, unacc) | 0.767442 | 0.021208 | 17 |
| 18 | {(0, 3)} - > (class, unacc) | 0.746269 | 0.192802 | 18 |
| 19 | {(1, 2), (2, 2)} - > (class, unacc) | 0.731959 | 0.04563 | 19 |
| 20 | {(3, 3), (4, 2), (5, high)} - > (class, acc) | 0.557377 | 0.021851 | 20 |
| 21 | {(3, 3), (5, high)} - > (class, acc) | 0.556757 | 0.066195 | 21 |

The candidate rules are ordered in the descending order before the pruning process begins. The rule ranking criteria is described below (Kliegr 2017):

For any two rules r1 and r2,

(a) r1 > r2, if confidence of r1 is higher than that of r2.
(b) r1 > r2, if confidence is same and r1 has higher support than r2.
(c) r1 > r2, if confidence and support are same and r1 has less number of items in its antecedents than r2.
(d) r1 > r2, if confidence, support and cardinality are same and r1 represents more classes than r2.

The rule ranking is needed to proceed for rule pruning to enhance classification accuracy. A new ranking criterion is added based on the frequency of the class name in the training dataset to prioritize among the remaining rules. From the above sample dataset, it can be found that the class unacc has the higher frequency than any other class. If in case, the rule's confidence value and support value are same then the rule will be ranked according to the rule length on the left hand side and if all the three parameters are same then the rule with highest class frequency would be given the higher rank.

# 3   The Proposed Rule Pruning Procedures

The rule pruning procedures are defined based on two criterions:

Firstly the type of rule matching with the test case, i.e. whether the rule matches the test case partially or fully. Secondly, the rule's class correctness with training case class. Both the criterions play significant role in identifying the effect of pruning on the classifiers.

A. Full Matching Pruning Method (FM)

The basic concept of this pruning method is that the rule is considered significant if the body of the rule completely matches with the training data case without any interest in class matching of the rule with that of the training case. The procedure below describes the FM pruning method (Fig. 1):

> Input: Ranked Rule set R1, Training Dataset D
> Output: Classifier C_FM
> Step1: R2=sort (R1)
> Step2: For each rule ri in R2
> Step3: Search training cases in D that matches ri's condition fully
> Step4: Add ri to the classifier C_FM
> Step5: Delete all the training cases in D, to which ri is applicable
> Step6: else
> Step7: Remove $r_i$ from R1
> Step8: end
> Step9: Next $r_i$

**Fig. 1.** Full Matching Pruning Method (FM)

Consider the classification rules obtained in Table 5, the rule r1 {(5, low/0– > (class, unacc)} in the FM pruning method, it can be seen that it covers case 19, so it is considered significant and the corresponding datacases are discarded from the training dataset. The process continues with next ranked rule r2, it covers cases 17, 18, hence it is also marked significant, and the corresponding data cases in the similar manner are discarded from the training data set. Similarly the next rule r3 does not cover any case so it is not marked as significant. Finally, proceeding in this manner the pruning process ends after the evaluation of all the rules or after the coverage of all the training data cases. This pruning method generates 7 rules (r1, r2, r9, r10, r17, r20, r21) found significant for the classifier. Hence the process was stopped after evaluating all 21 rules (Fig. 2).

## B. Partial Matching Pruning Method (PM)

The concept underlying this method considers rule significant if at least one of the item in the training data case is contained in the rule with no botheration about the correct class. The method works according to the algorithm.

Input: Ranked Rule set R1, Training Dataset D
Output: Classifier C_PM
Step1: R2=sort (R1)
Step2: For each rule $r_i$ in R2
Step3: Search training cases in D such that rule items in ri  matches with at least one of the item in the training data case di.
Step4: Add $r_i$ to the classifier C_PM
Step5: Delete all the training cases in D, to which $r_i$ is applicable
Step6: else
Step7: Remove $r_i$ from R1
Step8: end
Step9: Next $r_i$

**Fig. 2.**  Partial Matching Pruning Method (PM)

Let's reconsider Table 3 and implement partial matching concept to generate the rules in classifier. Beginning with high ranked rule r1, the training cases covered are (2, 6, 9, 12, 15, 19). So it is marked significant and added to the classifier. The corresponding training cases covered are discarded. Repeating the similar process with next ranked rule r2 cases 17, 18 are found covered. So the rule r2 is marked significant and added to the classifier. The corresponding cases 17, 18 are discarded. The process is continued till all the cases are covered or all the rules are evaluated. In this pruning methods after evaluating all rules 4 rules (r1, r2, r3, r6) are marked significant and added to the classifier. The process terminates after all training cases covered and the remaining rules are considered insignificant.

## C. Full and Partial Matching (FPM)

This is hybrid approach combining the Full Matching and Partial Matching pruning methods. In this method rule is checked, if the training cases fully matches the rule, if not whether training case partially matches the rule. If the rule fully or partially matches items of the training data case it will be considered significant and added to the classifier with no botheration towards class similarity, else will be discarded (Fig. 3).

In this method, in cases for which the rule items do not match fully with that of training data case then the partial matching pruning method is invoked which include the rules partially matching with the training case. While implementing this method on Table 1, the rule r1 covers the cases (2, 6, 9, 12, 15, 19), these data cases will be discarded and the rule is appended to the classifier. The further ranked rule r2 covered cases 17,18, so marked significant and added to the classifier. Similarly the other

Input: Ranked Rule set R1, Training Dataset D

Output: Classifier C_FPM

Step1: R2=sort (R1)

Step2: For each rule $r_i$ in R2

Step3: Search training cases in D that matches $r_i$'s condition completely

Step4: Add $r_i$ to the classifier C_FPM

Step5: Delete all the training cases in D, to which $r_i$ is applicable

Step6: else, if a rule $r_i$ partially matches the training case then

Step7: Append the rule to C_FPM

Step8: Delete all the training cases in D, to which $r_i$ is applicable

Step9: else

Step10: Remove $r_i$ from R1

Step11: end

Step12: Next $r_i$

**Fig. 3.** Full and Partial Matching (FPM)

ranked rules are applied. This process is stopped after evaluating all the rules and the rules r4, r5, r7 to r21 are found insignificant as they do not cover any training case and hence discarded. this process covered all training cases and the classifier consist of 4 rules (r1, r2, r3, r6).

## D. Partial Matching with Correct Class (PMCC)

This method considers a rule to be significant to be the part of classifier if it contains at least one item of the training case and the class of training data case is also similar to the class of the rule (Fig. 4).

This method considers a rule to be significant to be the part of classifier if it contains at least one item of the training case and the class of training data case is also similar to the class of the rule.

Input: Ranked Rule set R1, Training Dataset D

Output: Classifier C_PM

Step1: R2=sort (R1)

Step2: For each rule $r_i$ in R2

Step3: Search training cases in D such that any one item in ti matches rule items in $r_i$ and class of di is same as class of $r_i$

Step4: Add $r_i$ to the classifier C_PM

Step5: Delete all the training cases in D, to which $r_i$ is applicable

Step6: else

Step7: Remove $r_i$ from R1

Step8: end

Step9: Next $r_i$

**Fig. 4.** Partial Matching with Correct Class (PMCC)

Again consider the Table 1 and the Table 5. Lets begin with top ranked rule r1 {(5, low) – > (class, unacc)} covered (2, 6, 9, 12, 19, 19) training cases with similar class labels. Thus it is marked significant to be the part of the classifier and the corresponding training data cases are discarded. Then applied next ranked rule r2. It covered training cases 17, 18 added to the classifier. Similarly next ranked ruler r3 cover the training case 3 with similar class. Proceeding in the similar manner, this pruning process is stopped after evaluating all the rules. Some rules were not tested as their cases were already covered by higher ranked rules. Thus the classifier obtained consist of 5 rules (r1, r2, r3, r6, r20).

### E. Full and Partial Matching with Correct Class (FPMCC)

This hybrid approach consider a rule if the training case fully matches the rule, if not then the partial matching procedure is invoked with class correctness criteria (Fig. 5).

Input: Ranked Rule set R1, Training Dataset D
Output: Classifier C_FPMCC
Step1: R2=sort (R1)
Step2: For each rule $r_i$ in R2
Step3: if ri matches a training case fully with correct class then
Step4: Add $r_i$ to the classifier C_FPMCC
Step5: Delete all the training cases in D, to which $r_i$ is applicable
Step6: else if
Step7: Search training cases in D such that any one item in ti matches rule items in $r_i$ and class of di is same as class of $r_i$
Step8: Add $r_i$ to the classifier C_FPMCC
Step9: Delete all the training cases in D, to which $r_i$ is applicable
Step10: else
Step11: Remove $r_i$ from R1
Step12: end
Step13:Next $r_i$

**Fig. 5.** Full and Partial Matching with Correct Class (FPMCC)

In this method when rule r is tested, it checks if r can cover any training case whose body fully matches with the training case along with class correctness, it is marked significant and the process is repeated till the next ranked rule. The process continues till all the rules are tested for full or partial matching with class correctness, or if the training set becomes empty when all the cases are covered. Thus applying this method produces the resulting classifier containing 5 rules with class similarity as an important aspect.

Table 6 shows all the rules participating in the classification of training cases and added to the resultant classifier.

**Table 6.** Rule participation in classification

| S.No | Rule ID | FM | PM | FPM | PMCC | FPMCC |
|---|---|---|---|---|---|---|
| 1 | 1 | 19 | 2, 6, 9, 12, 15, 19 | 2, 6, 9, 12, 15, 19 | 2, 6, 9, 12, 15, 19 | 2, 6, 9, 12, 15, 19 |
| 2 | 2 | 17, 18 | 17, 18 | 17, 18 | 17, 18 | 17, 18 |
| 3 | 3 | Not Applicable | 1, 3, 4 | 1, 3, 4 | 3 | 3 |
| 4 | 4 | Not Applicable | Not Applicable | Not Applicable | Not Applicable | Not Applicable |
| 5 | 5 | Not Applicable | Not Applicable | Not Applicable | Not Applicable | Not Applicable |
| 6 | 6 | Not Applicable | 5, 7, 8, 10, 11, 13, 14, 16, 20 | 5, 7, 8, 10, 11, 13, 14, 16, 20 | 10, 11 | 10, 11 |
| 7 | 7 | Not Applicable | Not Applicable | Not Applicable | Not Applicable | Not Applicable |
| 8 | 8 | Not Applicable | Not Applicable | Not Applicable | Not Applicable | Not Applicable |
| 9 | 9 | 1, 2, 3, 4 | Not Applicable | Not Applicable | Not Applicable | Not Applicable |
| 10 | 10 | 8, 9, 10, 11 | Not Applicable | Not Applicable | Not Applicable | Not Applicable |
| 11 | 11 | Not Applicable | Not Applicable | Not Applicable | Not Applicable | Not Applicable |
| 12 | 12 | Not Applicable | Not Applicable | Not Applicable | Not Applicable | Not Applicable |
| 13 | 13 | Not Applicable | Not Applicable | Not Applicable | Not Applicable | Not Applicable |
| 14 | 14 | Not Applicable | Not Applicable | Not Applicable | Not Applicable | Not Applicable |
| 15 | 15 | Not Applicable | Not Applicable | Not Applicable | Not Applicable | Not Applicable |
| 16 | 16 | Not Applicable | Not Applicable | Not Applicable | Not Applicable | Not Applicable |
| 17 | 17 | 13 | Not Applicable | Not Applicable | Not Applicable | Not Applicable |
| 18 | 18 | Not Applicable | Not Applicable | Not Applicable | Not Applicable | Not Applicable |
| 19 | 19 | Not Applicable | Not Applicable | Not Applicable | Not Applicable | Not Applicable |
| 20 | 20 | 15 | Not Applicable | Not Applicable | 1, 4, 20 | 1, 4, 20 |
| 21 | 21 | 5 | Not Applicable | Not Applicable | Not Applicable | Not Applicable |
| Total | 21 rules | 7 rules | 4 rules | 4 rules | 5 rules | 5 rules |

## 4   Rule Pruning Impact on Classification Accuracy

Association Classification approach begins by generating association rules using Association rule mining techniques as discussed in literature. Association Rule mining techniques tend to discover frequent item sets in transactional databases then, followed by association rules which are often large in number. Handling such large data without setting constraints on rule generation process and appropriate pruning methods may result in production of very large number of rule. Different associative classification algorithm employ different pruning approaches such as CBA employing database coverage pruning method prefer general rules for classification over specific rules and tend to produce small classifiers (with less number of rules participation in the classification) as compared to Lazy pruning based algorithms such as L3 (Garza, et al. 2007) that often produce big sized classifier (huge number of rules as compared to database coverage method).

Consider the experimental results in Table 6 showing the results obtained by employing CBA based on database coverage and pessimistic error pruning approaches and our proposed method of pruning based on full and partial matching with class correctness. Results are obtained using minsup 2% and minconf of 50%. The number of rules produced on the datasets indicate that pruning approaches based on database coverage or pessimistic error pruning methods attempt to obtain general rules to form small sized classifiers with less computing cost but are relatively less accurate (Thabtah 2006).

In general small classifiers are preferable due to their less maintenance and easy understanding. However, small sized classifiers generally suffer from problems such as handling the datasets containing conflicting and missing data values. Secondly, the ability of database coverage being low. The Lazy pruning method is however capable of enhancing the accuracy but producing large number of rules leave it as a complex method of handling and maintenance due to the fact that LP method preserve the spare rules in order to cover the remaining test data cases. Our proposed pruning method take care of both aspects along with tile of execution which is also a parameter underlying the performance of the classification process. Our proposed pruning method called FPMCC tend to select specific rule and try to produce smaller classifier as compared to other pruning methods such as lazy pruning discussed in literature. However FPMCC produce slightly larger size classifier as compared to database coverage or pessimistic error but show increased classification accuracy.

## 5   The Prediction Approach Proposed

The Association rules based classification approach in data mining aims at classifying the unknown data by predicting the classes. The algorithms such as CBA and MCAR uses the prediction concept based on single rule that work on the following procedure: first rule that is applicable to the test case will classify it. CMAR and CPAR (Zhao et al. 2009) employs more than one rules for predicting the labels of test data cases. Such predictions are based on the concept that using more than one rule for prediction allows the contribution of more than one rules in the decision making process. However such algorithms relying on multiple rules for prediction also not take into consideration the independence of the rules (Zaïane and Antonie 2005) as in the training phase, the cases can be covered by more than one rules which differ in class names, it may lead to interdependence of rules creating conflicts during decision making. Thus the algorithms utilizing single rule for prediction of class names generally produce classifiers with high accuracy rate reducing the chances of dependency and conflicts but priority should be given to the rule with high confidence and high support.

The proposed prediction method (Fig. 6) chooses the frequent class among the set of high confidence rules from the set of rules R. It counts the class names of the rules that fully match the items of the test datacase first and then selects the class with more count and assign to the test datacase. If no rule is applicable then the default class is allocated to the test case. Consider again, the rule set in Table 6 derived by CBA_Optimized with minconf 50% and minsup 2% (Table 7).

**Input: R** -Classifier, T- test data set , A -an array
**Output**: The prediction error rate P_e
For any test data set (T), the class prediction method proceeds as follows:
  1.   For all test data case t do
  2.        Status = f
  3.        For every rule r in R (ranked rules) do
  4.                 Identify the possible rules matching with the  body of t and add it to  A
  5.                 If A is unempty do
  6.                      If a rule r fully matches t's conditions
  7.                           count_class+=1
  8.                      else
  9.                           allocate t, the default class
  10.                         set status=t
  11.       end
  12.      if status = f
  13.          empty A
  14.    end
  15.  Calculate total errors in T

**Fig. 6.** Frequent class based prediction approach example

**Table 7.** Example rule set

| Rule ID | Rule | Confidence |
|---------|------|------------|
| 1 | {(5, low)} - > (class, unacc) | 1 |
| 2 | {(3, 1)} - > (class, unacc) | 1 |
| 3 | {(1, 1)} - > (class, unacc) | 1 |
| 4 | {(0, 1)} - > (class, unacc) | 1 |
| 5 | {(0, 3)} - > (class, unacc) | 1 |
| 6 | {(0, 1), (4, 1)} - > (class, unacc) | 0.909091 |
| 9 | {(1, 1), (4, 1)} - > (class, unacc) | 0.873684 |
| 10 | {(3, 3), (5, high)} - > (class, acc) | 0.87218 |
| 14 | {(0, 1), (2, 1)} - > (class, unacc) | 0.833773 |
| 16 | {(0, 3), (2, 3)} - > (class, unacc) | 0.795918 |

**Table 8.** Test case

| t | 2 (0, med) | 3 (1, low) | 1 (2, 2) | 2 (3, more) | 2 (4, med) | high (5, high) |
|---|------------|------------|----------|-------------|------------|----------------|

To illustrate the method, let us work on the test data case given in above Table 8. The Table 9 given below presents the set of possible rules applicable to case t:

**Table 9.** Rules applicable to given test case

| Rule ID | Rule | Confidence |
|---------|------|------------|
| 10 | {(3, 3), (5, high)} - > (class, acc) | 0.87218 |
| 14 | {(0, 1), (2, 1)} - > (class, unacc) | 0.833773 |
| 16 | {(0, 3), (2, 3)} - > (class, unacc) | 0.795918 |

To classify t, the applicable rules per class are counted, it is found class "unacc" has the highest count, so class "unacc" is predicted for test case t.

## 6 Evaluation and Experimental Results

This section presents the comparison of different pruning methods with our proposed pruning methods based on two parameters i.e., no of rules in classifier and the accuracy rate. Many datasets are taken from UCI machine learning data repository are for performing the experiment. The commonly used pruning techniques (Database Coverage, Pessimistic error and Lazy Pruning) are compared with our proposed pruning methods in Tables 10, 11, 12 and 13 indicates the time of execution for different datasets with average execution time.

**Table 10.** Number of rules

| S.No. | Dataset | Database coverage, pessimistic error | Lazy pruning | FPM | FPMCC |
|-------|---------|--------------------------------------|--------------|-----|-------|
| 1 | Iris | 9 | 190 | 7 | 8 |
| 2 | Car Evaluation | 21 | 2275 | 16 | 17 |
| 3 | Banknote Authentication | 24 | 38475 | 22 | 23 |
| 4 | Facebook | 27 | 39475 | 24 | 25 |
| 5 | Forest | 30 | 41252 | 27 | 28 |
| 6 | Horse Colic | 33 | 41678 | 28 | 30 |
| 7 | Indian Liver patient | 34 | 4232 | 29 | 30 |
| 8 | Monks | 21 | 926 | 18 | 19 |
| 9 | Wine | 25 | 40775 | 20 | 21 |
| 10 | Vertebra | 14 | 876 | 11 | 12 |
| | **Number of Rules** | 238 | 210154 | 202 | 213 |

**Table 11.** Average error rate

| S.No. | Datasets | Database coverage, pessimistic error | Lazy prunning | FPM | FPMCC |
|-------|----------|--------------------------------------|---------------|-----|-------|
| 1 | Iris | 0.2 | 0.4 | 0.02 | 0.011 |
| 2 | Car | 3.25 | 3.32 | 3.31 | 2.202 |
| 3 | Banknote Authentication | 0.3 | 1.34 | 3.33 | 2.228 |
| 4 | Facebook | 7.6 | 7.81 | 7.8 | 6.09 |
| 5 | Forest | 1.95 | 2.12 | 2.11 | 1.005 |
| 6 | Horse Colic | 0.33 | 0.51 | 0.501 | 0.38 |
| 7 | Indian Liver Patient | 0.24 | 0.23 | 0.221 | 0.11 |
| 8 | Monks | 0.32 | 0.11 | 0.102 | 0.01 |
| 9 | Wine | 0.61 | 0.62 | 0.61 | 0.48 |
| 10 | Vertebra | 0.1 | 0.12 | 0.1 | 0.02 |
| | **Total** | 14.9 | 16.58 | 18.104 | 12.536 |
| | **Average Error Rate** | 1.49 | 1.658 | 1.8104 | 1.2536 |

**Table 12.** Average accuracy rate

| S.No. | Dataset | Database coverage, pessimistic error | Lazy pruning | FPM | FPMCC |
|---|---|---|---|---|---|
| 1 | Iris | 99.8 | 99.6 | 99.98 | 99.989 |
| 2 | Car Evaluation | 96.75 | 96.68 | 96.69 | 97.798 |
| 3 | Banknote Authentication | 99.7 | 98.66 | 96.67 | 97.772 |
| 4 | Facebook | 92.4 | 92.19 | 92.2 | 93.91 |
| 5 | Forest | 98.05 | 97.88 | 97.89 | 98.995 |
| 6 | Horse Colic | 99.67 | 99.49 | 99.499 | 99.62 |
| 7 | Indian Liver Patient | 99.76 | 99.77 | 99.779 | 99.89 |
| 8 | Monks | 99.68 | 99.89 | 99.898 | 99.99 |
| 9 | Wine | 99.39 | 99.38 | 99.39 | 99.52 |
| 10 | Vertebra | 99.9 | 99.88 | 99.9 | 99.98 |
| | **Total** | 985.1 | 983.42 | 981.896 | 987.464 |
| | **Average Accuracy Rate** | 98.51 | 98.342 | 98.1896 | 98.7464 |

**Table 13.** Average execution Time(s)

| S.No. | Dataset | Database coverage, pessimistic error | Lazy pruning | FPM | FPMCC |
|---|---|---|---|---|---|
| 1 | Iris | 0.02 | 0.89 | 0.03 | 0.03 |
| 2 | Car Evaluation | 0.05 | 0.78 | 0.04 | 0.05 |
| 3 | Banknote Authentication | 0.04 | 0.62 | 0.03 | 0.04 |
| 4 | Facebook | 0.03 | 0.32 | 0.02 | 0.03 |
| 5 | Forest | 0.02 | 0.61 | 0.03 | 0.03 |
| 6 | Horse Colic | 0.03 | 0.82 | 0.02 | 0.04 |
| 7 | Indian Liver Patient | 0.06 | 0.34 | 0.042 | 0.05 |
| 8 | Monks | 0.01 | 0.12 | 0.01 | 0.02 |
| 9 | Wine | 0.03 | 0.71 | 0.02 | 0.04 |
| 10 | Vertebra | 0.01 | 0.15 | 0.01 | 0.02 |
| | **Total** | 0.3 | 5.36 | 0.252 | 0.35 |
| | **Average Execution Time** | 0.03 | 0.536 | 0.0252 | 0.035 |

## 7 Results and Discussion

The results are derived during new enhanced version of CBA. The Table 10 indicates that the algorithm that employ Lazy Pruning (L3) produce large number of rule. The database coverage method eliminate the space rules and thus produce moderate sized classifier. Specifically CBA algorithm as compared to other algorithm MCAR, CMAR produce moderate sized classifier which enables the user to manage, understand the rules and get benefitted from them. The pruning techniques proposed in this study takes into consideration mainly two aspects, the size of the classifier produced and the rate of classification accuracy, that attempts to choose only specific rules and produce moderate sized classifier as compared to other pruning methods such as Lazy Pruning. The classifier produced by the proposed pruning methods with CBA algorithm often produce slightly larger in size then that produced by CBA using database coverage and pessimistic error pruning methods. The Table 12 presents the classification accuracy rate of different pruning methods along with our proposed pruning methods (FPM, FPCC) for different datasets taken from UCI Machine Learning data repository. As per the results, the proposed rule pruning methods produced competitive classification accuracy.

## Number of Rules

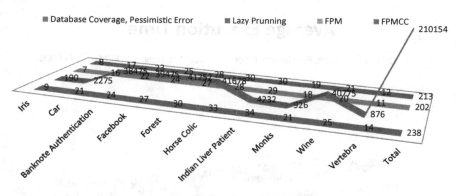

**Fig. 7.** Number of rules

Figure 7 represents number of rules produced. The number of rules produced by the algorithms FPM and FPMCC are minimum and approximately similar, however the rules produced by database coverage algorithm are larger and lazy pruning approach produced far large number of rules compared to others.

Figure 8 represents the accuracy results of the different pruning methods. The pruning methods FPM and FPCC shown a competitive accuracy rates while comparing to others.

Also their execution time is also competitive as compared to CBA with database coverage, pessimistic error pruning method in Fig. 9.

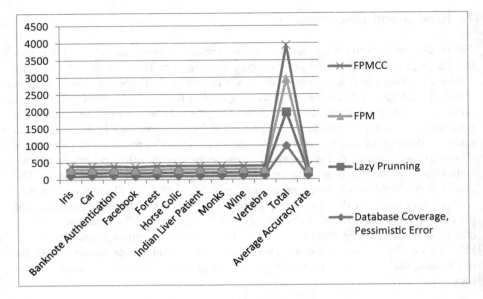

**Fig. 8.** Average classification accuracy

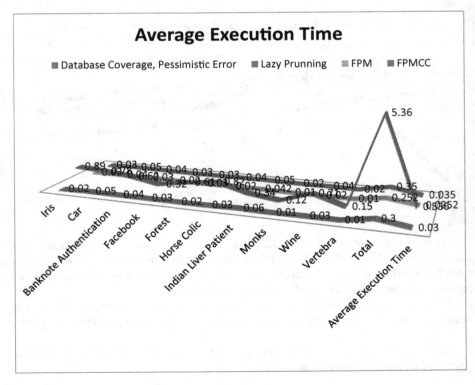

**Fig. 9.** Average execution time

Thus the results produced in Table 12, the accuracy rate indicate that FPM, FPCC methods enhance the classification accuracy than other pruning methods and attempt to build competitive classifier in terms of classification accuracy rate and the size.

## 8 Conclusion

The large number of rules produced during the training phase of AC algorithms may result in difficulty in interpretation and maintenance of the data. These two issues need to be addressed in building the classifier in order to eliminate the insufficient rules containing the noisy information leading to wrong classification. Also the execution time increases in managing such large number of rules. This chapter presented different and efficient rule pruning methods in order to build moderate sized and accurate AC classifier. Further, a new class prediction method is also proposed to handle more than one rules if any and to enhance the classification accuracy rate using a single rule.

The experiments are conducted on different datasets taken from UCI repository based on two parameters, the number of rules and the accuracy rate using improved CBA algorithm. According to the results, the algorithm FPCC pruning method shown outstanding performance as compared to other methods.

## References

Kliegr, T. (2017): Quantitative CBA: Small and Comprehensible Association Rule Classification Models arXiv:1711.10166. (eprint arXiv:1711.10166)

Liu, B., Hsu, W., Ma, Y.: Integrating classification and association rule mining. In: Proceedings of the KDD, pp. 80–86. New York (1998)

Merz, C., Murphy, P.: UCI Repository of Machine Learning Databases. University of California, Department of Information and Computer Science, Irvine (1996)

Mittal, K., Aggarwal, G., Mahajan, P.: Int. J. Inf. Tecnol. 1–6 (2018). https://doi.org/10.1007/s41870-018-0233-x

Zaïane, O.R., Antonie, M.-L.: On pruning and tuning rules for associative classifiers. KES **3**, 966–973 (2005)

Garza, P., Chiusano, S., Baralis, E.: A lazy approach to associative classification. In: IEEE Transactions on Knowledge & Data Engineering, vol. 20, pp. 156–171 (2007). https://doi.org/10.1109/tkde.2007.190677

Sahu, S.K., Kumar, P., Singh, A.P.: Modified K-NN algorithm problems for classification with improved accuracy. Int. J. Inf. Technol. **10**, 65 (2018). https://doi.org/10.1007/s41870-017-0058

Thabtah, F.: Pruning techniques in associative classification: survey and comparison. JDIM **4**, 197–202 (2006)

Zhao, M., Cheng, X., He, Q.: An algorithm of mining class association rules. In: Cai, Z., Li, Z., Kang, Z., Liu, Y. (eds.) Advances in Computation and Intelligence, vol. 5821. Springer, Berlin (2009). https://doi.org/10.1007/978-3-642-04843-2_29

# Author Index

Printed in the United States
By Bookmasters